W9-AVF-135

SENSING,

INTELLIGENCE,

MOTION

SENSING, INTELLIGENCE, MOTION

HOW ROBOTS AND HUMANS MOVE IN AN UNSTRUCTURED WORLD

Vladimir J. Lumelsky

A JOHN WILEY & SONS, INC., PUBLICATION

Published by John Wiley & Sons, Inc., Hoboken, New Jersey.
Published simultaneously in Canada.

For general information on our other products and services or for technical support, please contact
our Customer Care Department within the United States at (800) 762-2974, outside the United
States at (317) 572-3993 or fax (317) 572-4002.

Wiley also publishes its books in a variety of electronic formats. Some content that appears in print
may not be available in electronic formats. For more information about Wiley products, visit our
web site at www.wiley.com.

Library of Congress Cataloging-in-Publication Data:

Lumelsky, Vladimir.
  Sensing, intelligence motion : how robots and humans move in an unstructured world /
Vladimir L. Lumelsky.
    p. cm.
  "A Wiley-Interscience publication."
  Includes bibliographical references and index.
  ISBN-13 978-0-471-70740-0
  ISBN-10 0-471-70740-6
    1. Robots—Motion. 2. Manipulators (Mechanism) I. Title.

TJ211.L85 2005
629.8'92—dc22

                                                                        2005041748

Printed in the United States of America.

10 9 8 7 6 5 4 3 2 1

To Rakhil, Nadya, Michael, and Anna

CONTENTS

We humans are good at moving around in this world of ours. If we are serious about the ubiquity of robots' help to humankind, we must pass this skill to our robots. It also turns out that in some tasks, robots can find their way better than humans. This suggests that it is time for humans and robots to join forces.

Imagine you arrive at a party. You are a bit late. The big room is teeming with voices and movement. People talk, drink, dance, walk. As you look around, you notice a friend waving to you from the opposite side of the room. You fill two glasses with wine, glance quickly across the room, and start on your journey. You maneuver between people, bend your body this way and that way to avoid collision or when shoved from the side, you raise your hands and squeeze your shoulders, you step over objects on the floor. A scientifically minded observer would say that you react to minute disruptions on your path while also keeping in mind your global goal; that you probably make dozens of decisions per second, and a great many sensors are likely involved in this process; that you react not only to what you see, but also to what you sense at your sides, your back, your feet. In a minute's time you happily greet your friend and hand him a glass of wine.

You may be surprised to hear that in your trip across the room you planned and executed a complex motion planning strategy whose emulation in technology is a yet unachieved dream of scientists and engineers. Providing a robot with a seemingly modest skill that you just demonstrated, an ability to move safely among surrounding objects using incomplete sensing information about them would be a breakthrough in science and technology whose consequences for society is hard to overestimate. This would be the beginning of a new era, with a great number of machines of unimaginable variety moving quietly and productively in the world around us.

The main reason that we desire such technology is not, of course, the convenience of a wine-serving automatic maid. A machine's ability to safely operate in a reasonably arbitrary environment will lead to our automating a wide span of tasks that have eluded automation so far—from the delivery of drugs and food to patients in hospitals and nursing homes, to a robot "nurse" in the homes of elderly people, and to such indispensable tasks as cleaning chemical and nuclear waste sites, demining of old and new mine fields, planetary exploration, repair of faraway space satellites, and a great number of other tasks in agriculture, undersea, deep space, and so on. Equipped with this skill, the recent Mars rovers Spirit and Opportunity would have accomplished in hours what took them weeks.

We do not have such automation today. Today, humans are not even allowed to share space with serious robots, though a good number of the tasks above would require this. The only reason for this constraint is that today's robot bodies are too insensitive, too oblivious to their surroundings, and hence too dangerous to themselves and to objects and people around them.

Looking ahead to the near future, however, there are at least three good reasons for optimism. One is social: The problem will not go away and so the pressure on scientists and engineers will stay strong. The need for machines capable of working in our midst or far away with little or no supervision will only grow with time. The value of human life and the increasing costs of human labor combined with ever riskier undertakings in space, undersea, and in rough places on Earth will continue the push for more automation. A very good example of this trend is the recent unique "attempt for on attempt" for a robot mission to save the ailing Hubble Telescope.

One may say that having a painful problem is not enough to find a solution. True, but then there are the other two reasons. The second reason for optimism is the successes of robot systems in recent years. Almost 1,000,000 highly reliable industrial robots are doing useful, sometimes quite complex, work worldwide. True, almost none of these robots can operate outside of their highly specialized man-made environment, and those few that do are too simplistic to be taken seriously. Hence the third reason for optimism: Research laboratories around the world report more and more sophistication in robot systems operating outside the "sanitized" factory environment. Robots have been shown to be as good as or better than humans in some tasks that require spatial reasoning and motion planning. Systems have been demonstrated where synergistic human–robot teams operate better, even smarter, than each of them separately. This trend is bound to continue.

It is the ability to plan its own motion that makes a robot qualitatively different from other machines. After all, the mechanical parts, electronics, computers, some functional abilities, and sophistication that robots possess are present in many other digitally controlled machines. Thus the half-humorous debates of the 1960s and 1970s when designers of digitally controlled factory machinery were accusing specialists in robotics of inflating the prestige of their field by calling their machines robots—aren't these just slightly modified digitally controlled machines? There is truth to it. Now we are approaching a time when the field of robotics will be able to say that it is the ability to plan its own motion that makes a robot a robot.

Doesn't such technology already exist? Haven't we read about robots that paint and weld and do assembly in automotive and computer manufacturing factories? For factories, yes, but for tasks outside the factory floor—hospitals and outer space and mine fields—no, not really, except perhaps in a few simplistic cases. What is the difference?

For you and me, the success of, say, returning a bottle to the refrigerator depends little on whether at this very instant the arrangement of objects in the refrigerator differs from what it was half hour ago when the bottle was taken out. This is not so for today's robots.

If the required motion is to be repeated over and over again and if all the objects in the robot workspace can be described precisely—as they are, for example, on the car assembly line or in an automatic painting booth—using robots to automate the task presents no principal difficulties today. Designing the required trajectories for the tool in the robot hand is a purely geometric problem, fully solvable by computer. (Depending on the task specifics, it may of course require an unrealistically large amount of computation time, but this is another matter.) Once the car model changes next year, the new data are fed into the computer, and the required motion is recalculated. This is an example of a *structured task*, and it takes place in a *structured environment*. The word "structured" is roughly equivalent to "well-organized," "known precisely," "man-made." Objects in a structured environment can be safely assumed fully known in space and time.

As a rule, a structured environment is designed, carefully and often at great cost, by highly qualified professionals. From the standpoint of motion planning, the input information that the robot needs in order to generate the desired motion is available before the motion starts. What is needed is appropriate algorithms for transforming this information into proper motion trajectories. Today there are plenty of such algorithms. This setup represents the *Intelligence–Motion* planning paradigm.

This algorithmic paradigm was formulated right at the beginning of robotics as a field of science and technology, around the mid-1960s. Today the Intelligence–Motion paradigm boasts a large literature, appearing under such names as *motion planning with complete information*, or *model-based motion planning*, or the *Piano Mover's model*. The symbolism behind the latter term is that when movers set out to move a piano, they can first sit down and figure out the whole sequence of moves and turns and raisings and lowerings, before they start the actual motion. After all, the physical setting that encompasses this information is right there before them. (Except, one might comment, "Who in this world would ever do it this way?" More likely the movers just say, "Let's do it!", and they discuss every move as they get to it—thereby losing an opportunity to contribute to a great theory.)

On the theoretical level, the problem of motion planning with complete information is more or less closed: remarkably complete and enlightening studies of the problem have provided computational complexity bounds, motion planning algorithms, and deep insights into the problem. Which is not to say that all problems in this area are solved. Most of today's work in this area is devoted to special cases and to struggling with computational issues in realistic settings. Somewhat ironically, applications where such techniques are used today relate not so much to robotics as to other areas: computer-aided design (CAD, e.g., to design an aircraft engine such as to allow quick removal or replacement of a given unit), models of protein folding in biology, and a few others. The major property of such tasks is that the required motion is designed in a database rather than in a physical setting. Given the wealth of published work in this area, this book reviews the Piano Mover's paradigm only cursorily.

The focus of this book is on *unstructured tasks*—tasks that unfold in an *unstructured environment*, an environment that is not predesigned and has to be taken as is. Most of the motion planning examples above (homes, outdoors, deep space, etc.) refer to unstructured tasks. Until recently, robotics practitioners have either ignored this area or have limited their efforts to grossly simplified tasks with robot hands or with mobile robots. Even in the latter cases the operation is mostly limited to a tight human teleoperation, with a minimum of robot autonomy (as in the case of recent Mars rovers). All kinds of helpful "artificial" measures—for example, an extremely slow operation—are taken to allow the operator to precede commands with a careful analysis.

Automating motion planning for mobile robots will be considered in the first sections of this text. We will also see later that teaching a robot arm manipulator to safely move in an unstructured environment is a much taller order than the same request for a mobile robot. This is unfortunate because a large number of pressing applications require manipulators. Today people use a great deal more arm manipulators than mobile robot vehicles. An arm manipulator is a device similar to a human arm. If the task is to just move around and sense data or take pictures, that is a job for a mobile robot. But if the task requires "doing things"—welding, painting, putting things together or taking them apart—one needs an arm manipulator. Interestingly, while collision avoidance is a major bottleneck in the use of robot manipulators, there is minuscule literature on the subject. This book attempts to fill the gap.

Objects in an unstructured robot workspace cannot be described fully—either because of their unyielding shape, or because of lack of knowledge about them, or because one doesn't know which object is going to be where and when, or because of all three. In dealing with an environment that has to be taken as is, our robots have a good example to follow: The evolution has taught us humans how to move around in our messy unstructured world. We want our robots to leap-frog this process.

And then there are tasks—especially, as we will see, with motion planning for arm manipulators—where human skills and intuition are not as enviable. In fact, not enviable at all. Then not only do we need to enter unchartered territories and synthesize new robot motion planning strategies that are way beyond human spatial reasoning skills, but also we must built a solid theoretical foundation behind them, because human experience and heuristics cannot help ascertain their validity.

If the input information about one's surroundings is not available beforehand, one cannot of course calculate the whole motion at once, or even in large pieces. What do we humans and animals do in such cases? We compensate by real-time *sensing* and *sensor data processing*: We look, touch, listen, smell, and continuously use the sensing information to plan, execute, and replan our motion. Even when one thinks one knows by heart how to move from point A to point B—say, to drive from home to one's office—the actual execution still involves a large amount of continuous sensor-based motion planning.

Hence the names of approaches to motion planning in an unstructured environment that one finds in the literature are: *motion planning with incomplete information*, or *sensor-based motion planning*. Another good name comes from the crucial role that this paradigm assigns to sensing: Similar to the phrase *Intelligence–Motion* for motion planning with complete information, we will use the name *Sensing–Intelligence–Motion* (SIM) for motion planning with incomplete information. The SIM approach will help open the door for robotics into automation of unstructured tasks. (Recall "Open door, Simsim!" in the Arabian tale "Ali Baba and the Forty Thieves.")

The described differences in how input information appears in the Piano Mover's and SIM paradigms affect their approach to motion planning in crucial ways—so much so that attempted symbiosis of some useful features of "structured" and "unstructured" approaches have been so far of little theoretical interest and little practical use.

While techniques for motion planning with complete information started in earnest in the first years of robotics, sometime in early 1960s, the work on SIM approaches started later, in the late 1980s, and has proceeded more slowly. The slow pace is partly due to the fact that the field of robotics in general and the area of motion planning in particular have been initiated primarily by computer scientists. The combinatoric–computational professional inclinations of these visionaries made them more enthusiastic about geometric and computational issues in robotics than about real-time control and the algorithmic role of sensing. Another important reason is the tight connection between algorithms and hardware that the SIM approach espouses. As we will see later, some of this (sensing) hardware has only started appearing recently. Finally, a quick look at this book's table of contents will show that the work on SIM approaches requires from its practitioners a somewhat unusual combination of background: topology, computational complexity, control theory, and a rather strange sensing hardware.

Whatever the reasons, in spite of its great theoretical interest and an immense practical potential, the literature on the sensor-based motion planning paradigm is small, especially for arm manipulators. In fact, today there are no textbooks devoted to it.

Our goals in this book are as follows:

(a) Formulate the problem of sensor-based motion planning. We want to explore why the relevant issues are so hard—so much so that in spite of hard work and some glorious successes of robotics, there is no robot today that can be left to its own devices, without supervision, outdoors or in one's home. Build a theoretical foundation for sensor-based motion planning strategies.

(b) Study in depth a variety of particular algorithmic strategies for mobile robots and robot arm manipulators, and try to identify promising directions for conquering the general problem.

(c) Given the similarity of underlying tasks and requirements, compare robot performance and human performance in sensor-based motion planning.

The hope is that by doing so we can get a better insight into the nature of the problem, and can help build synergistic human–robot teams for tele-operation tasks.

(d) Review sensing hardware that is necessary to realize the SIM paradigm.

The book is intended to serve three purposes: (1) as a course textbook; (2) as a research text covering in depth one particular area of robotics; (3) as a program of research and development in robotic automation of unstructured tasks.

As a Textbook. A good portion of this book grew out of graduate and senior undergraduate courses on robot motion planning taught by the author at Yale University and the University of Wisconsin—Madison. As often happens with research-oriented courses, the course kept changing as more research material appeared and our knowledge of the subject expanded.

The text assumes a basic college background in mathematics and computer science. A prior introductory course in robotics and some knowledge in topology will be helpful but are not required. Some more exposure to topology is advised for mastering the analysis that appears in Section 5.8 (Chapter 5) and the first two pages of Section 6.2.4 (Chapter 6). Conclusions from this analysis, in particular the formulation of algorithms, are written at the level compatible with the rest of the book, though. The instructor is advised to glance through the chapters beforehand to decide which level of what background a given chapter or section requires.

Homework examples are provided as needed. In my view, a good homework structure for an advanced course like this one includes two components: (a) ordinary homework assignments that dig deeper in the student's knowledge, are modest in number, and require a week or two to complete each assignment; and (b) a course project that is initiated in the course's first few weeks, goes in parallel with it, and is defended at the end of the course, with the defense treated as the final exam. The weights of those components in the student final grade can be, say, 50% for the homework, 20% for the midterm assessment of the project, and 30% for the final text-plus-presentation-before-class of the project. A list of ideas for course projects is provided in Chapter 9.

Assuming a conventional two-semester school year, this book has about two semesters worth of material. A one-semester course hence calls for choices. A typical structure that covers ideas and computational schemes of the sensor-based motion planning paradigm will include Chapters 1, 2, 3, 5, and 6 (Motion Planning—Introduction, A Quick Sketch of Major Issues in Robotics, Motion Planning for a Mobile Robot, Motion Planning for Two-Dimensional Arm Manipulators, Motion Planning for Three-Dimensional Arm Manipulators). Let us call this sequence the *core course*. The sequence contains no control theory or electronics, and it allows for the widest audience in terms of students' majors.

For a strictly engineering class where students have already had courses in controls and electronics, the instructor may want to sharply contract the time for Chapter 2 and provide instead a deeper understanding of the effects of robot dynamics on motion planning, covered in Chapter 4, plus a cursorial review of principles of design of sensing devices necessary for realizing sensor-based motion planning strategies, Chapter 8. Any group can benefit from Chapter 7, which is devoted to human performance in motion planning and spatial reasoning tasks. A two-semester sequence will comfortably cover all those chapters (with the danger of one's noticing some repetitions necessitated by the foreseen different uses of the book).

The decision to include in the course the topics covered in Chapters 4, 7, and 8, as well as the time devoted to the introductory Chapters 1 and 2 will depend much on the mixture of students in class, in particular their prior exposure to robotics, control theory, and electronics. Mandating prior courses on these topics may introduce interesting difficulties. In my experience, a significant percentage of graduate students attracted to this course come from disciplines outside of engineering, computer science, physics, and mathematics—such as business administration, psychology, and even medicine. This is not surprising since the course material touches upon the future of their disciplines rather deeply. Students from some areas, especially the latter three above, are usually interested in ideas and cognitive underpinnings of the subject. These students are often extremely good, quick, and knowledgeable and have a reasonably good background in mathematics. Often such students do well in homework assignments, bring in new ideas, and come up with wonderful course projects in their appropriate areas. Denying their participation would be a pity, in my view—after all, robotics is a wide and widely connected field.

With such students in class, the instructor may choose to spend a bit more time on the introductory sections, in order to bring up to speed students who have had no past exposure to the robotics field. The instructor may also want to complement introductory material with a relevant textbook (some such textbooks are mentioned in Chapters 1 and 2). Students' grades in the homework at the end of Chapter 2 will give the instructor a good indication of how prepared they are for the core course.

As a Research Text. This book is targeted to people who are interested in or are directly involved in research and development of robot and human–robot interaction systems. If one's goal is to understand the underlying issues or design a system capable of purposeful motion in an unstructured environment while protecting the robot's whole body—in streets, homes, undersea, deep space, agriculture, and so on—today SIM is the only consistent approach one can count on. This is not to say that the book contains answers to all questions. It provides some constructive answers, and it calls for continuation.

The book should also be of interest to people working in areas that are tangentially connected to robotics, such as sensor development and design of tele-operated systems. And finally, the book will hopefully appeal to people interested in the wide complex of underlying issues in robotics and human–robot interaction, from mathematical and algorithmic questions to cognitive science to advanced robot applications.

As a Program for Continued Research and Development. To repeat the statement above, today the Sensing–Intelligence–Motion (SIM) approach seems to be the only paradigm that holds promise to bring about robot automation of unstructured tasks. This is not because of some special sophistication of SIM techniques, but simply because only SIM techniques take care of the necessary whole body awareness of the robot and do it "on the fly," in real time, making it possible to handle a high level of uncertainty. And only this approach guarantees results in this area when human intuition breaks down.

And yet, as one will see later, only a limited number of SIM algorithms and sensing schemes for real-world robot systems have been explored so far. Much of the theory and of algorithmic and hardware machinery that is necessary to bring the SIM approach to full fruition lies ahead of us. The book starts on the misty route that lies ahead and that has to be traversed if we are serious about bringing automation into unstructured tasks. With the risk of being seen less than balanced, I suggest that not many areas of computer science and engineering can compete with the excitement, the required breadth of knowledge, and the potential impact on society of the topics covered in this book.

Professional and commercial importance of robotics aside, robots have been always of immense interest to the general public. Isaac Asimov's robot heros are household names. Crowds invariably surround fake robots (controlled by humans from nearby buildings) on the Disneyland streets. Robot exploits on Mars or on the Space Shuttle or in a minefield disarming operation make front pages of newspapers. What excites laymen is a human-like behavior potential of a robot. This book takes the reader further in this same direction by providing a solid foundation behind one human-like ability of robots that was so far assumed to be an inherent monopoly of humans—namely, the ability to think of and plan one's motion in an unstructured world.

Robots are often referred to derisively: "He moves like a robot," "Yours is a robot reaction," "Hey, don't behave like a robot." What is meant is crude, unintelligent, and mechanical; even the word "mechanical" signifies here crude and unintelligent. Many mimes entertain the crowd on the street corners by moving "like a robot"—that is, switching sharply from one movement to the other and being oblivious to the surroundings.

That is not what robots should be and even are today. Examples in Chapter 8 will show that when equipped with means for self-awareness and with strategies to use it, robots become sensitive to their surroundings, "pensive," and even gentle in how they "mind" their movement.[1] A nonprofessional reader curious about the possibilities of intelligent robots will find long layman-level passages in

---

[1]Sharp "robot-like" movements have been a persistent science fiction-maintained myth. Many robot applications—car painting is a good example—require smooth motion and simply cannot tolerate sharp turns. Today's industrial robots can generate a motion that is so smooth and delicate that it may be the envy of "Swan Lake" ballerinas. For those who know calculus, what dancer can promise, for example, a motion so smooth that both its derivatives have guaranteed continuity!

the Introduction, introductory sections to other chapters, discussions, examples, and simplified explanations of the underlying ideas throughout the text.

Designing a whole-sensitive robot is almost like designing a friend. One day you move your hand in a stroking movement along the robot's skin, and it responds with a gentle appreciative movement. This gives you a strange feeling: We humans are totally unprepared to see a machine exhibit a behavior that we fully expect from a cat or a dog. I hope that both professional and layman readers will share this gratifying feeling. And, of course, I hope the book will further our attempts toward populating our environment with helpful and loyal robot friends.

VLADIMIR J. LUMELSKY

Madison, Wisconsin
Washington D.C.
April 2005

ACKNOWLEDGMENTS

When pieces of a large multiyear project start falling into place, a sign that it functions right is that the pieces "know by themselves" what to do and when to do it. A product of one section logically invites and defines the other; theory calls for the experiment to confirm its correctness; experiments beg for turning theory into useful products. The project then operates as a leisurely human walk: As the right foot is thrown forward, the left foot knows it should stay behind on the ground, the body bends slightly forward as if ready to fall, the left arm moves forward, and the right arm heads back—all at once, seemingly effortlessly, and then they switch, one-two, one-two, a pleasure to watch, so hard to emulate, one-two, one-two.

A piece of science or new technology cannot be like this, not that perfect, simply because there is always more unknown and yet undiscovered than known and understood. But the feeling is similar: All of a sudden, things fall into place. This picture fully applies to this book. While the knowledge that it treats will be always incomplete, a moment came when individual smaller projects started looking as parts of a tightly coordinated organism.

This would not be possible without my graduate students. Much of today's science is produced this way. It is the graduate students' sleepless nights, enthusiasm, and unwavering commitment to science that help cover the skeleton of ideas with flesh and blood of details of design and proofs and tests and computer simulations. They help turn the skeleton's jerky squeakiness into smooth and coordinated and pleasing to the professional eye elegant whole.

"What if" is rarely a reliable game. There is no way of knowing what this book would look like if I had different students, not those I was privileged to have. I do think that some pieces would have been quite different, because the personalities and prior background of my students invariably left a strong trace on my choice of projects for them and hence the joint papers that became the foundation of this book. I am grateful to them for sharing with me the joy of doing science. With all those different personalities, there was also something in common that emerged in them as the work progressed–perhaps the desire for dry precision, for doing things right. In thanking them for sharing with me our life in the lab and discussions in seminars and at the blackboard, I am mentioning here only those whose work was pivotal for this book: Kang Sun, Timothy Skewis, Edward Cheung, Susan Hert, Andrei Shkel, Fei Liu, Dugan Um. Other students helped as well, but their main work centered on topics that are beyond our subject here.

From the beginning of this research in the late 1980s, the National Science Foundation was incredibly generous to me, funding in parallel the theory/software and the hardware/sensing lines of this work. I am also indebted to the Sandia Laboratories and Hitachi Corporation for providing necessary resources.

Every book has to be started, and that moment calls for an appropriate setting. My thanks go to the Rockefeller Foundation, whose invitation to spend a month at the incomparable Villa Serbelloni in the village of Bellagio, Lake Como, Italy, made the start of this book quick and easy. Putting in a day of work, along with a couple more hours in the evening, was tiring but easy, in anticipation of the game of bocce on the lake by 5 o'clock and then dressing up for drinks and dinner with the Villa's guest artists and writers and scientists, among the seventeenth-century rugs and furniture. It is not for nothing that the Villa Serbelloni's library is crammed with books authored by many of its visitors from all over the world.

V. J. L.

CHAPTER 1

Motion Planning — Introduction

Midway along the journey ... I woke to find myself in a dark wood, for I had wandered off from the straight path.

— Dante Alighieri, The Divine Comedy, "Inferno"

1.1 INTRODUCTION

In a number of Slavic languages the noun "robota" means "work"; its derivative "robotnik" means a worker. The equivalent of "I go to robota" is a standard morning sentence in many East European homes. When in 1921 the Czech writer Karel Capek needed a new noun for his play R.U.R. (Rossum's Universal Robots), which featured a machine that could work like a human, though in a somewhat mechanical manner, he needed only to follow Slavic grammar: Chopping off "a" at the end of "robota" not only produced a new noun with a similar meaning but moved it from feminine to masculine. It was just what he wanted for his aggressive machines that eventually rebelled against the humankind and ran amok. The word *robot* has stuck far beyond Capek's wildest expectations— while, interestingly, still keeping his original narrow meaning.

Among the misconceptions that society attaches to different technologies, robotics is perhaps the most unlucky one. It is universally believed that a robot is almost like a human but not quite, with the extent of "not quite" being the pet project of science fiction writers and philosophers alike. The pictures of real-life robots in the media, in which they look as close to a human as, say, a refrigerator, seem to only insult the public's insistence on how a robot should look.

How much of "not quite"-ness is or ever will be there is the subject of sometimes fierce arguments. It is usually agreed upon that high intelligence is a must for a robot, as is a somewhat wooden personality. And, of course, the public refuses to take into account the tender age of the robotics field. One standard way of expressing the "not quite"-ness is a jerky motion sold as robot motion in Hollywood movies and by young people imitating a robot on street corners. Whatever future improvements the public is willing to grant the field, a smooth motion and a less-than- wooden personality are not among them. A robotics professional will likely give up when hearing from friends or school audiences that the best robots are found in Disneyworld. ("What do you mean? Last week I myself talked to one in Disneyworld in Orlando." Don't try to tell him he actually spoke to an operator in the nearby building.)

It would not be fair to blame Karel Capek, or Disney, or Hollywood for the one-dimensional view of robotics. The notion of a robotic machine goes far back in time. People have always dreamt of robots, seeing them as human-like machines that can serve, fascinate, protect, or scare them. In Egyptian temples, large figurines moved when touched by the morning sun rays. In medieval European cities, bronze figures in large tower clocks moved (and some still move) on the hour, with bells ringing.

Calling on human imagination has been even easier and more effective than relying on physical impersonation. Jewish mysticism, with its Cabbala teachings and literary imagery, has also favored robots. Hence the image of Golem in Cabbala, a form that is given life through magic. In the Hebrew Bible (Psalms 139:16) and in the Talmud, Golem is a substance without form. Later in the Middle Ages the idea took the form; it was said that a wise man can instill life in an effigy, thus creating a Golem with legs and arms and a head and mighty muscles. A "typical" Golem became a human-like automaton, a robot.

Perhaps the best-known such story is of Rabbi Loew of sixteenth-century Prague, in Czechia. (The Rabbi's somewhat scary gravestone still greets the visitor in the Jewish cemetery at Prague's center.) Rabbi Loew created his Golem from clay, to serve as his servant and to help protect the Jews of Prague. Though the creature was doing just that, saving Jews of Prague from many calamities by using its great strength and other supernatural skills, with time it became clear that it was getting out of hand and becoming dangerous to its creator and to other Jews. Rabbi Loew thus decided to return the Golem back to its clay immobility, which he achieved using a secret Cabbalistic formula. He then exiled the figure to the attic of his Prague synagogue, where it presumably still is, within two blocks from Loew's grave. This story became popular through the well-written 1915 novel called *Der Golem*, by a German writer Gustav Meyrink, and the 1920 movie under the same title by the German director Paul Wegener (one can still find it in some video shops). The Golem, played by Wegener himself, is an impressive figure complete with stiff "robotic" movement and scary square-cut hairdo.

We still want the helpful version of that robot—in fact, we never wanted it more. The last 40 years have seen billions of dollars, poured by the United States, European, Japanese, and other governments, universities, and giant companies into development of robots. As it often happens with new technologies, slow progress would breed frustration and gaps in funding; companies would lose faith in quick return and switch loyalties to other technologies. Overall, however, since 1960 the amount of resources poured by the international community into robotics has been steadily going up. For what it's worth, even the dream of an anthropomorphic likeness is well and alive, even among professionals and not only for toy robots. Justifications given—like "people feel comfortable with a human-looking robot," as if people would feel less comfortable with a dishwasher-shaped robot—may sound somewhat slim; nevertheless, the work on anthropomorphic robots still goes on, especially in Japan and from time to time in the United States and Europe.

The reasons behind the strong interest in robotics technology have little to do with Hollywood dreams. Producing a machine that can operate in a reasonably arbitrary environment will allow us to automate a wide span of tasks.

If some of us feel that we have more than enough automation already, this feeling is not necessarily due to our ambivalence about machines. It is hard to feel a need for something that does not exist. Think, for example, of such modern-day necessities as paper towels and paper napkins. Who would think of "needing" them back in the nineteenth century, before they became available?

To have a sense of what is the "right" amount of automation, consider the extent of automation in today's industrialized world, and then consider the kind of automation we may have if the right technology becomes available. Wouldn't we welcome it if our dishwashers knew how to collect dirty dishes from the table, drop the solid waste into the waste basket, slightly rinse the dishes under the faucet, put them into the dishwashing basin—and only then proceed to what today's dishwashers do—and later of course put the clean dishes and silverware where they belong? More seriously, wouldn't we embrace a machine capable of helping an old person prolong her independent living by assisting with simple household chores such as answering the doorbell, serving food, and bringing from the closet clothing to wear? How about a driverless security car patrolling the streets and passing along information to the police control room; automatic waste collection and mail delivery trucks; driverless tractors and crops picking machines in farms? There is no end to this list.

Then there are tasks in which human presence is not feasible or highly undesired, and for which no expense would be too big: demining of minefields in countries after war (there is no shortage of these in recent years); deep-sea oil exploration; automatic "repairmen" of satellites and planet exploration vehicles; and so on. For example, unlike the spectacular repairs of the Hubble Space Telescope by astronauts, no human help will be feasible to its one-million-miles-away replacement, the James Webb Space Telescope—only because the right robots do not exist today. Continuing our list in this fashion and safely assuming the related automation will be feasible at some point, observe that only a small fraction, perhaps 5% or so, of tasks that *could and should* be automated *have been* automated today. Robotics is the field we turn to when thinking about such missing automation. So, why don't we have it? What has been preventing this automation from becoming a reality?

It may sound surprising, but by and large the technology of today is already functionally ready for many of the applications mentioned above. After all, many factory automation machines have more complex actuators—which translates into an ability to generate complex motion—than some applications above require. They boast complex digital control schemes and complex software that guides their operation, among other things. There is no reason why the same or similar schemes could not be successful in designing, say, a robot helper for the homes of elderly individuals. So, why don't we have it? What is missing? The answer is, yes, something is missing, but often it is not sophistication and not functional abilities.

What is missing are two skills. One absolutely mandatory, is a local nature and is a seemingly trivial "secondary" ability in a machine not to bump into unexpected objects while performing its main task—be it walking toward a person in a room with people and furniture, helping someone to dress, replacing a book on the shelf, or "scuba-diving" in an undersea cave. Without this ability the robot is dangerous to the environment and the environment is dangerous to the robot—which for an engineer simply means that the robot cannot perform tasks that require this ability. We can call this ability *collision avoidance in an uncertain environment*.

The other skill, which we can call *motion planning*, or navigation, is of a global nature and refers to the robot ability to guarantee arrival at the destination. The importance of this skill may vary depending on a number of circumstances.

For humans and animals, passing successfully around a chair or a rock does not depend on whether the chair or the rock is in a position that we "agreed" upon before we started. The same should be true for a robot—but it is not.

Let us call the space in which the robot operates the *robot workspace*, or the *robot environment*. If all objects present in the robot workspace could be described precisely, to the smallest detail, automating the necessary motion would present no principal difficulties. We would then be in the realm of what we call *the paradigm of motion planning with complete information*. Though, depending on details, the problem may require an inordinate computation time, this is a purely geometric problem, and the relevant software tools are already there. Algorithmic solutions for this problem started appearing in the late 1970s and were perfected in the following decades.

A right application for such a strategy is, for example, one where the motion has to be repeated over and over again in exactly the same workspace, precisely as it happens on the car assembly line or in a car body painting booth. Here complete information about all objects in the robot environment is collected beforehand and passed to the motion planning software. The computed motion is then tried and optimized via special software or/and via many trial-and-error improvements, and only then used. Operators daily make sure that nothing on the line changes; if it does purposely, the machine's software is updated accordingly. Advantages of this strategy are obvious: It delivers high accuracy and repeatability, consistent quality, with no coffee breaks. If the product changes, say, in the next model year, a similar "retraining" procedure is applied.

We will call tasks and environments where this approach is feasible *structured tasks* and *structured environments*, which signifies the fact that objects in the robot environment are fully known and predictable in space and time. Such environments are, as a rule, man-made.

An automotive assembly line is a perfect example of a structured environment: Its work cells are designed with great care, and usually at a great cost, so as to respect the design constraints of robots and other machinery. A robot in such a line always "knows" beforehand what to expect and when. Today the use of robotics on such lines is an extremely successful and cost-effective proposition, in spite of their high cost.

Unfortunately, some tasks—in fact, the great majority of tasks we face every day—differ in some fundamental ways from those on the automotive assembly line. We live in the world of uncertainty. We deal with *unstructured tasks*, tasks that take place in an *unstructured environment*. Because of unpredictable or changing nature of this environment, motions that are needed to do the job are not amenable to once-and-for-all calculation or to honing via direct iterative improvement. Although some robots in the structured automotive environment are of great complexity, and functionally could be of much use in unstructured tasks, their use in an unstructured environment is out of the question without profound changes in their design and abilities. Analyzing this fact and finding ways of dealing with it is the topic of this book.

Sometime in the late 1950s John McCarthy, from Stanford University [who is often cited as father of the field of artificial intelligence (AI)], was quoted as saying that if the AI researchers had as much funding as NASA was given at the time to put a man on the moon, then within 10 years robot taxi cabs would roam the streets of San Francisco. McCarthy continued talking about "automatic chauffeurs" until at least the late 1990s. Such loyalty to the topic should certainly pay off eventually because the automatic cab drivers will someday surely appear.

Today, over 40 years since the first pronouncement, we know that such a robot cannot be built yet—at any cost. This statement is far from trivial—so it is not surprising that many professional and nonprofessional optimists disagree with it. Not only it is hard to quantify the difficulties that prevent us from building such machines, but these difficulties have been consistently underestimated. As another example, in 1987, when preparing an editorial article for the special issue on robot motion planning for the *IEEE Transactions on Robotics and Automation*, this author was suggested to take off from the Foreword a small paragraph saying that in the next 10 years—that is, between 1987 and 1997—we should not expect a robot capable of, say, tying one's shoelaces or a necktie. The text went on to suggest that the main bottleneck had less to do with lacking finger kinematics and more with required continuous sensing and accompanying continuous sensor data processing. "This sounds too pessimistic; ten years is a long time; science and technology move fast these days," the author was told. Today, almost two decades later, we still don't have robots of this level of sophistication—and not for a lack of trying or research funding. In fact, we can confidently move the arrival of such robots by at least another decade.

One way to avoid the issue is to say that a task should be "well engineered." This is fine except that no task can be likely "well engineered" unless a technician has a physical access to it once or twice a day, as in any automotive assembly line. Go use this recipe with a robot designed to build a large telescope way out in deep space!

Is the situation equally bleak in other areas of robotics? Not at all. In recent years robotics has claimed many inroads in factory automation, including tasks that require motion planning. Robots in automotive industry are today among the most successful, most cost-effective, and most reliable machines. Robot motion planning algorithms have penetrated areas far from robotics, from designing quick-to-disassemble aircraft engines (for part replacement at the airport gate) to studies of folding mechanisms of DNA molecules.

It is the unstructured environment where our success stops. We have difficulty moving robots into our messy world with its unending uncertainty. That is where the situation is bleak indeed—and that is where robotics is needed badly.

The situation is not black and white but rather continuous. The closer a task is to that in a fully structured environment, the better the chance that today's approaches with complete information will apply to it. This is good news. When considering a robot mission to replace the batteries, gyroscopes, and some scientific instruments of the aging Hubble Space Telescope, NASA engineers were gratified to know that, with the telescope being a fully man-made creature, its repair presents an almost fully structured task. The word "almost" is not to be overlooked here—once in a while, things may not be exactly as planned: The robot may encounter an unscrewed or bent bolt, a broken cover, or a shifted cable. Unlike an automotive plant, where operators check out the setup once or twice a day, no such luxury would exist for the Hubble ground operators. Although, luckily, the amount of "unstructuredness" is small in the Hubble repair task, it calls for serious attention to sensing hardware and to its intimate relation to robot motion planning. Remarkably, even the "unstructuredness" that small led to the project's cancellation.

A one-dimensional picture showing the effect of increase in uncertainty on the task difficulty, as one moves from a fully structured environment to a fully unstructured environment, is shown in Figure 1.1. An automotive assembly line (the extreme left in the figure) is an example of a fully structured environment: Line operators make sure that nothing unexpected happens; today's motion planning strategies with complete information can be confidently used for tasks like robot welding or car body painting.

As explained above, the robot repair of the Hubble Telescope is slightly to the right of this extreme. Just about all information that the robot will need is known beforehand. But surprises—including some that may be hard to see from the ground—cannot be ruled out and must be built in the mission system

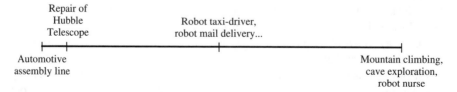

Figure 1.1 An increase in uncertainty, from a fully structured environment to a fully unstructured environment, spells an increase in difficulty when attempting to automate a task using robots.

design. In comparison with this task, designing a robot taxi driver carries much more uncertainty and hence more difficulty. Though the robot driver will have electronic maps of the city, and frequent remote updates of the map will help decrease the uncertainty due to construction sites or street accidents, there will still be a tremendous amount of uncertainty caused by less than ideally careful human car drivers, bicyclists, children running after balls, cats and dogs and squirrels crossing the road, potholes, slippery road, and so on. These will require millions of motion planning decisions done on the fly. Still, a great many objects that surround the robot are man-made and well known and can be preprocessed.

Not so with mountain climbing—this task seems to present the extreme in unstructured environment. While the robot climber would know exactly where its goal is, its every step is unlike the step before, and every spike driven in the wall may be the last one—solely due to the lack of complete input information. A tremendous amount of sensing and appropriate intelligence would be needed to compensate for this uncertainty. While seemingly a world apart and certainly not as dangerous, the job of a robot nurse would carry no less uncertainty. Similar examples can be easily found for automating tasks in agriculture, undersea exploration, at a construction site on Earth or on the moon, in a kindergarten, and so on.[1]

In terms of Figure 1.1, this book can be seen as an attempt to push the envelope of what is possible in robotics further to the right along the uncertainty line. We will see, in particular, that the technology that we will consider allows the robot to operate at the extreme right in Figure 1.1 in one specific sense—it makes a robot safe to itself and to its environment under a very high level of uncertainty. Given the importance of this feature and the fact that practically all robots today operate at the line's extreme left, this is no small progress. Much, but certainly not everything, will also become possible for robot motion planning under uncertainty.

What kind of input information and what kind of reasoning do we humans use to plan our motion? Is this an easy or is it a difficult skill to formalize and pass along to robots? What is the role of sensing—seeing, touching, hearing—in this process? There must be some role for it—we know, for instance, that when a myopic person takes off his glasses, his movement becomes more tentative and careful. What is the role of dynamics, of our mass and speed and accelerations relative to the surrounding objects? Again, there must be some role for it—we slow down and plan a round cornering when approaching a street corner. Are we humans universally good in motion planning tasks, or are some tasks more difficult for us than others? How is it for robots? For human–robot teams?

Understanding the issues behind those questions took time, and not everything is clear today. For a long time, researchers thought that the difficulties with motion planning are solely about good algorithms. After all, if any not-so-smart animal can successfully move in the unstructured world, we got to be able to teach our robots to do the same. True, we use our eyes and ears and skin to sense the

---

[1] The last example brings in still another important dimension: The allowed uncertainty depends much on what is at stake.

environment around us—but with today's technology, don't we have more than enough sensor gadgetry to do the job?

The purpose of this book is to identify those difficulties, see why they are so hard, attempt solutions, and try to identify directions that will lead us to conquering the general problem. A few points that will be at the center of our work should be noted. First, we will spend much effort designing motion planning algorithms. This being the area that humans deal with all the time, it is tempting to try to use human strategies. Unfortunately, as often happens with attempts for intelligent automation, asking humans how they do it is not a gratifying experience. Similar to some other tasks that humans do well (say, medical diagnostics), we humans cannot explain well how we do it. Why did I decide to walk around a table this way and not some other way, and how did this decision fit into my plan to get to the door? I can hardly answer. This means that robot motion planning strategies will not likely come from learning and analysis of human strategies. The other side of it is, as we will see, that often humans are not as good in motion planning as one may think.

Second, the above example with moving in the dark underlines the importance of sensing hardware. Strategies that humans and animals use to realize safe motion in an unstructured environment are intimately tied to the sensing machinery a species possesses. When coming from the outside into a dark room, your movement suddenly changes from brisk and confident to slow and hesitant. Your eyes are of no use now: Touching and listening are suddenly at the center of the motor control chain. Your whole posture and gait change. If audio sources disappear, your gait and behavior may change again. This points to a strong connection between motion planning algorithms and sensing hardware. The same has to be true for robots.

We will see that today's sensing technology is far from being adequate for the task in hand. In an unstructured environment, a trouble may come from any direction and affect any point of the robot body. Robot sensing thus has to be adequate to protect the robot's whole body. This calls for a special sensing hardware and specialized sensor data processing. One side effect of this circumstance is that algorithms and sensing hardware are to be addressed in the same book—which is not how a typical textbook in robotics is structured. Hence we hope that a reader knowledgeable in the theory of algorithms will be tolerant of the material on electronics, and we also hope that a reader comfortable with electronics will be willing to delve into algorithms.

Third, human and animals' motion planning is tied to the individual's kinematics. When bending to avoid hitting a low door opening, one invokes multiple sequences of commands to dozens of muscles and joints, all realized in a complex sequence that unfolds in real time. Someone with a different kinematics due to an impaired leg will negotiate the same door as skillfully though perhaps very differently. Expect the same in robots: Sensor-based motion planning algorithms will differ depending on the robot kinematics.

Aside from raising the level of robot functional sophistication, providing a robot with an ability to operate in an unstructured world amounts to a jump in its universality. This is not to say that a robot capable of moving dirty dishes from the table to a dishwasher will be as skillful in cutting dead limbs from trees. The higher universality applies only to the fact that the problem of handling uncertainty is quite generic in different applications. That is, different robots will likely use very similar mechanisms for collision avoidance. A robot that collects dishes from the table can use the same basic mechanism for collision avoidance as a robot that cuts dead limbs from trees.

As said above, we are not there yet with commercial machines of this kind. The last 40 years of robotics witnessed a slow and rather painful progress—much slower, for example, than the progress in computers. Things turned out to be much harder than many of us expected. Still, today's robots in automation-intensive industries are highly sophisticated. What is needed is supplying them with an ability to survive in an unstructured world. There are obvious examples showing what this can give. We would not doubt, for example, that, other issues aside, a robot can move a scalpel inside a patient's skull with more precision than a human surgeon, thus allowing a smaller hole in the skull compared to a conventional operation. But, an operating room is a highly unstructured environment. To be useful rather than to be a nuisance or a danger, the robot has to be "environment-hardened."

There is another interesting side to robot motion planning. Some intriguing examples suggest that it is not always true that robots are worse than people in space reasoning and motion planning. Observations show that human operators whose task is to plan and control complex motion—for example, guide the Space Shuttle arm manipulator—make mistakes that translate into costly repairs. Attempts to avoid such mistakes lead to a very slow, for some tasks unacceptably slow, operation. Difficulties grow when three-dimensional motion and whole-body collision avoidance are required. Operators are confused with simultaneous choices—say, taking care of the arm's end effector motion while avoiding collision at the arm's elbow. Or, when moving a complex-shaped body in a crowded space, especially if facing simultaneous potential collisions at different points of the body, operators miss good options. It is known that losing a sense of direction is detrimental to humans; for example, during deep dives the so-called Diver's Anxiety Syndrome interferes with the ability of professional divers to distinguish up from down, leading to psychological stress and loss in performance.

Furthermore, training helps little: As discussed in much detail in Chapter 7, humans are not particularly good in learning complex spatial reasoning tasks. These problems, which tend to be explained away as artifacts of poor teleoperation system design or insufficient training or inadequate input information, can now be traced to the human's inherent relatively poor ability for spatial reasoning.

We will learn in Chapter 7 that in some tasks that involve space reasoning, robots can think better than humans. Note the emphasis: We are *not* saying that robots can *think faster* or *compute more accurately* or *memorize more data* than humans—we *are* saying that robots can *think better* under the same conditions.

This suggests a good potential for a synergism: In tasks that require extensive spatial reasoning and where human and robot thinking/planning abilities are complementary, human–robot teams may be more successful than each of them separately and more successful than today's typical master–slave human–robot teleoperation systems are. When contributing skills that the other partner lacks, each partner in the team will fully rely on the other. For example, a surgeon may pass to a robot the subtask of inserting the cutting instrument and bringing it to a specific location in the brain.

There are a number of generic tasks that require motion planning. Here we are interested in a class of tasks that is perhaps the most common for people and animals, as well as for robots: One is simply requested to go from location A to location B, typically in an environment filled with obstacles. Positions A and B can be points in space, as in mobile robot applications, or, in the case of robot manipulators, they may include positions of every limb.

Limiting our attention to the go-from-A-to-B task leaves out a number of other motion planning problems—for example, terrain coverage, map-making, lawn mowing [1]; manipulation of objects, such as using the fingers of one's hand to turn a page or to move a fork between fingers; so-called power grips, as when holding an apple in one's hand; tasks that require a compressed representation of space, such as constructing a Voronoi diagram of a given terrain [2]; and so on. These are more specialized though by no means less interesting problems.

The above division of approaches to the go-from-A-to-B problem into two complementary groups—(1) motion planning with complete information and (2) motion planning with incomplete information—is tied in a one-to-one fashion to still another classification, along the scientific tools in the foundation of those approaches. Namely, strategies for motion planning with complete information rely exclusively on *geometric tools*, whereas strategies for motion planning with incomplete information rely exclusively on *topological tools*. Without going into details, let us summarize both briefly.

1. *Geometric Approaches.* These rely, first, on geometric properties of space and, second, on complete knowledge about the robot itself and obstacles in the robot workspace. All those objects are first represented in some kind of database, typically each object presented by the set of its simpler components, such as a number of edges and sides in a polyhedral object. According to this approach, then, passing around a hexagonal table is easier than passing around an octagonal table, and much easier than passing around a curved table, because of these three the curved table's description is the most complex.

Then there is an issue of information completeness. We can hear sometimes, "I can do it with my eyes shut." Note that this feat is possible only if the objects involved are fully known beforehand and the task in hand has been tried many times. A factory assembly line or the list of disassembly of an aircraft engine are examples of such structured tasks. Objects can be represented fully only if they allow a final size (practical) description. If an object is an arbitrary rock, then only its finite approximation will do—which not only introduces an error, but is in itself a nontrivial computational task.

If the task warrants a geometric approach to motion planning, this will likely offer distinctive advantages. One big advantage is that since everything is known, one should be able to execute the task in an optimal way. Also, while an increased dimensionality raises computational difficulties—say, when going from two-dimensional to three-dimensional space or increasing the robot or its workspace complexity—in principle the solution is still feasible using the same motion planning algorithm.

On the negative side, realizing a geometric approach typically carries a high, not rarely unrealistic, computational cost. Since we don't know beforehand what information is important and what is not for motion planning, everything should be in. As we humans never ask for "complete knowledge" when moving around, it is not obvious how big that knowledge can be even in simple cases. For example, to move in a room, the database will have to include literally every nut and bolt in the room walls, every screw holding a seat in every chair in the room, small indentations and extensions on the robot surface etc. Usually this comes to a staggering amount of information. The number of those details becomes a measure of complexity of the task in hand.

Attempts have been made to connect geometric approaches with incomplete sources of information, such as sensing. The inherent need of this class of approaches in a full representation of geometric data results in somewhat artificial constructs (such as "continuous" or "X-ray" or "seeing-through" sensors) and often leads to specialized and hard-to-ascertain heuristics (see some such ideas in Ref. 3).

With even the most economical computational procedures in this class, many tasks of practical interest remain beyond the reach of today's fastest computers. Then the only way to keep the problem manageable is to sacrifice the guarantee of solution. One can, for example, reduce the computational effort by approximating original objects with "artificial" objects of lower complexity. Or one can try to use some beforehand knowledge to prune nonpromising path options on the connectivity graph. Or one can attempt a random or pseudorandom search, checking only a fraction of the connectivity graph edges. Such simplification schemes leave little room for directed decision-making or for human intuition. If it works, it works. Otherwise, a path that has been left out in an attempt to simplify the problem may have been the only feasible path. The ever-increasing power of today's computer make manageable more and more applications where having complete information is feasible.

The properties of geometric approaches can be summarized as follows (see also Section 2.8):

(a) They are applicable primarily to situations where complete information about the task is available.

(b) They rely on geometric properties (dimensions and shapes) of objects.

(c) They can, in principle, deliver the best (optimal) solution.

(d) They can, in principle, handle tasks of arbitrary dimensionality.

(e) They are exceedingly complex computationally in more or less complex practical tasks.

2. *Topological Approaches.* Humans and animals rarely face situations where one can approach the motion planning problem based on complete information about the scene. Our world is messy: It includes shapeless hard-to-describe objects, previously unseen settings, and continuously changing scenes. Even if faced with a "geometric"-looking problem, say, finding a path from point A to point B in a room with 10 octagonal tables, we would never think of computing first the whole path. We take a look at the room, and off we go. We are tuned to dealing with partial information coming from our sensors. If we want our robots to handle unstructured tasks, they will be thrown in a similar situation.

In a number of ways, topological approaches are an exact opposite of the geometrical approaches. What is difficult for one will be likely easy for the other.

Consider the above example of finding a path from point A to point B in a room with a few tables. The tables may be of the same or of differing shapes; we do not know their number, dimensions, and locations. A common human strategy may look something like this: While at A, you glance at the room layout in the direction of point B and start walking toward it. If a table appears on your way, you walk around it and continue toward point B. The words "walking around" mean that during this operation the table is on the same side from you (say, on the left). The table's shape is of no importance: While your path may repeat the table's shape, "algorithmically" it is immaterial for your walk around it whether the table is circular or rectangular or altogether highly nonconvex. Why does this strategy represent a topological, rather than geometric, approach? Because it relies implicitly on the topological properties of the table—for example, the fact that the table's boundary is a simple closed curve—rather than on its geometric properties, such as the table's dimensions and geometry.

We will see in Chapter 3 that the aforementioned rather simplistic strategy is not that bad—especially given how little information about the scene it requires and how elegantly simple is the connection between sensing and decision-making. We will see that with a few details added, this strategy can guarantee success in an arbitrarily complex scene; using this strategy, the robot will find a path if one exists, or will conclude "there is no path" if such is the case.

On the negative side, since no full information is available in this process, no optimality of the resulting path can be guaranteed. Another minus, as we will see, is that generalizations of such strategies to arm manipulators are dependent on the robot kinematics. Let us summarize the properties of topological approaches to motion planning:

(a) They are suited to unstructured tasks, where information about the robot surroundings appears in time, usually from sensors, and is never complete.

(b) They rely on topological, rather than geometrical, properties of space.

(c) They cannot in principle deliver an optimal solution.

(d) They cannot in principle handle tasks of arbitrary dimensionality, and they require specialized algorithms for each type of robot kinematics.

(e) They are usually simple computationally: If a technique is applicable to the problem in hand, it will likely be computationally easy.

1.2   BASIC CONCEPTS

This section summarizes terminology, definitions, and basic concepts that are common to the field of robotics. While some of these are outside of this book's scope, they do relate to it in one way or another, and knowing this relation is useful. In the next chapter this material will be used to expand on common technical issues in robotics.

1.2.1   Robot? What Robot?

Defining *what a robot is* is not an easy job. As mentioned above, not only scientists and engineers have labored here, but also Hollywood and fiction writers and professionals in humanities have helped much in diffusing the concept. While this fact will not stand in our way when dealing with our topic, starting with a decent definition is an old tradition, so let us try.

There exist numerous definitions of a robot. *Webster's Dictionary* defines it as follows:

> A robot is an automatic apparatus or device that performs functions ordinarily ascribed to humans, or operates with what appears to be almost human intelligence.

Half of the definition by *Encyclopaedia Britannica* is devoted to stressing that a robot does not have to look like a human:

> Robot: Any automatically operated machine that replaces human effort, though it may not resemble human beings in appearance or perform functions in a humanlike manner.

These definitions are a bit vague, and they are a bit presumptuous as to what is and is not "almost human intelligence" or "a humanlike manner." One senses that a chess-playing machine may likely qualify, but a machine that automatically digs a trench in the street may not. As if the latter does not require a serious intelligence. (By the way, we do already have champion-level chess-playing machines, but are still far from having an automatic trench-digging machine.) And what about a laundry washing machine? This function has been certainly "ordinarily ascribed to humans" for centuries. The emphatic "automatic" is also bothersome: Isn't what is usually called an operator-guided teleoperation robot system a robot in spite of not being fully automatic?

The Robotics Institute of America adds some engineering jargon and emphasizes the robot's ability to shift from one task to another:

A robot is a reprogrammable multifunctional manipulator designed to move material, parts, tools, or specialized devices through variable programmed motions for the performance of a variety of tasks.

Somehow this definition also leaves a sense of dissatisfaction. Insisting solely on "manipulators" is probably an omission: Who doubts that mobile vehicles like Mars rovers are robots? But "multifunctional"? Can't a robot be designed solely for welding of automobile parts? And then, is it good that the definition easily qualifies our familiar home dishwasher as a robot? It "moves material" "through variable programmed motions," and the user reprograms it when choosing an appropriate cycle.

These and other definitions of a robot point to dangers that the business of definitions entails: Appealing definition candidates will likely invite undesired corollaries.

In desperation, some robotics professionals have embraced the following definition:

I don't know what a robot is but will recognize it when I see one.

This one is certainly crisp and stops further discussion, but it suffers from the lack of specificity. (Try, for example, to replace "robot" by "grizzly bear"—it works.)

A good definition tends to avoid explicitly citing material components necessary to make the device work. That should be implicit and should leave enough room for innovation within the defined function. Implicit in the definitions above is that a robot must include *mechanics* (body and motors) and a *computing device*. Combining mechanics and computing helps distinguish a robot from a computer: Both carry out large amounts of calculations, but a computer has information at its input and information at its output, whereas a robot has information at its input and motion at its output.

Explicitly or implicitly, it is clear that *sensing* should be added as the third necessary component. Here one may want to distinguish *external sensing* that the machine uses to acquire information about the surrounding world (say, vision, touch, proximity sensing, force sensing) from *internal sensing* used to acquire information about the machine's own well-being (temperature sensors, pressure sensors, etc.). This addition would help disqualify automobiles and dishwashers as robots (though even that is not entirely foolproof).

Perhaps more ominously, adding "external" sensing as a necessary component may cause devastation in the ranks of robots. If the robot uses sensing to obtain information about its surroundings, it would be logical to suggest that it must be using it to react to changes in the surrounding world. The trouble is that this innocent logic disqualifies a good 95–98% of today's robots as robots, for the simple reason that all those robots are designed to work in a highly structured environment of a factory floor, which assumes no unpredictable changes.

With an eye on the primary subject of this book—robots capable of handling tasks in an unstructured environment—we accept that reacting to sensing data is essential for a robot's being a robot. The definition of a robot accepted in this text is as follows:

A robot is an automatic or semiautomatic machine capable of purposeful motion in response to its surroundings in an unstructured environment.

Added in parentheses or seen as unavoidably tied to the defined ability is a clause that a robot must include mechanical, computing, and sensing components.

While this definition disqualifies many of today's robots as robots, it satisfies what for centuries people intuitively meant by robots—which is not a bad thing. Purists may still point to the vagueness of some concepts, like "purposeful" (intelligent) and "unstructured." This is true of all other attempts above and of human definitions in general. Be it as it may, for the purpose of this book this is a working definition, and we will leave it at that.

1.2.2 Space. Objects

A robot operates in its *environment* (*workspace, work cell*). The real-world robot environment appears either in *two-dimensional space* (2D), as, for example, with a mobile robot moving on the hospital floor, or in *three-dimensional space* (3D), as with an arm manipulator doing car body painting.

Robot workspace is physical continuous space. Depending on approaches to motion planning, one can model the robot workspace as *continuous* or *discrete*. Robotics deals with moving or still *objects*. Each object may be

- A point—for example, an abstract robot automaton used for algorithm development
- A rigid body—for example, boxes in a warehouse, autonomous vehicles, arm links
- A hinged body made of rigid bodies—for example, a robot arm manipulator

The robot environment may includes *obstacles*. Obstacles are objects; depending on the model used and space dimensionality, obstacles can be

- Points
- Polygonal (polyhedral) objects, which can be rigid or hinged bodies
- Other analytically described objects
- Arbitrarily shaped (physically realizable) objects

1.2.3 Input Information. Sensing

Similar to humans and animals, robots need *input information* in order to plan their motion. As discussed above, there may be two situations: (a) Complete information about all objects in the robot environment is available. (b) There is *uncertainty* involved; then the input information is, by definition, incomplete and is likely obtained in real time from robot's sensors.

Note the algorithmic consequences of this distinction. If complete information about the workspace is available, a reasonable method to proceed is to build a model of the robot and its workspace and use this model for motion planning. The significant effort that is likely needed to build the model will be fully justified by the path computed from this model. If, however, nothing or little is known beforehand, it makes little sense to spend an effort on building a model that is of doubtful relevance to reality.

In the above situation (b), the robot hence needs to "think" differently. From its limited sensing data, it may be able to infer some topological properties of space. It may be able to infer, for example, whether what it sees from its current position as two objects are actually parts of the same object. If the conclusion is "yes," the robot will not be trying to pass between these two "objects." If the conclusion is "no," the robot will know that it deals with separate objects and may choose to pass between them. The objects' actual shapes will be of little concern to the robot.

What type of sensing is suitable for a competent motion planning? It turns out that just about any sensing is fine: tactile, sonar, vision, laser ranger, infrared proximity, and so on. We will learn a remarkable result that says that even the simplest tactile sensing, when used with proper motion planning algorithms, can guarantee that the robot will reach its target (provided that the target is reachable). In fact, we will consistently prefer tactile sensing when developing algorithms, before attempting to use some richer sensing media; this will allow us to clarify the issues involved. This is not to say that one should prefer tactile sensors in real tasks: As a blind person will likely produce a more circuitous route than a person with vision, the same will be true for a robot.

Being serious about collision avoidance means that robot motion planning algorithms must protect the whole robot body, every one of its points. Accordingly, robot sensors must provide sufficient input information. Intuitively, this requirement is not hard to understand for mobile robots. Existing mobile robots typically have a camera or a range finder that rotates as needed, or sonar sensors that cover the whole robot's circumference.

Intuition is less helpful when talking about arm manipulators. Again, sensors can be of any type: tactile, proximal, vision, and so on. What is harder to grasp but is absolutely necessary is a guarantee that the arm has sensing data regarding all points of its body. No blind spots are allowed.

We tend not to notice how strictly this requirement is followed in humans and animals. We often tie our ability to move around solely with our vision. True, when I walk, my vision is typically the sole source of input information. I may not be aware of, and not interested in, objects on my sides or behind me. If something worthwhile appears on the sides, I can turn my head and look there.

However, if I attempt to sit down and the seat will happen to have a nail sticking out of it, I will be quickly made aware of this fact and will plan my ensuing motions quickly and efficiently. If a small rock finds its way into my shoe, I will react equally efficiently. The sensor that I use in these cases is not vision, but is the great many tactile sensors that cover my whole skin. Vision alone would never be able to become a whole-body sensor.

Think of this: If among millions and millions of spots inside and outside of our bodies at least one point would not be "protected" by sensing, in our unstructured messy world sooner or later that very point would be assaulted by some hostile object. Evolution has worked hard on making sure that such situations do not occur. Those of our forbears millions of years ago who did not have a whole-body sensing have no offspring among us.

The fact that losing some type of sensing is a heavy blow to one's lifestyle is a witness to how important all our sensing systems are. Blind people have to make special precautions and go through special training to be able to lead a productive life. People suffering from diabetes may incur a loss of tactile facilities, and then they are warned by their doctors to be extra careful when handling objects: A small cut may become a life-threatening wound if one's sensors sound no alarm.

People tend to think that vision is more essential for one's survival than tactile sensing. Surprisingly, the reality is the other way around. While many blind people around us have productive lives, the human ability to function decreases much more dramatically if their tactile system is seriously damaged. This has been shown experimentally [4]. Today's knowledge suggests that a person losing his or her tactile facilities completely will not be able to survive, period.

Animals are similarly vulnerable. Some are able to overcome the deficit, but only at a high cost. If a cat loses its tactile sensing (say, if the nerve channel that brings tactile information to the brain is severed), the cat can relearn some operational skills, but its locomotion, gait, behavior, and interaction with the surrounding world will change dramatically [5].[2]

One can speculate that the reason for a higher importance of tactile sensing over vision for one's survival is that tactile sensing tends to have no "blind spots" whereas vision does by definition have blind spots. Vision improves the efficiency of one's interaction with the environment; tactile sensing is important for one's very survival. In other words, our requirement of a whole-body robot sensing is much in line with live nature.

Similarly, it is not uncommon to hear that for robot arm manipulators "vision should be enough." Vision is not enough. Sooner or later, some object occluded from the arm's cameras by its own links or by power cables will succeed in coming through, and a painful collision will occur. Whole-body sensing will prevent this from happening. This suggests that our robots need a *sensitive skin* similar to human skin, densely populated with many sensors.

Whether those sensors are tactile or proximal, like infrared sensors, is a matter of efficiency, not survival. Be it as it may, motion planning algorithms developed for simpler sensing can then be expanded to more sophisticated sensing. This point is worth repeating, because misunderstanding is not uncommon:

---

[2]To be sure, nature has developed means to substitute for an incomplete sensing system. A turtle's shell makes tactile sensing at its back unnecessary. Such examples are rare, and they look more like exceptions confirming the rule.

When we develop our motion planning algorithms based on tactile sensing, this does not mean we suggest tactile sensing as a preferable sensing media, nor does it mean that the algorithms are applicable solely to tactile sensing. As we will see, expanding algorithms to more complex sensing is usually relatively easy, and usually results in higher efficiency.

1.2.4 Degrees of Freedom. Coordinate Systems

It is known from mechanics that depending on space dimensionality and object complexity, there is a minimum number of independent variables one needs to define the object's position and orientation in a unique way. These variables are called the object's *degrees of freedom (DOF)*. The reference (coordinate) system expressed in terms of object's DOF is called the *configuration space (C-space)*. *C*-space is hence a special representation of the robot *workspace (W-space)*. From a textbook on mechanics, the minimum number of DOF that a rigid body needs for an arbitrary motion is

In 2D, if only translation is allowed:	2
In 2D, translation plus orientation allowed:	3
In 3D, if only translation is allowed:	3
In 3D, translation plus orientation allowed:	6

For example, for a planar (2D) case with a rigid object free to translate and rotate, the object is defined by three DOF (x, y, θ): two Cartesian coordinates (x, y) that define the object's position, plus its orientation angle θ.

A robot arm manipulator's DOF also determine its ability to move around. Specific values of all robot's DOF signify the specific *arm configuration* of its links and joints. Shown in Figure 1.2a is a revolute planar (2D) arm with two links. Its two DOF, two rotation angles, allow an arbitrary position of its endpoint in the robot workspace, but not an arbitrary orientation. The 3-link 3-DOF planar arm manipulator shown in Figure 1.2b can provide an arbitrary position and an arbitrary orientation at its endpoint.

The DOF that a robot arm possesses are usually realized via independent control means, such as *actuators* (motors), located in the arm's *joints*. Joints connect together the arm's *links*. Links and joints can be designed in different configurations: The most common are the *sequential linkage*, which is similar to the kinematics of a human arm, like in Figure 1.2, and the *parallel linkage*, where links form a parallel structure. The latter is used in some spatial applications, such as a universal positioner for various platforms. We will be interested in only sequential linkages.

The most popular types of joints are *revolute joints*, where one link rotates relative to the other (like in the human elbow joint), and *sliding joints* (also called *prismatic joints*), where one link slides relative to the other. The arm shown in Figure 1.2a has two revolute joints, of which the first joint is located in the arm's fixed *base*. The freely moving distal link or links on a typical arm manipulator is called the *end effector*. The end effector can carry a tool for doing the robot's

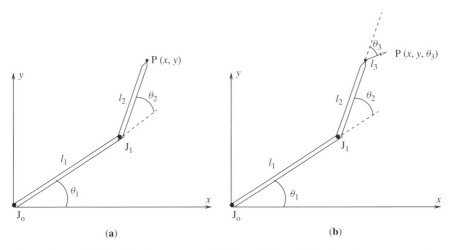

Figure 1.2 (a) A simple planar arm manipulator with two links (l_1, l_2), and two revolute joints (J_0, J_1). The robot base coincides with joint J_0 and is fixed in 2D space. This 2 DOF manipulator can bring its end effector to any point $P(x, y)$ within its workspace; the end effector's orientation at each such point will then be defined by the orientation of link l_2. (b) Same arm with added third revolute link (hand), l_3; now the end effector can be put in any arbitrary position of workspace with arbitrary orientation, $P(x, y, \theta_3)$.

job; this can be some kind of a gripper, a screwdriver, a paint or welding gun, and so on. The arm end effector may have its own DOF.

To use tools, the end effector needs to be put not only in a particular position but also at a particular orientation: For example, not only the screwdriver's blade has to be in the screw's slot, but its axis has to be perpendicular to the surface into which a screw is driven. In the human arm, it is of course the hand that handles the tool orientation, a relatively compact device, compared with the rest of the arm, with its own (many) DOF.

A similar organization of kinematics is common to robot arm manipulators: The robot's DOF are divided into the *major linkage*, relatively long links whose function is to bring the end effector to the vicinity of the job to be done, and the more compact *minor linkage* (called the *wrist* or *hand*), which is the end effector proper. Although from the theoretical standpoint such division is not necessary, it is useful from the design considerations. Often a real arm's major linkage consists of three DOF, and its minor linkage presents a 3-DOF hand.

The values of all robot arm DOF define its coordinates in physical space. A given position plus orientation can be described via two coordinate (reference) systems: *Cartesian coordinates* and *joint (joint space) coordinates*. The latter are also called *configuration space coordinates*. As long as both systems describe the robot configuration—that is, position plus orientation—in a unique way, they are equivalent. For the 3-link 3-DOF arm manipulator shown in Figure 1.2b, its

Cartesian coordinates are (x, y, θ_3), and its configuration space coordinates are $(\theta_1, \theta_2, \theta_3)$.

Typically, the user is interested in defining the robot's paths in terms of Cartesian coordinates. Robot's motors are controlled, however, in terms of joint values (that is, configuration space coordinates). Hence a standard task in robot motion planning and control is translation from one reference system to the other. This process gives rise to two problems: (a) *direct kinematics*—given joint values, find the corresponding Cartesian coordinates—and (b) *inverse kinematics*—given the robot's Cartesian coordinates, find the corresponding joint values. As we will see in the next chapter, calculation of inverse kinematics is usually significantly more difficult than the calculation of direct kinematics.

1.2.5 Motion Control

The robot's *path* is a curve that the robot's end effector (or possibly its some other part) moves along in the robot workspace. To be physically realizable, each point of the path must be associated with the joint values that fully describe the robot position and orientation in the respective configuration. The term *trajectory* is used sometimes to designate a path geometry plus timing, velocity, and acceleration information along the path.[3]

As used in robotics, the term *motion control* or *motion control system* refers to the lower-level control functions, such as algorithmic and electronic and mechanical means that direct individual motors, as opposed to *motion planning*, which signifies the upper-level control—that is, control that requires some intelligence. This is not to say that motion control is a simple matter—robot controllers are often quite sophisticated. The control means are used to realize a given path or trajectory with required fidelity. While control means are beyond the scope of this book, for completeness we will review them briefly in the next chapter.

Depending on the number of DOF available for motion planning, we distinguish between three types of systems:

- *Holonomic Systems.* These have enough DOF for an arbitrary motion. The minimum number of those is equal to the dimensionality of the corresponding *C*-space: For example, 6 is the minimum number of DOF a 3D arm manipulator needs to realize an arbitrary motion in space without obstacles.
- *Nonholonomic Systems.* These are systems with constraints on their motion. For example, a car is a nonholonomic system: with its 2-DOF control—forward motion and steering—it cannot execute a lateral motion; this creates a well-known difficulty in parallel car parking. Note that a car's *C*-space is 3D, with its axes being two position variables (x, y) plus the orientation angle.
- *Redundant Systems.* Those with the number of DOF well above the minimum necessary for holonomic motion. Humans, animals, and some complex robots present redundant systems.

---

[3]In some books, and also here, terms "trajectory" and "path" are used interchangeably.

A serious analysis of holonomic and nonholonomic systems requires more rigorous mathematical definitions. These will be introduced later as needed.

1.2.6   Robot Programming

A robot executes a given motion because it is *programmed* to do so. The meaning of the words *robot motion programming* is not dissimilar to what an adult does when teaching a child how to walk.

One can distinguish between two basic approaches to robot motion programming:

- *Explicit robot motion programming*—when every robot configuration along the path is prescribed explicitly. One variation of this is when a set of configurations is given explicitly beforehand, and the robot interpolates configurations between the set points using some rule.
- *Task-level robot motion programming*—in which contents-based subtasks are given, such as "Grasp a part" or "Insert a peg in a hole," and the robot figures out further details on its own.

The "subtask" above can be a complex motion or procedure that has been programmed separately beforehand. For example, details of the task "Grasp a part" may differ from one task instantiation to the other, depending on the sensing data. Clearly, task-level programming is, in general, preferable to explicit programming. It is also significantly more difficult to realize because it requires much beforehand knowledge on the part of the robot.

The "programming" of a dancer or a gymnast is clearly closer to the task-level approach than to the explicit programming. The choreographer can, for example, say to the ballerina, "Here you do a pirouette followed by an arabesque." A pirouette is a rather complex combination of little motions that the ballerina learned while at school. The motions have been "programmed" into her, so just naming it is sufficient for her to know what to do. The same is so for an arabesque. On the other hand, this does not mean the pirouette will be exactly the same at all times: For instance, the ballerina may slightly deviate from her usual pirouette when seeing another dancer backing up toward her.

Another classification of robot motion programming is given by different types of robot teaching systems. A robot teaching system is a specific technique for robot programming. The following list applies primarily to robot arm manipulators; the corresponding analogues for mobile robots are simpler and present a subset of this list:

- *Manually guiding* a robot through the path in real time
- *Point-to-point guidance*, with an automatic generation of the time pattern
- *Teach pendant*
- *Off-line programming*, procedural languages
- *Automatic programming* using the task database

In *manual guidance* systems, a specially trained technician grabs the end tool of the robot and performs the actual operation by moving it along the required motion. The system automatically records all robot configurations along the path, which can then be reproduced faithfully. The arm may be specially mechanically balanced for easier motion, or be even completely replaced by a mockup arm that includes all electronic means to document the motion in system's memory. This is similar to the situation when the choreographer physically moves a dancer's arm through the air, "That's how you do it." For example, in preparation for a robot painting of a car body, an experienced human painter moves the arm with the attached painting gun through the necessary motion, actually painting a car body; the recorded motion is then used to paint a batch of car bodies.

While looking attractively simple, manual guidance systems are hard to realize. For example, during actual job execution the robot must produce a perfectly painted surface after a single motion, whereas a human painter can usually use his powerful visual feedback to detect mistakes along the way and then touch the paint here and there if needed. Various techniques have been designed to mathematically "massage" the technician-taught motion to perfection—for example, to assure the path curve smoothness or the robot end effector uniform speed.

The *point-to-point teaching* is a variation of the previous technique that disposes with the real-time teaching. Here the human "teacher" brings the robot end tool into the right position, pushes a button to save the corresponding robot configuration in the robot memory, and goes to the next point, and so on, until a set of points representing the whole path is accumulated. The set must then be "massaged" in the above fashion, which is more difficult than in a manual guidance system because the teaching session was removed from the real-time operation and hence likely misses some important dynamic characteristics.

The *teach pendant* is a hardware accessory for point-to-point teaching. The pendant is a small box connected with the robot by a cable, with a variety of buttons for the operator to generate robot configurations. By intermittently giving increments in robot joint values—or alternatively, in Cartesian positions and orientations of the robot tool—the operator brings the end effector to the desired position, pushes a button to save it, and goes to the next position. This is a tedious process: A reasonably complex path—say, painting a car engine compartment—may require 150–200 or more points, each requiring 40–60 button pushes to produce it. The resulting path will likely need a considerable preprocessing by a special software before being ready for actual use. Most of today's industrial robot programming systems are of this kind.

The *off-line programming* method is a logical and rather dramatic departure from the techniques above, in that it tries to address their shortcomings by delegating the whole robot programming work to software. After all, each robot configuration along the path is a function of the required path, which is in turn a function of the task to which the motion applies. This motion can in principle be coded in some specialized programming language, the way we write computer programs. Hundreds of robot programming languages have been developed in the last three decades. For a while, some of them became "widely known in narrow circles" of robotics engineers; today almost none of them are remembered. Why so, especially given the remarkable success of computer programming languages?

The main reason for this is, one might say, a linguistic inadequacy of such languages to the problem of describing a motion. The product of a human oral or written speech, or of a computer programming language, is a linear, one-dimensional, discreet set of signals—sounds if spoken and symbols if written. A motion, on the other hand, happens in two- or three-dimensional space and is a continuous phenomenon. It is very difficult to describe in words, or in terms of a computer program, a reasonably complex two- or three-dimensional curve (unless it has a mathematical representation). Try to show your friend a motion—say, try to wave your hand goodbye. Then ask your friend to repeat this motion; he will likely do it quite close to your original motion. Now try to describe this same motion on paper with words. Take your time. Once ready, give your description to another friend, who did not see your motion, and ask him to reproduce the motion from this description. (Writing "Please wave your hand goodbye" is, of course, not allowed). The result will likely be far from the original.

This is undoubtedly the reason why we will never know how people danced in ancient Egypt and Greece and Rome, and even in Europe at the end of the XIX century, until the appearance of moving motion cameras. Unlike the millennia-old alphabets for recording human speech, alphabets for describing motion have been slow to come. *Labanotation*, the first system for recording an arbitrary (but only human) motion, appeared only in the mid-1920s and is rather clumsy and far from perfect.

The *automatic programming* technique is a further development in robot teaching techniques, and it is even further removed from using real motion in teaching. Take an example of painting the car engine compartment for a given car model. The argument goes as follows. By the time the robot painting operation is being designed, complete description of the car body is usually in a special database, as a result of the prior design process. Using this database and the painting system parameters (such as dimensions of the paint spray), a special software package can develop a path for the painting gun, and hence for the robot that holds the gun, such that playing that path would result in a complete and uniform paint coverage of the engine compartment. There is no need to involve humans in the actual motion teaching. Only a sufficiently sophisticated manufacturing environment can benefit from this system: Even with the right robot, one will have hard time producing a database necessary to paint one's backyard fence.

While showing an increasing sophistication from the first to the last robot teaching techniques above, from manual guidance to automatic programming of robot motion, each of these techniques has its advantages and its shortcomings. For example, no other techniques can match the ingenious teaching-by-showing ability of the first, manual guidance, method. This has led some researchers to attempt combination techniques from the list above. For example, first a vision system would record the human manually guided motion, and then a special software would try to reproduce it. In conclusion, in spite of a long history (in relative terms of the robotics field), the robot teaching methods can still be said to be in their infancy.

The data "massaging" techniques mentioned above are widely used in manufacturing robot systems. These include, for example,

- *Path smoothing*
- *Straight line interpolation*
- *Achieving a uniform velocity path*
- *Manipulating velocity/acceleration profiles* along the path

Path smoothing is usually done to improve the system performance. Smoothing the first and second derivatives of the robot path will help avoid jerky motion and sharp turns.

Straight-line interpolation is something different. Many applications—for example, welding two straight line beams along their length—require a straight-line path. Arms with revolute joints, such as in Figure 1.2, tend to move along curved path segments, so approximating a straight-line path takes special care. This is a tedious job, and we will consider it further in the next chapter. The human arm has a similar problem, though we often are not aware of this: Humans are not good in producing straight lines, even with the powerful feedback control help of one's vision.

A *uniform velocity path* may be needed for various purposes. For a quality weld in continuous welding, the robot has to move the gun with the uniform speed. In the example above with the painting robot, a nonuniform velocity of the painting gun will produce streaks of thinner and thicker paint on the surface that is being painted. Furthermore, note that the meaning of "uniform velocity" in this example must refer not to the velocity at the painting gun endpoint, but to the velocity at an imaginary point in space where paint meets the painted surface. That is the gun aiming point (say, 20 cm away from the gun's endpoint) that has to move with the uniform velocity. This may coincide with the gun endpoint moving sometimes faster and sometimes slower, and sometimes even stopping, with the gun rotating in space.

Manipulating velocity/acceleration profiles presents an extension of the velocity control. Some tasks may require control of the robot linear or angular acceleration—for example, to ascertain a certain pattern of starting and finishing a motion. A good robot system will likely include software that allows creating various profiles of robot velocity and acceleration.

1.2.7 Motion Planning

Motion planning is the single unique defining core of the field of robotics—same as computation is the single unique defining core of the field of computers. Many components and disciplines contribute to producing a good robot—the same is true for a good computer—but it is motion planning that makes a robot a robot.

There are different criteria of quality of robot paths. We may want the robot to do one or more of these:

- Execute a predefined path.
- Find an optimal path (the shortest, or fastest, or one requiring a minimum energy, etc.).
- Plan a "reasonable" path.
- Plan a path that respects some constraints—say, a path that would not make the robot bang into the walls of an automotive painting booth.

Robots in the factory environment tend to follow predefined paths, sometimes with deviations allowed by their programs. Robot car painting operation is a good example. Such tasks often put a premium on path optimization: After all, in a mass production environment, shaving 1–2 sec out of a 50-sec cycle can translate into large savings. On the other hand, for a robot operating in an environment with uncertainty, optimality is ruled out and more often than not is of little concern anyway. For instance, a mobile robot that is used for food and drug delivery in a hospital is expected to go along more or less reasonable, not necessarily optimal, paths. Either of these systems can also be subject to additional constraints: For example, an arm manipulator may need to work in a narrow space between two walls.

CHAPTER 2

A Quick Sketch of Major Issues in Robotics

Personally I'm always ready to learn, although I do not always like being taught.

— Winston Churchill

The material in this chapter is given primarily as a review and is structured quite similar to such reviews elsewhere (see, for example, Ref. 6). Some sections—in particular, Sections 2.5 and 2.6—are only tangentially relevant to our main topic of sensor-based motion planning, so they can be just glanced through or skipped by those who have had some introductory course in robotics. Those who have not are suggested to go through this chapter more carefully.

Robotics is a multidisciplinary field. It deals with a multiplicity of issues, and its tools relate to various disciplines, from mechanical engineering to computer science to mathematics to human factors. The issues covered in this chapter relate to generating a desired motion—that is, motion that would bring the robot to the right destination, with acceptable dynamics and collision-free. Addressing the reader who knows little about robotics, our goal is to give at least a perfunctory understanding of areas that relate to motion planning. Besides those, other areas may be as essential to a designer of robotic systems: object manipulation (e.g., design of hands and appropriate intelligence); grasping, which in turn divides into precision grasping and power grasping (think of the difference between holding a pen or an apple); robot (computer) vision, and so on. The list of issues that we are about to review is as follows:

- Kinematics
- Statics
- Dynamics
- Feedback control
- Compliant motion
- Trajectory modification
- Motion planning and collision avoidance; navigation

Sensing, Intelligence, Motion, by Vladimir J. Lumelsky

Of these and other issues mentioned above, the last one, motion planning and collision avoidance, is the central problem in robotics—first, because it appears in just about any robotic task and application, and, second, because it appears to be the most "robotic" issue in robotics. Indeed, the other areas above have been developed in, and are of importance to, other engineering fields, not only to robotics, whereas the subject of motion planning and collision avoidance is unique to robotics. For example, kinematics, statics, and dynamics are central to the design of an immense variety of machines (of which robots are only a small part); feedback control is the central issue in control theory and control engineering; and so on.

Readers interested in deeper understanding of those and other issues are referred to other sources; some such will be cited in the sequel.

Consider a simple planar two-link arm (shown in Figure 2.1) that we will use in a few sections of this chapter. Here is some notation that we will use:

θ_1—shoulder angle

θ_2—elbow angle

J_0, J_1—arm joints

l_1, l_2—arm links

m_1, m_2—link masses

R—link radius

We will assume that link masses are distributed uniformly within each link. Besides axes x and y shown in the figure, imagine also an axis z perpendicular to the plane of the figure. Assume that axes of joints J_0 and J_1 are parallel to the axis z. (In this chapter we will not need axis z; it is mentioned here only to define the joint axes.)

The workspace of the arm in Figure 2.1 is a disk of radius $(l_1 + l_2)$. Because link l_1 is longer than link l_2, centered at the arm's base J_0 there is a dead zone of radius $|l_1 - l_2|$, no point of which can be reached by the arm endpoint b. Note that if l_1 happened to be shorter than link l_2, there would still be a dead zone of exactly the same radius $|l_1 - l_2|$. The arm's workspace is therefore the area sandwiched between the circles of radii $(l_1 + l_2)$ and $|l_1 - l_2|$. In case the arm links are of equal lengths, the arm's workspace is a circle of radius $(l_1 + l_2)$. If one or both arm joints are subject to constraints on their values, the workspace will change accordingly. This does happen with real arms; for example, the arm's joint angle θ_1 may be limited to the range $\pm 120°$.

The arm's endpoint b can occupy any point in the arm's workspace. When the arm is fully outstretched, its endpoint b is at the workspace outer circle boundary; when it is fully folded, its endpoint b is at the workspace inner circle boundary. Only one arm configuration corresponds to any such point. Any other position of endpoint b in the arm workspace corresponds to two arm configurations. The second arm configuration is shown by dashed lines in Figure 2.1. If $l_1 = l_2$, an infinite number of configurations can place the arm endpoint b at the base J_0, with $\theta_2 = \pi$.

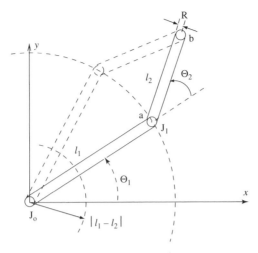

Figure 2.1 A planar two-link arm manipulator: l_1 and l_2 are links, with their respective endpoints a and b; J_0 and J_1 are two revolute joints; θ_1 and θ_2 are joint angles. Both links are of the same thickness $2R$. The robot's base coincides with the joint J_0 and is fixed.

Position of the endpoint b can be defined in the arm workspace either by two coordinates x and y in the Cartesian plane (x, y) or by two joint angles θ_1 and θ_1. Hence we distinguish two representations of an arm configuration, in *Cartesian space* (x, y) and in *joint space* (θ_1, θ_1).

An arm with more degrees of freedom or a three-dimensional arm will result in a higher complexity of those representations.

2.1 KINEMATICS

Kinematics describes the relationship between positions, velocities, and accelerations of a set of bodies—in our case, of the robot arm links.

While we are at it, let us also define the concepts of *statics* and *dynamics*, which often go together with kinematics when describing a body's motion, and which we will address in the following sections:

Statics describes (a) the relationship between forces and torques that, say, an arm manipulator exerts on the surrounding objects and (b) the relationship between internal forces and torques at the arm links.

Dynamics describes the relationship between kinematics and statics. For example, the relationship between torques at the arm joints and link positions represents the arm's dynamics.

For trajectory planning of robot arm manipulators, kinematics is especially important. Here is one reason for that. More often than not, people prefer to command arm's positions in terms of Cartesian coordinates—in our case the two coordinates (x, y)—whereas the arm control system expects them in terms of arm's joint values—in our case the angular joint values (θ_1, θ_1). Inversely, if the arm somehow—say, acting upon the sensor data—arrives at some position, from the arm's joints we obtain its joint angles, and we would like to know which position (x, y) in Cartesian space they correspond to (Figure 2.1). Hence there is a need to translate from one coordinate system to the other.

Accordingly, there are two relationships between these two coordinate representations:

Direct Kinematics. Given the values (θ_1, θ_1), find the corresponding Cartesian coordinates (x, y) of the arm endpoint.

Inverse Kinematics. Given Cartesian coordinates (x, y) of the arm endpoint, find the corresponding joint values (θ_1, θ_1).

Note that if p_i^* is the vector from the proximal to the distal joint of link i (Figure 2.2), $i = 1, 2$, then

$$p_1^* = l_1 \begin{bmatrix} \cos \theta_1 \\ \sin \theta_1 \end{bmatrix}$$
$$p_2^* = l_2 \begin{bmatrix} \cos(\theta_1 + \theta_2) \\ \sin(\theta_1 + \theta_2) \end{bmatrix} \tag{2.1}$$

Direct Transformation (Direct Kinematics). From Figure 2.2, it is not hard to derive equations for the joint position, and by taking their derivatives to find equations for velocity and accelerations of the arm endpoint in terms of the arm joint angles:

Position:

$$X = \begin{bmatrix} x \\ y \end{bmatrix} = \begin{bmatrix} l_1 \cos \theta_1 + l_2 \cos(\theta_1 + \theta_2) \\ l_1 \sin \theta_1 + l_2 \sin(\theta_1 + \theta_2) \end{bmatrix} \tag{2.2}$$

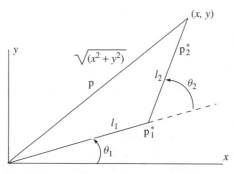

Figure 2.2   A sketch for deriving the two-link arm's kinematic transformations.

Velocity:

$$\dot{X} = \begin{bmatrix} \dot{x} \\ \dot{y} \end{bmatrix} = \begin{bmatrix} -l_1 \sin \theta_1 - l_2 \sin(\theta_1 + \theta_2) & -l_2 \sin(\theta_1 + \theta_2) \\ l_1 \cos \theta_1 + l_2 \cos(\theta_1 + \theta_2) & l_2 \cos(\theta_1 + \theta_2) \end{bmatrix} \begin{bmatrix} \dot{\theta}_1 \\ \dot{\theta}_2 \end{bmatrix} \quad (2.3)$$

or, in vector form, $\dot{X} = J\dot{\theta}$, where the 2×2 matrix J is called the system's *Jacobian* (see, e.g., Refs. 6 and 7).

Acceleration:

$$\begin{bmatrix} \ddot{x} \\ \ddot{y} \end{bmatrix} = \begin{bmatrix} -l_1 \sin \theta_1 & -l_2 \sin(\theta_1 + \theta_2) \\ l_1 \cos \theta_1 & l_2 \cos(\theta_1 + \theta_2) \end{bmatrix} \begin{bmatrix} \ddot{\theta}_1 \\ \ddot{\theta}_1 + \ddot{\theta}_2 \end{bmatrix}$$

$$- \begin{bmatrix} l_1 \cos \theta_1 & l_2 \cos(\theta_1 + \theta_2) \\ l_1 \sin \theta_1 & l_2 \sin(\theta_1 + \theta_2) \end{bmatrix} \begin{bmatrix} \dot{\theta}_1^{\,2} \\ (\dot{\theta}_1 + \dot{\theta}_2)^2 \end{bmatrix} \quad (2.4)$$

Inverse Transformation (Inverse Kinematics). From Figure 2.2, obtain the position and velocity of the arm joints as a function of the arm endpoint Cartesian coordinates:

Position:

$$\cos \theta_2 = \frac{x^2 + y^2 - l_1^2 - l_2^2}{2l_1 l_2}$$

$$\theta_1 = \tan^{-1} \frac{y}{x} - \tan^{-1} \frac{l_2 \sin \theta_2}{l_1 + l_2 \cos \theta_2} \quad (2.5)$$

Velocity:

$$\begin{bmatrix} \dot{\theta}_1 \\ \dot{\theta}_2 \end{bmatrix} = \frac{1}{l_1 l_2 \sin \theta_2} \begin{bmatrix} l_2 \cos(\theta_1 + \theta_2) & l_2 \sin(\theta_1 + \theta_2) \\ -l_1 \cos \theta_1 - l_2 \cos(\theta_1 + \theta_2) & -l_1 \sin \theta_1 - l_2 \sin(\theta_1 + \theta_2) \end{bmatrix}$$

$$\times \begin{bmatrix} \dot{x} \\ \dot{y} \end{bmatrix} \quad (2.6)$$

Obtaining equations for acceleration takes a bit more effort; for these and for other details on equations above, one is referred, for example, to Ref. 8. In general, for each point (x, y) in the arm workspace there are two (θ_1, θ_2) solutions: One can be called "elbow up," while the other can be called "elbow down" (Figure 2.3a). This is not always so—one should remember special cases and degeneracies:

- Any point on the workspace boundaries—that is, when $\theta_2 = 0$ or $\theta_2 = \pi$—has only one solution (Figure 2.3b).

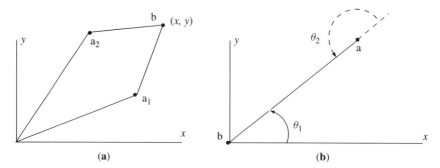

Figure 2.3 **(a)** In general, for each (x, y) point in the arm workspace there are two (θ_1, θ_2) solutions: One can be called "elbow up," and the other can be called "elbow down." **(b)** The arm's degeneracies.

- If the given coordinates (x, y) of the arm endpoint happen to lie outside of the arm workspace, then the inverse kinematics will have no solution for (θ_1, θ_2).
- If the arm links are of equal length, $l_1 = l_2$, then, when $\theta_2 = \pi$, the arm endpoint b falls into the arm origin, $x = y = 0$, and $\tan^{-1}(y/x)$ is undefined, resulting in an infinite number of solutions for the angle θ_1 (Figure 2.3b).

For an arbitrary rigid object in two-dimensional space, such as the rectangular object in Figure 2.4, the set of three values (x, y, θ) describe fully its *configuration*. The object's configuration information includes its position and its orientation. These three values are the object's three *degrees of freedom (DOF)*, each of which can be manipulated independently. All such sets form the object's three-dimensional *configuration space*, or *C-space*.

Our two-link arm of Figure 2.1 consists of two bodies, two links, and so in principle it should have $(3 + 3) = 6$ DOF. But, since the arm is bound in its motion by *kinematic constraints*–its fixed base and the connection at its joints—it

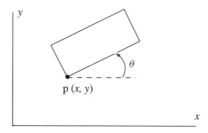

Figure 2.4 To describe the configuration of this rectangular object, an arbitrary point p is chosen on it first. Now the object's configuration is described by three values: x, y, and θ, where x and y are the Cartesian coordinates of point p.

has only two degrees of freedom, angles θ_1 and θ_2, two DOF that fully define its configuration in space. All such sets (θ_1, θ_2) define the arm's two-dimensional C-space.

2.2 STATICS

As formulated in Section 2.1, *statics* describes (a) the relationships between forces and torques that the arm exerts on the surrounding objects and (b) the relationships between internal forces and torques at the links. Statics analysis is done on an isolated link, taking into account the forces and torques contributed by neighboring links. For link i, the result of analysis is the net force $\mathbf{f}_i$, net torque $\mathbf{n}_i$, and

$$
\begin{aligned}
&m_i\mathbf{g} &&\text{Force of gravity}\\
&\mathbf{f}_{i-1,i} &&\text{Force exerted on link } i \text{ by link } i-1\\
&\mathbf{n}_{i-1,i} &&\text{Torque exerted on link } i \text{ by link } i-1
\end{aligned}
$$

The force balance is

$$\mathbf{f}_i = \mathbf{f}_{i-1,i} - \mathbf{f}_{i,i+1} + m_i\mathbf{g} \tag{2.7}$$

The minus in front of the second term above is due to the changed direction of the exerted force. In Figure 2.5, we have

$$
\begin{aligned}
&\mathbf{p}_i^* - \mathbf{r}_i^* &&\text{Vector from the link } i \text{ center of gravity to the joint } i\\
&\mathbf{r}_i^* &&\text{Vector from the link } i \text{ center of gravity to the joint } i+1
\end{aligned}
$$

The torque balance is (see Figure 2.5)

$$\mathbf{n}_i = \mathbf{n}_{i-1,i} - \mathbf{n}_{i,i+1} - (\mathbf{p}_i^* + \mathbf{r}_i^*) \times \mathbf{f}_{i-1,i} + \mathbf{r}_i^* \times \mathbf{f}_{i,i+1} \tag{2.8}$$

2.3 DYNAMICS

As defined above, *dynamics* describes relationships between kinematics and statics. For example, the relation between torques at the arm joints and link positions represents the arm's dynamics. Dynamics is typically a final step in deriving joint torques. Given various forces acting on the arm, one needs joint torques to realize the desired trajectory. Then by Newton–Euler equations we relate the D'Alamber force $\mathbf{f}_i$ and torque $\mathbf{n}_i$ (from the static equations) to the acceleration of link i [7].

Let $\mathbf{r}_i$ be the vector from the arm base to the center of mass of link i (Figure 2.5). Take two links, link 1 and link 2, of masses m_1 and m_2, respectively. Then the net forces $\mathbf{f}_1$ and $\mathbf{f}_2$ acting upon the links 1 and 2 relate to

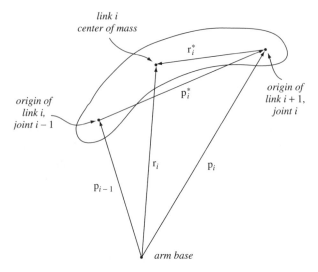

Figure 2.5   Balance of forces and torques acting on a single link.

the accelerations $\ddot{\mathbf{r}}_1$ and $\ddot{\mathbf{r}}_2$ of the links' centers of mass by Newton's second law,

$$\begin{aligned}\mathbf{f}_1 &= m_1\ddot{\mathbf{r}}_1 \\ \mathbf{f}_2 &= m_2\ddot{\mathbf{r}}_2\end{aligned} \tag{2.9}$$

From these equations, accelerations $\ddot{\mathbf{r}}_i$ of the centers of mass can be derived.

Let ω_i be the angular velocity vector of the center of mass of link i.
Let $\dot{\omega}_i$ be the corresponding angular acceleration.
Let $\mathbf{I}_i$ be the inertia matrix of link i.

Then torques are related to angular velocities and accelerations by Euler's equations,

$$\begin{aligned}\mathbf{n}_1 &= \mathbf{I}_1\dot{\omega}_1 + \omega_1 \times \mathbf{I}_1\omega_1 \\ \mathbf{n}_2 &= \mathbf{I}_2\dot{\omega}_2 + \omega_2 \times \mathbf{I}_2\omega_2\end{aligned} \tag{2.10}$$

For our planar two-link manipulator shown in Figure 2.1, the torque is normal to the arm's plane. Rotary inertia through the centers of mass of links 1 and 2 are [7]

$$\begin{aligned}I_1 &= m_1 l_1^2/12 + m_1 R^2/4 \\ I_2 &= m_2 l_2^2/12 + m_2 R^2/4\end{aligned} \tag{2.11}$$

Angular velocities and accelerations are

$$
\begin{aligned}
\omega_1 &= \dot{\theta}_1 \\
\omega_2 &= \dot{\theta}_1 + \dot{\theta}_2 \\
\dot{\omega}_1 &= \ddot{\theta}_1 \\
\dot{\omega}_2 &= \ddot{\theta}_1 + \ddot{\theta}_2
\end{aligned}
\tag{2.12}
$$

Substituting those into Euler's equations (and taking into the account that $\omega_i \times \mathbf{I}_i \omega_i = 0$), we obtain

$$
\begin{aligned}
n_1 &= I_1 \ddot{\theta}_1 \\
n_2 &= I_2 (\ddot{\theta}_1 + \ddot{\theta}_2)
\end{aligned}
\tag{2.13}
$$

Finally, Newton–Euler equations are combined with static equations [Eq. (2.8)] to produce the torques at arm joints—that is, to do inverse dynamics. After simplifications, these become (details can be found in Refs. 7 and 8)

$$
\begin{aligned}
n_{1,2} = &\; \ddot{\theta}_1 \left(I_2 + \frac{m_2 l_1 l_2}{2} \cos \theta_2 + \frac{m_2 l_2^2}{4} \right) + \ddot{\theta}_2 \left(I_2 + \frac{m_2 l_2^2}{4} \right) \\
&+ \frac{m_2 l_1 l_2}{2} \dot{\theta}_1^2 \sin \theta_2 + \frac{m_2 l_2 g_2}{2} \cos (\theta_1 + \theta_2) \\
&- l_2 \sin (\theta_1 + \theta_2) f_{2,3x} - l_2 \cos (\theta_1 + \theta_2) f_{2,3y} + n_{2,3}
\end{aligned}
\tag{2.14}
$$

$$
\begin{aligned}
n_{0,1} = &\; \ddot{\theta}_1 \left(I_1 + I_2 + m_2 l_1 l_2 \cos \theta_2 + \frac{m_1 l_1^2 + m_2 l_2^2}{4} + m_2 l_1^2 \right) \\
&+ \ddot{\theta}_2 \left(I_2 + \frac{m_2 l_2^2}{4} + \frac{m_2 l_1 l_2}{2} \cos \theta_2 \right) \\
&- \frac{m_2 l_1 l_2}{2} \dot{\theta}_2^2 \sin \theta_2 - m_2 l_1 l_2 \dot{\theta}_1 \dot{\theta}_2 \sin \theta_2 \\
&+ \left(\frac{m_2 l_2}{2} \cos (\theta_1 + \theta_2) + l_1 \left(\frac{m_1}{2} + m_2 \right) \cos \theta_1 \right) g_2 \\
&- (l_1 \sin \theta_1 + l_2 \sin(\theta_1 + \theta_2)) f_{2,3y} + n_{2,3}
\end{aligned}
$$

There are three types of terms that appear in such equations. Taking as an example the above equation for $n_{1,2}$, these are:

Dynamic Torques (Terms 1, 2, and 3). These arise from the arm movement, and depend on velocities and accelerations.

Gravity Torques (Term 4). These are due to the (vertical) gravity force.

External Torques (Terms 5, 6, and 7). These are due to external forces and torques that come from the arm's interaction with other objects; they appear when the arm physically touches objects, such as in assembly or cleaning. During a free arm motion these torques are zeros.

Then there is another classification of dynamic torques, also with three types:

Inertial Torques. These are proportional to accelerations in the arm joints, and they arise from normal action/reaction forces of an accelerating body.

Centripetal Torques. These torques arise from a constrained rotation about a point, and they are proportional to the squares of joints' velocities. For example, the arm's forearm must rotate about the arm's shoulder joint, and so the centripetal acceleration is aimed at the shoulder joint along link l_1 (Figure 2.1).

Coriolis Torques. These torques arise from vortical forces, as a result of interaction between two rotating systems (in our case two arm links), and they are proportional to the product of joint velocities of two different links.

Notice the remarkable growth in the complexity of equations as we proceed from kinematics to statics to dynamics equations. Then there is another natural source of complexity—the arm complexity, measured by the number of robot degrees of freedom. The reader is reminded that in our example, Figure 2.1, we are dealing with the simplest two-link planar manipulator: In its analysis we started with modest kinematic equations (2.2) and (2.3) and arrived at rather complex dynamic equations in (2.14). Will the equations complexity grow as quickly with the growth in the number of robot DOF?

Indeed they will. As an example, if we write only the acceleration-related coefficients for an arm with six DOF, they form this 6×6 matrix (note that a great many of today's industrial robot arm manipulators have six or more DOF):

$$\mathbf{D} = \begin{bmatrix} D_{11} & D_{12} & D_{13} & D_{14} & D_{15} & D_{16} \\ D_{12} & D_{22} & D_{23} & D_{24} & D_{25} & D_{26} \\ D_{13} & D_{23} & D_{33} & D_{34} & D_{35} & D_{36} \\ D_{14} & D_{24} & D_{34} & D_{44} & D_{45} & D_{46} \\ D_{15} & D_{25} & D_{35} & D_{45} & D_{55} & D_{56} \\ D_{16} & D_{26} & D_{36} & D_{46} & D_{56} & D_{66} \end{bmatrix} \quad (2.15)$$

The diagonal terms in this matrix represent uncoupled terms—that is, terms caused by a single joint—and off-diagonal terms represent pairwise interaction effects for all six joints. Each such term is itself a rather complex relationship. Out of curiosity, if the very first term in the matrix above, D_{11}, is written in full, it looks as shown in Figure 2.6 [9]. As you glance at this formula, try to imagine what the whole matrix $\mathbf{D}$ must look like, and imagine what kind of complexity a control system based on such expressions must involve. Consequently, all kind of simplifications are done in real-world systems when designing robot control schemes. Simplifications in equations bring, of course, imprecision, and so the design process involves carefully studied trade-offs.

$$D_{11} = m_1 k_{122}^2$$

$$+ m_2 \left[k_{211}^2 s^2\theta_2 + k_{233}^2 c^2\theta_2 + r_2(2\bar{y}_2 + r_2) \right]$$

$$+ m_3 \left[k_{322}^2 s^2\theta_2 + k_{333}^2 c^2\theta_2 + r_3(2\bar{z}_3 + r_3)s^2\theta_2 + r_2^2 \right]$$

$$+ m_4 \left\{ \tfrac{1}{2}k_{411}^2 \left[s^2\theta_2(2s^2\theta_4 - 1) + s^2\theta_4 \right] + \tfrac{1}{2}k_{422}^2(1 + c^2\theta_2 + s^2\theta_4) \right.$$

$$\left. + \tfrac{1}{2}k_{433}^2 \left[s^2\theta_2(1 - 2\,s^2\theta_4) - s^2\theta_4 \right] + r_3^2 s^2\theta_2 + r_2^2 - 2\bar{y}_4 r_3 s^2\theta_2 + 2\bar{z}_4(r_2 s\theta_4 + r_3 s\theta_2 c\theta_2 c\theta_4) \right\}$$

$$+ m_5 \left\{ \tfrac{1}{2}(-k_{511}^2 + k_{522}^2 + k_{533}^2) \left[(s\theta_2 s\theta_5 - c\theta_2 s\theta_4 c\theta_5)^2 + c^2\theta_4 c^2\theta_5 \right] \right.$$

$$+ \tfrac{1}{2}(k_{511}^2 - k_{522}^2 + k_{533}^2)(s^2\theta_4 + c^2\theta_2 c^2\theta_4)$$

$$+ \tfrac{1}{2}(k_{511}^2 + k_{522}^2 - k_{533}^2) \left[(s\theta_2 c\theta_5 + c\theta_2 s\theta_4 s\theta_5)^2 + c^2\theta_4 s^2\theta_5 \right] + r_3^2 s^2\theta_2 - r_2^2$$

$$\left. + 2\bar{z}_5 \left[r_3(s^2\theta_2 c\theta_5 + s\theta_2 s\theta_4 c\theta_4 s\theta_5) - r_2 c\theta_4 s\theta_5 \right] \right\}$$

$$+ m_6 \left\{ \tfrac{1}{2}(-k_{611}^2 + k_{622}^2 + k_{633}^2) \left[(s\theta_2 s\theta_5 c\theta_6 - c\theta_2 s\theta_4 c\theta_5 c\theta_6 - c\theta_2 c\theta_4 s\theta_6)^2 + (c\theta_4 c\theta_5 c\theta_6 - s\theta_4 s\theta_6)^2 \right] \right.$$

$$+ \tfrac{1}{2}(k_{611}^2 - k_{622}^2 + k_{633}^2) \left[(c\theta_2 s\theta_4 c\theta_5 s\theta_6 - s\theta_2 s\theta_5 s\theta_6 - c\theta_2 c\theta_4 c\theta_6)^2 + (c\theta_4 c\theta_5 s\theta_6 + s\theta_4 c\theta_6)^2 \right]$$

$$+ \tfrac{1}{2}(k_{611}^2 + k_{622}^2 - k_{633}^2) \left[(c\theta_2 s\theta_4 s\theta_5 + s\theta_2 c\theta_5)^2 + c^2\theta_4 s^2\theta_5 \right]$$

$$- \left[r_6 c\theta_2 s\theta_4 s\theta_5 + (r_6 c\theta_5 + r_3)s\theta_2 \right]^2 + (r_6 c\theta_4 s\theta_5 - r_2)^2$$

$$+ 2\bar{z}_6 \left[r_6(s^2\theta_2 c^2\theta_5 + c^2\theta_4 s^2\theta_5 - c^2\theta_2 s^2\theta_4 s^2\theta_5 + 2\,s\theta_2 c\theta_2 s\theta_4 s\theta_5 c\theta_5) \right.$$

$$\left. \left. - r_3(s\theta_2 c\theta_2 s\theta_4 s\theta_5 + s^2\theta_2 c\theta_5) - r_2 c\theta_4 s\theta_5 \right] \right\}$$

Figure 2.6   The full expression for one term, D_{11}, of the matrix (2.15).

2.4 FEEDBACK CONTROL

In the jargon of control theory, the system under control is called a *plant*. In our case the plant is a robot arm manipulator. In doing its job of controlling the motion of arm motors, the robot control system realizes an appropriate *control law*, which is the relationship between the system's input and output. For the arm control, the input is the arm's desired position(s), and the output is the arm's corresponding actual position. One can distinguish three types of control:

- *Open-loop control*, where the control action is applied regardless the system errors
- *Linear control*, in which the control law is a linear relationship
- *Nonlinear control*, in which the control law is a nonlinear relationship

Because there are a great number of nonlinear relationships, the term "nonlinear control" calls for further precision. Typically, the nonlinear control is more complex to realize than linear control, so a need for nonlinear control suggests that the system in question is quite complex. One finds in literature many terms related to different methods of nonlinear control: switching controls (which are further divided into a bang-bang control, duty-cycle modulation, logic control), globally nonlinear feedback mapping (e.g., saturating controls), adaptive control (with its own division into, for example, model-based reference control and self-tuning control), and so on. Both linear and nonlinear control are typically realized as a *feedback*-based control, as opposed to open-loop control. In a feedback control system the control law becomes a *feedback control law*, and is calculated based on the desired system behavior and the contemplated error (i.e. difference between the system's input and output), forming a *feedback loop* (Figure 2.7).

Assume that each of the two joints of our arm (Figure 2.1) has a torque motor with a unity input-to-torque conversion coefficient. As before, denote the motor torques at links l_1 and l_2 as $n_{0,1}$ and $n_{1,2}$, respectively. To demonstrate how the control system is synthesized, we will use an *independent joint controller*, called also a *linear decentralized feedback law*. Assume that the desired joint angles are given.

The control law we choose is of *proportional-integral-derivative (PID)* type, a widely used type of controller. In Figure 2.8, K is position (error) gain, L is integral (error) gain, and H is derivative (error) gain. In simple terms, the controller's position feedback component improves the speed of response of the control system; the integral feedback component ensures a steady-state tracking

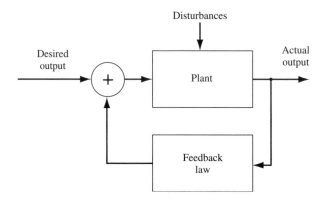

Figure 2.7   A sketch of a feedback-based control system.

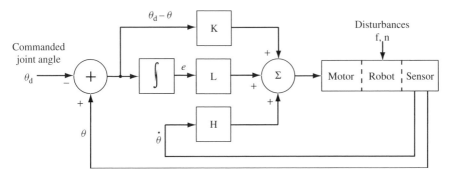

Figure 2.8   A PID control system.

of the desired output; and the derivative feedback component enhances the system stability and reduces oscillation due to control.

Equations characterizing the control law here will come out as (see Ref. 7),

$$n_{0,1} = K_{01}(\theta_1 - \theta_{1d}) + L_{01}e_1 + H_{01}\theta_1$$
$$\dot{e}_1 = \theta_1 - \theta_{1d} \tag{2.16}$$

and

$$n_{1,2} = K_{12}(\theta_2 - \theta_{2d}) + L_{12}e_2 + H_{12}\theta_2$$
$$\dot{e}_2 = \theta_2 - \theta_{2d} \tag{2.17}$$

Steady-State Analysis. Assume that the commanded angles $(\theta_{1d}, \theta_{2d})$ are constant, the disturbances are constant, and the control system is stable. System stability implies that eventually all time derivatives approach zero, which, using Eq. (2.14), gives

$$\theta_1 = \theta_{1d}$$
$$\theta_2 = \theta_{2d}$$
$$e_2 = \frac{1}{L_{12}}\left[\frac{1}{2}m_2l_2g_2\cos(\theta_{1d} + \theta_{2d}) + n_{2,3}\right]$$
$$\quad + \frac{1}{L_{12}}\left[l_2\cos(\theta_{1d} + \theta_{2d})\,f_{2,3y} - l_2\,\sin(\theta_{1d} + \theta_{2d})f_{2,3x}\right] \tag{2.18}$$
$$e_1 = \frac{1}{L_{01}}\left[L_{12}e_2 + \left(\frac{1}{2}m_1 + m_2\right)l_1g_2\cos\theta_{1d}\right]$$
$$\quad + \frac{1}{L_{01}}\left[l_1\cos\theta_{1d}\,f_{2,3y} - l_1\sin\theta_{1d}f_{2,3x}\right]$$

Analysis of these equations shows that steady-state tracking will occur regardless of disturbances. Note that the value of error e_1 must compensate for the value of e_2 but not vice versa, which is understandable given the links' sequential connection. It can be shown that if the gains L_{01} and L_{12} are zeroes, then joint values θ_1 and θ_2 differ from their desired values by amounts proportional to the disturbance values and inversely proportional to the feedback gains K_{01} and K_{12}. If, in turn, both gains K are zero, then the system's steady state is independent of the commands and reflects only the balance of gravitational and disturbance forces. This of course indicates the importance of the feedback.

Dynamic Stability Analysis. This is done to verify the stability hypothesis. Suppose that $n_{2,3} = 0$, $f_{2,3x} = 0$, and $f_{2,3y} = 0$. Assume a small initial error $\delta\theta_1(t)$,

$$\delta\theta_1(0) = \theta_1 - \theta_{1d}$$

$$\delta\theta_1(0) = \theta_1 - \theta_{1d}$$

and constant θ_{1d} and θ_{2d} for $t > 0$. Then, assuming that stability can be achieved via feedback, linearized equations in terms of $\delta\theta_1(t)$ and $\delta\theta_2(t)$ are written. Stability of those equations can be assessed by applying to them the Laplace transform and studying the characteristic polynomial [7]. Tests for stability are in general difficult to apply, so simpler necessary conditions are used, followed by a detail experimental verification.

2.5 COMPLIANT MOTION

When the robot is expected to physically interact with other objects, additional care has to be taken to ensure a smooth operation. Imagine, for example, that the robot has to move its hand along a straight line, on a flat surface, say a table. It is easy to program such motion, but what if the table has a tiny bump right along the robot's path? The robot will attempt to produce a straight line, effectively trying to cut through the bump. Serious forces will develop, with a likely unfortunate outcome. What is needed is some mechanism for the robot to "comply" with deviations of the table's surface from the expected surface. Two types of motion are considered in such cases:

- *Guarded motion*, when the arm is still moving in free space, before it contacts an object. Position control similar to the one above is used.
- *Compliant motion*, when the arm is in continuous contact with the object's surface. Position control and force control are then used simultaneously.

Consider an example: Let us say that our task requires the robot to grasp an object A (see Figure 2.9) that is initially positioned on top of object B_1, move it first into contact with the surface of table T, then slide it along T until it contacts an object B_2, and stop there. Let us assume that the grasping operation itself presents no difficulty and that the grasp is a rigid grasp; that is, for all

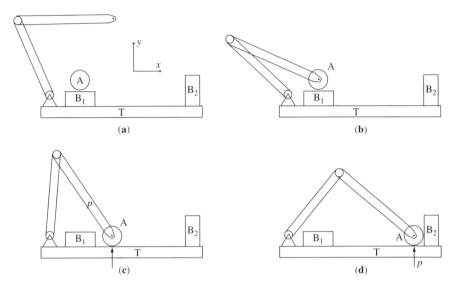

Figure 2.9   In this task, as object A is moved by the robot arm along the surface of table T, between positions **(c)** and **(d)**, the control system has to maintain a compliant contact between A and T.

practical purposes the grasped object A becomes the rigid continuation of the robot hand. The contact relation between object A and objects B_1, T, and B_2 is each a point relation and easy to accomplish. Our primary concern is the contact relation between object A and surface T during the motion of A along T. This contact relation is an external constraint on the arm's position control during this part of the path. The PID controller in our example in Figure 2.8 realizes positioning control; that is, it concerns itself only with the robot arm endpoint position (θ_1, θ_2) [or, equivalently, (x, y)]. In the example in Figure 2.9, such a controller would work as long as the arm moves in free space. But when the arm attempts to move its endpoint along the surface of table T, a contact constraint comes in, affecting the y component of the arm motion and acting as another "control." With two simultaneous constraints affecting the same motion components, one from the control system and the other from a contact constraint, the system is *overspecified*.

In other words, we have a potential conflict. Imagine that at some moment a little bump on the table will require the arm to move for an instant up and then down again. As the robot control will "insist" on moving along the straight line, the arm's positioning system will likely fail because of inconsistency of the simultaneous constraints. Besides deviations of surface T from its model, there may be a conflict due to modeling uncertainty, controller errors, and so on. We need to resolve this conflict.

The solution is to (a) add a force control at this part of the path and (b) limit the force and positioning controls so that each type never creates an overspecified system. In the task in Figure 2.9 the motion will be planned in two stages:

1. A guarded motion in the y direction will be used for the part of the path in free space, namely from the arm initial position (Figure 2.9a) to grasping the object A (Figure 2.9b) to the position when object A contacts surface T (Figure 2.9c). Only position control will be used at this stage. (Since object A is immobile during the grasping operation, let us assume that such control will suffice for grasping.)
2. Compliant motion control will be done during the part of motion where object A is in continuous contact with table T, between positions shown in Figures 2.9c and 2.9d. Both position and force control will be used at this stage: position control in the direction x and force control in direction y.

Here is why this control is called compliant. During this part of the path the control system will only attempt to maintain a set force pressure in the y direction. If a little bump is encountered on the table, the arm's attempt to maintain the same y coordinate as before will instantaneously develop a stronger reaction force from the table. As the arm's control measures and reacts to forces in this y direction, it will then *comply*, gently raising the arm enough to keep the same action/reaction forces in the y direction. As the bump is passed, the reaction force will quickly decrease, and the arm's control will move the arm endpoint a notch down, just to maintain the force at the set value.

This *hybrid controller* therefore has two feedback loops (see Figure 2.10): one for position control and one for force control. (Each loop may of course have its own complications; for example, each can be built as a PID controller shown in Figure 2.8.)

Remember that the controller shown in Figure 2.10 can provide a successful compliance control only specifically along the y axis, which is what is needed for the task in Figure 2.9. In reality the direction of the compliance line may differ from case to case, so for the general case a more general scheme is needed. The controller shown in Figure 2.11 can handle such cases. Its main difference from the controller in Figure 2.10 is that instead of specific matrices M_1 and M_2 in Figure 2.10, a generalized *constraint frame* 2×2 rotation matrix $\mathbf{Q}$ is used. Matrix $\mathbf{Q}$ describes orientation of the constraint axes. Other inputs in the scheme are as follows:

Axis s specifies the position versus control differentiation of axes,

$$s_i = \begin{cases} 1, & \text{where } s_i = 1 \text{ if axis } i \text{ of constraint frame is position-controlled} \\ 0, & \text{where } s_i = 0 \text{ if axis } i \text{ of constraint frame is force-controlled} \end{cases}$$

$\mathbf{p}_d = (x_d, y_d)$ is the desired position vector.
$\mathbf{f}_d = (f_{xd}, f_{yd})$ is the desired force vector.
$\mathbf{R}$ is the coordinate transformation of the force control loop.
$\mathbf{T}$ is the coordinate transformation of the position control loop.

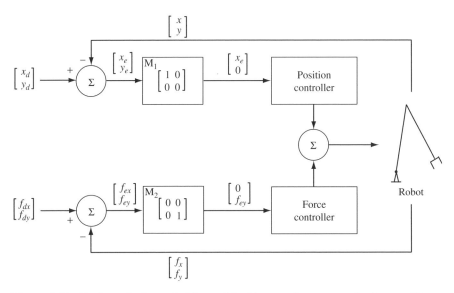

Figure 2.10 In the task shown in Figure 2.9, this compliance control scheme will provide position control during the motion between the positions shown in Figures 2.9a and 2.9c and will then provide force control between the positions shown in Figures 2.9c and 2.9d.

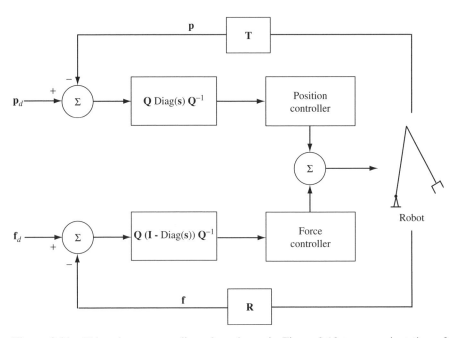

Figure 2.11 This scheme generalizes the scheme in Figure 2.10 to any orientation of the compliance line.

Suppose that the line of compliant motion forms an angle δ with the horizontal. Define a unit vector along that line as $\mathbf{u}(\delta)$. Then:

- The position loop projection becomes $\mathbf{u}(\delta)(\mathbf{p}_e \cdot \mathbf{u}(\delta))$.
- The force loop projection becomes $\mathbf{u}(\delta + \frac{\pi}{2})(\mathbf{f}_e \cdot \mathbf{u}(\delta + \frac{\pi}{2}))$.

These operations will be implemented properly if matrix $\mathbf{Q}$ is defined to align the constraint frame with the known compliance line, and vector $\mathbf{s}$ differentiates the directions of control loop actions:

$$\mathbf{Q} = \begin{bmatrix} \cos\ \delta & -\sin\ \delta \\ \sin\ \delta & -\cos\ \delta \end{bmatrix}$$

$$\mathbf{s} = \begin{bmatrix} 1 \\ 0 \end{bmatrix}$$

(2.19)

2.6 TRAJECTORY MODIFICATION

Robot trajectories (equivalently, robot paths) are generated in many ways. For example, as explained in Section 1.2, not rarely a path is obtained manually: A technician brings the arm manipulator to one point at the time, he or she presses a button, and the point goes into the trajectory database. A sufficient number of those points makes for a path. Or, the path can be obtained automatically via some application-specific software. Either way, if the robot goes through the obtained path, it is very possible that the motion would be less than perfect; for example, it may be jerky or make corners that are too sharp. For some applications, path smoothness may be very critical. Then, the set of collected path points has to be further processed into a path that satisfies additional requirements, such as smoothness.

Depending on the application, more requirements to the path quality may appear: a continuity of its second and even third derivatives (which relates to the path smoothness), precision of its straight line segments, and so on. That is, techniques used for modifying the robot path often emphasize appropriate mathematical properties of the path curves. The path preprocessing will likely include both position and orientation information of the path. If, for example, such work is to be done for a six-DOF arm manipulator, the desired properties of the path are expected from all DOF curve components.

Common trajectory modification techniques are *polynomial trajectories*, which amount to the satisfaction of appropriate constraints, and *straight-line interpolation*.

Polynomial Trajectories (Satisfaction of Constraints). Consider an example in Figure 2.12. We want to obtain a mathematical expression for a simple path that would bring this two-link planar arm from its initial point (position) p_a to the destination point p_b. Positional constraints are defined by the joint angle vectors

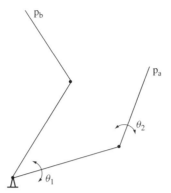

Figure 2.12 By constraining the path to only starting and finishing points (p_a, p_b), a simple straight-line path between p_a and p_b can be used. Adding additional constraints on velocities along the path requires going to more complex path shapes.

$\boldsymbol{\theta}_a$ and $\boldsymbol{\theta}_b$ at positions p_a and p_b, respectively. Here and below, $\boldsymbol{\theta}$, $\boldsymbol{\omega}$, and so on, are vectors whose dimensionality is equal to the number of system DOF; for a two-DOF robot arm, $\boldsymbol{\theta} = (\theta_1, \theta_2)$, and so on. The constraints can be met by a trajectory

$$\boldsymbol{\theta}(t) = (1 - f(t))\boldsymbol{\theta}_a + f(t)\boldsymbol{\theta}_b$$

where function f is defined on the segment [0,1] and converts the arm's path into a trajectory; $f(a) = 0$, $f(b) = 1$. In its simplest, the function $f(t) = t$ produces a linear trajectory,

$$\boldsymbol{\theta}(t) = (1 - t)\boldsymbol{\theta}_a + t\boldsymbol{\theta}_b \qquad (2.20)$$

While geometrically a straight line is a nice trajectory, it has drawbacks: (a) A simple differentiation shows that the angular velocity of this trajectory is a constant, which means that if we want to connect together two such path segments, large (formally, infinite) accelerations will develop at the joint point of those segments. (b) From our path expression it is hard to know if the whole trajectory lies in the robot workspace. Recall that the workspace of our arm may have a circular dead zone around its base (Figure 2.1): If our straight line passes through that zone, it would mean that the path is not physically realizable.

This suggests that we may want to add other constraints to the path. Let us add constraints on velocity, $\dot{\boldsymbol{\theta}}(a) = \boldsymbol{\omega}_a$, $\dot{\boldsymbol{\theta}}(b) = \boldsymbol{\omega}_b$. Two constraints can be met by a cubic trajectory,

$$\boldsymbol{\theta}(t) = (1 - t)^2 [\boldsymbol{\theta}_a + (2\boldsymbol{\theta}_a + \boldsymbol{\omega}_a)t] + t^2 [\boldsymbol{\theta}_b + (2\boldsymbol{\theta}_b - \boldsymbol{\omega}_b)(1 - t)] \qquad (2.21)$$

In terms of its execution, this trajectory is significantly more realistic than the one in (2.20). But, it still has drawbacks: (a) The trajectory does not take into account the fact of maximum attainable velocity. (b) Accelerations (and hence torques) cannot be independently specified at the ends of the trajectory. This significantly limits our freedom in specifying the pattern of robot motion. The problem can be fixed via additional constraints, namely by specifying accelerations α at the path's beginning and end, $\alpha(a)$ and $\alpha(b)$. Now we have a total of six constraints; to meet them, a minimum of fifth-order polynomial is needed:

$$\theta(t) = (1-t)^3 \left[\theta_a + (3\theta_a + \omega_a)t + (\alpha_a + 6\omega_a + 12\theta_a)\frac{t^2}{2} \right]$$
$$+ t^3 \left[\theta_b + (3\theta_b - \omega_b)(1-t) + (\alpha_b - 6\omega_b + 12\theta_b)\frac{(1-t)^2}{2} \right] \quad (2.22)$$

Straight-Line Interpolation. Achieving a straight-line motion with the arm shown in Figure 2.12 may be tricky. The reason for that is that the arm's rotating joints move links' endpoints on circular curves. This means that we cannot have an ideal straight line, and so we need to synthesize it from curves. The more those curves and the shorter they are, the better our straight-line approximation. To accomplish this, we will calculate a number of points along the desired straight line, and will force the arm endpoint through those points. Between those guaranteed points, the arm will move as it pleases; more precisely, the arm's own control will be linearly interpolating points between our specified points. The interpolation is done in terms of joint angles, or, as we call it, in the arm's *joint space* (or *configuration space*).

To summarize, the straight-line interpolation of added points is to be done in Cartesian space, and the arm's own interpolation between those given points takes place in joint space. Assume, for example, that the θ_1 and θ_2 angles of points p_a and p_b in Figure 2.12 are $p_a = (0, \frac{\pi}{6})$ and $p_b = (\frac{\pi}{3}, \frac{\pi}{2})$. This corresponds, respectively, to these Cartesian coordinates:

$$p_a = \left(\left(l_1 + \frac{\sqrt{3}}{2}l_2 \right), \frac{1}{2}l_2 \right)$$
$$p_b = \left(\left(\frac{1}{2}l_1 - \frac{\sqrt{3}}{2}l_2 \right), \left(\frac{\sqrt{3}}{2}l_1 + \frac{\sqrt{3}}{2}l_2 \right) \right)$$

If we provide the robot with only starting and ending points (p_a, p_b) of the desired straight-line path (p_a, p_c, p_b) (Figure 2.13), then, given the robot's joint space linear interpolation procedure, it will produce instead the curve (p_a, p_j, p_b). The midpoints of the joint and Cartesian paths indicate the extent of deviation of the actual path from the desired path: in Cartesian terms, $p_j = (\frac{\sqrt{3}}{2}l_1, (\frac{1}{2}l_1 + l_2))$ and $p_c = (\frac{3}{4}l_1, (\frac{\sqrt{3}}{4}l_1 + \frac{1}{2}l_2))$.

A reasonable idea then is to further approximate the desired straight-line path by forcing the robot endpoint through more intermediate points along the straight line. This is called the *bounded deviation paths* technique [10]; the added points are called *knots*. The process starts with two knots, the initial and ending points.

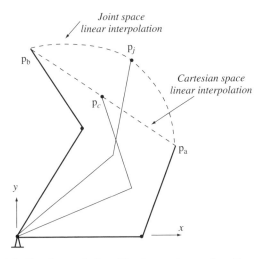

Figure 2.13 Straight-line interpolation. If only starting and ending points (p_a, p_b) are given, the arm's control will do the rest by interpolating points in between in joint space, producing as a result the dotted curve signified by a position p_j. To obtain a trajectory closer to a straight line, a set of points interpolated along the straight line, like point p_c, should be given.

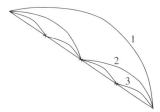

Figure 2.14 Forcing the robot to pass not only through the path endpoints, but also through the midpoint knot p_c, decreases the deviation from the straight-line path (path 2). Adding more knots further improves the path (path 3).

If the resulting deviation is too big compared to the specified threshold, more knot points are contemplated. Intuitively, adding the path midpoint (Figure 2.14) should decrease the deviation. Hence the robot will be guaranteed to pass through three points on the straight line: the initial point, the added midpoint knot, and the ending point. Between the knots, the path is interpolated in joint space, as shown. The resulting path consists of two curves. In the bounded deviation method, the deviation resulting from each added knot is calculated, and extra knots are iteratively added to each subsegment of the path until the resulting deviation from the straight line is below the specified threshold (see paths 1, 2, and 3 in Figure 2.14).

The reality is more complex than this simple scheme may suggest. First, depending on the path's initial and ending points, in general the deviation is not necessarily symmetrical on both sides of the midpoint knot. This means that obtaining the same deviation on both sides of the path midpoint may require uneven distribution of knot points. In fact, minimizing the number of knot points that satisfy a given threshold is a complex computational problem, and so is the optimal choice of locations for knot points. Second, as Figure 2.14 demonstrates, the tangents at endpoints of two curves that meet at a knot are not equal, causing unhealthy accelerations and jerky motion when the arm manipulator is passing through knot points. Third, every arm has degenerate points where various control and computational difficulties arise [7]. For example, as discussed above (Figure 2.3), if the arm links are of the same length, $l_1 = l_2$, the arm's joint values are ill-defined when its endpoint is at the arm base. If a knot point happened to be located in the base, the inverse kinematics procedure that turns out joint angles for every proposed knot point may give solutions that are in sharp contrast with the arm's prior and future motion.

Various schemes have been considered to address those complications, such as splines between the path curve segments or approximation of the curves with more complex polynomials featuring desired characteristics.

2.7 COLLISION AVOIDANCE

Whatever robot application is considered—assembly, welding, cleaning, exploring a new planet—for the robot's own sake and for its environment, it is paramount that the robot does not bump into surrounding objects. Of the issues in robotics that we set out to review in this chapter, *motion planning and collision avoidance* is perhaps the most universal robotic problem. It is also the most "robotic" robotic problem: Whereas other issues and techniques considered above are common to other areas of sciences and engineering, collision avoidance—especially its branch that deals with partial input information (such as from sensors)—is the monopoly of robotics. This is true for all robots and all variations of the collision avoidance problem, from mobile robots operating in a two-dimensional surface to multilink robot arm manipulators moving in three-dimensional space among three-dimensional objects. This monopoly does not imply, of course, that the problem of motion planning is harder or easier than those other issues, but it does imply two things: (a) that robotics is a distinct discipline, with its own problems and its own methodological apparatus, and (b) that solving this problem is our full responsibility—there will be no help from other disciplines.

To avoid collisions, the robot must know something about objects that it tries to avoid. Knowledge carries a price, either in terms of sensing that is necessary to acquire it, or in terms of the amount and speed of memory the robot needs to store it, or in terms of computational power it needs to process this knowledge. In fact, a complete information about the robot workspace is usually of tremendous volume and tremendous complexity and can tax the most advanced computers. The less knowledge about its surroundings the robot needs for successful collision avoidance, the more attractive the corresponding strategy. In this sense, *collision avoidance is an information-theoretical problem.*

Once the robot knows enough about objects in its surroundings, it has to figure out how to avoid those objects, while not jeopardizing its primary task. If moving my hand to replace a book on the shelf is about to cause my elbow to bump into a nearby file cabinet, there are great many ways to avoid the collision—I just need to think about this. I may think hard and slowly, or I may react instantly based on my instincts and experience; either way, I am using my intelligence to avoid collision. This example suggests that *collision avoidance is a problem of artificial intelligence.*

Collision avoidance relates to moving in space among objects; hence it is not surprising that *collision avoidance is heavily tied to concepts and techniques from geometry and topology.* Objects in the robot workspace that are to be avoided may be static, or they may be moving. Moving obstacles add to complexity of the collision avoidance problem. Some techniques are amenable to moving obstacles and some are not. While this book addresses static obstacles, we will stress the applicability of some strategies to moving obstacles. Most of the time we will limit the discussion to the effects of kinematics, leaving out the robot dynamics. Some collision avoidance problems with dynamics are considered in Chapter 4.

The information-theoretical base of the collision avoidance problem gives rise to one classification of motion planning strategies that turns out to be very productive. The classification divides all approaches into two groups, each presenting a distinct paradigm:

- *Motion planning with complete information*, also called in literature the *Piano Mover's model* or *off-line planning* approach. Here the path is computed all at once before the motion starts; in principle, an optimal path can be found in this way.

- *Motion planning with incomplete information*, also called *sensor-based motion planning* or *on-line motion planning*, or *path planning with uncertainty*, or the *Sensing–Intelligence–Motion (SIM)* paradigm. Here the decision-making is done continuously as the robot moves along, based on on-line information, such as from sensors. By its very nature, an optimal solution is ruled out in this formulation.[1]

A simple relation governs the choice of one or the other approach in robot applications. If all the information necessary to produce the desired path is available beforehand one would want to produce the path beforehand and would hence choose the Piano Movers approach. On the other hand, if the information

---

[1] The term "reactive planning" that is used sometimes in literature in reference to sensor-based motion planning is unfortunate: It emphasizes the operation's local nature, suggests that intelligence is not necessary, and hides the global component of motion planning, with its algorithmic connections to convergence and computational complexity.

happens to be coming in real time from robot sensors, and thus there is always uncertainty about the robot's surroundings, one is forced to turn to the second approach, SIM.

In other words, as a rule, only one approach applies to a given task. Consider, for example, a maze-searching task (called also a mouse-in-the-labyrinth problem). One starts at some starting point S inside the labyrinth and attempts to reach some target point T, also in the labyrinth.[2] Imagine we have in our possession complete information about the labyrinth. We can feed these data into the computer, produce the bird's-eye view of the maze, and study the problem in great detail using this map. We can investigate different paths between points S and T, figure out the optimal (shortest) path, and so on. This is planning with complete information, and the Piano Mover's model should be the preferred approach.

On the other hand, if all of a sudden you find yourself in a maze, at any given moment you would see only the surrounding walls of the maze and perhaps remember a few corridors that you have just passed. You do not know what is ahead; input information is scant; what you learn comes from your sensors. Any movement, including the unfortunate deviations into dead end corridor appendices, becomes a part of the path. Doing anything approaching an optimal path is of course out of the question. Here you deal with incomplete information and produce the path as you go. This is planning with incomplete information, and so you need to turn to the SIM techniques.

Since robot motion planning is the topic of this book, in Sections 2.8 and 2.9 we will further explore differences between these two paradigms for motion planning, the Piano Mover's model and the Sensing–Intelligence–Motion model.

Provable Versus Heuristic Algorithms. Another important distinction between algorithms is between *provable* (other terms: nonheuristic, exact, algorithmic) and *heuristic* approaches.

A provable motion planning algorithm is one for which there is a guarantee that if a path between the starting and target points exist, the algorithm will find one in finite time and without an exhaustive search—or else will conclude in finite time that there is no path if such is the case. We then say that the algorithm *converges*. To obtain such a guarantee, people go through the trouble of proving the algorithm convergence. An algorithm itself should allow such a proof; for example, the so-called "common sense" strategies—we call them heuristic algorithms—do not allow a proof of convergence and are not likely to be convergent.

Whereas for some applications, having a guarantee of convergence may be a moot point—as, for example, when the user's knowledge or intuition pretty much replaces it—for more complex cases, seeking convergence reflects more than a love for academic purity. As we will see in Chapter 7, in complex problems—most motion planning problems with robot arm manipulators fit this

---

[2]In other variations of this problem, one starts inside the maze and tries to find an exit from it; or, one starts outside the maze and tries to reach the location with a hidden treasure somewhere inside the labyrinth.

category—human intuition is not a good advisor. If, while operating under some reasonably sounding algorithm with unproven convergence, the robot fails to find a path, the failure may simply mean that feasible paths do exist but the algorithm has missed them. A guarantee of convergence then becomes a very practical issue.

2.8 MOTION PLANNING WITH COMPLETE INFORMATION

In this type of motion planning, input information is processed before the actual motion starts. This means that the input information must exist beforehand.

The model with complete information is formulated as follows.[3] Given a solid object (*robot*), or a combination of solid objects, in two- or three-dimensional space, whose size, shape, and initial and target position and orientation are fully described, and given a set of *obstacles* whose shapes, positions, and orientations in space are likewise known, the task is to find a continuous path for the object from the initial to the target position while avoiding collisions with obstacles along the way. An important assumption used in the model is that the surfaces of the moving object and the obstacles are algebraic or semialgebraic. This guarantees a final description of the input data. In some works a stricter requirement of planar surfaces is imposed.

Because complete information about the problem is assumed, the whole operation of path planning is a one-time, off-line operation. The main difficulty is not in proving existence of algorithms that would guarantee a solution (they obviously exist), but in assessing the problem complexity and obtaining a computationally efficient procedure. Reaching a solution means either finding a path or concluding in finite time that no path exists. Since a solution is always feasible, cases of arbitrary complexity can in principle be considered. Another apparent advantage of dealing with complete information is that various optimization criteria—finding the shortest path, or the minimum-time path, or the safest path, and so on—can be introduced easily.

Historically, Piano Mover's approach strategies were the first to come, starting in late 1960s. Most of the people who formulated the problem of robot collision avoidance were computer scientists. For them, collision avoidance was a purely computational problem, and the question of handling input information boiled down to a search in the database that contained that information. They often perceived sensing, partial information, uncertainty, control, and all such issues as small conceptual bumps that only interfered with the beautiful computational problem in hand. By the late 1980s, the area became one of the richest and popular areas in computational geometry. Hundreds of planning algorithms with complete information have been published; the problem's computational complexity has been studied in depth, and ingenuous ways of dealing with it were reported [11].

---

[3]A good survey of the work on provable algorithms for the Piano Mover's problem can be found in Ref. 11; specialized maze search algorithms are considered in Ref. 12.

By the late 1980s and early 1990s, it was slowly becoming clear that the domains to benefit from the Piano Mover's approach related not so much to robotics as to some other specialized areas where "clean" information would be available. One would read less about robots and more about a strategy for a quick extraction of an assembly unit from an aircraft engine without disassembly of the whole engine; or of optimizing the design of a car door opening so as to simplify the installation of car seats; or of finding the route that a protein molecule follows when folding into a complex shape during the DNA mapping of proteins. Note that in these cases the complete aircraft engine database, or a complete car body database, or a complete database of the protein geometry would be available beforehand.

The way the Piano Movers strategies proceed is as follows. Before the motion planning proper is attempted, the task's *configuration space (C-space)* is calculated. Assume for now that the robot—or, in general, an object for which the motion planning task is contemplated—is a rigid body moving in two-dimensional (2D) space (see object A, Figure 2.15a). In C-space the robot shrinks to a point, whereas the surrounding objects—we call them *obstacles*—are grown accordingly, to compensate for the shrinking robot.

Free subspace of C-space is the complement to the grown obstacles. Any path that lies in a continuous subset of free space is a physically realizable path for object A. In order to decide which areas of free space are connected and which are disconnected, and whether and how two patches of free space can be passed from one to another, an additional intermediate structure is built, the *connectivity graph* of C-space. A path is then declared to exist if the start and the destination nodes on the connectivity graph are connected [11, 13].

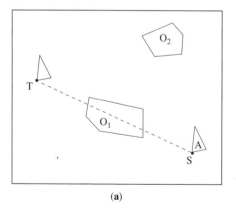
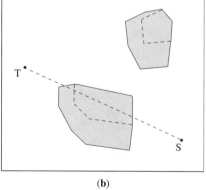

(a)   (b)

Figure 2.15 (a) The task is to move object A from its starting position S to the target position T, while keeping its orientation constant. (b) The C-space of task (a): Object A shrinks to a point, and obstacles O_1 and O_2 grow accordingly. Notice a simple fact: For A to be able to pass between O_1 and O_2 in the physical space (a), the grown obstacles in (b) must not overlap.

If the robot's orientation in space is maintained constant for the duration of its motion, the computation of C-space is relatively easy. If, in addition, the robot and obstacles are 2D polygons, then the grown obstacles are also polygons, though not necessarily with the same number of edges as in the original polygons (Figure 2.15b).

If, however, the robot orientation is allowed to change during the motion then, the C-space for a 2D workspace becomes a 3D space of parameters (x, y, θ), where θ is the robot orientation angle (Figure 2.16a). In this case, even the original polygonal robot and/or obstacles produce nonpolygonal and, in general, nonlinear grown 3D obstacles in C-space (Figure 2.16b). As complexity of the robot and obstacles increases—for example, in cases with 3D multilink arm manipulators and nonpolygonal obstacles—computation and proper representation of the C-space becomes an exceedingly difficult task.

The computational complexity of the problem is measured in the Piano Movers model in terms of the connectivity graph's structure. Overall, the computational price for dealing with perfect information and for the said advantages—optimal paths and, in principle, solutions for cases of arbitrary dimensionality—is high. Today many reasonable-size problems still cannot be approached.

From the application standpoint, unless there is a reason to believe that obstacle boundaries are algebraic (which would squarely mean that we are dealing with man-made objects), an appropriate approximation of the robot workspace has to be performed before the connectivity graph can be calculated. The approximation itself necessarily depends on considerations that are secondary to the path planning problem. One may specify, for example, the accuracy of presenting actual obstacles with polygons, or—a different criterion—one may put a limit on the computational cost of processing the resulting connectivity graph.

The approximation process can introduce significant computational costs of its own. John Reif has shown that approximation of nonlinear surfaces with linear constraints itself requires time exponential in the prescribed accuracy of approximation [14]. Also, the space of possible approximations is not continuous in the approximation accuracy; that is, a slight change in the specified accuracy

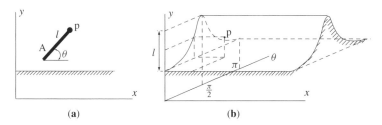

(a)   (b)

Figure 2.16 (a) When robot orientation is added to a 2D task, the (b) resulting 3D C-space of parameters (x, y, θ) is nonlinear, even if the original robot and obstacles are polygons. Here the "robot" A is a line segment of length l, and the obstacle is a horizontal "table" line.

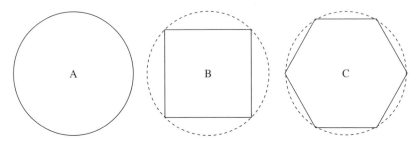

Figure 2.17   Which of the three obstacles, A, B, or C, would be easier to pass around?

of approximation can cause a dramatic change in the number and positions of nodes of the approximated surfaces and eventually in the generated paths.

Measuring the computational burden in terms of complexity of the connectivity graph may create peculiar situations where the derived computational complexity of a given task contradicts our intuitive notion of problem complexity. Consider, for example, a circular obstacle A in Figure 2.17. Assume that the motion planning algorithm that we plan to use requires polygonal obstacles. Then, obstacle A is first approximated—say, by one of the polygons B or C in Figure 2.17.

Now, according to Piano Mover's algorithms, planning a path around obstacle C is computationally more difficult than planning a path around the obstacle B, because of the greater number of nodes in C. Moreover, in the limit, increasing the accuracy of polygon approximation takes the computational burden to infinity. But, from the human and from the robotics control viewpoints, walking around the circle A is actually easier than walking around obstacles B or C, because the latter require special decisions at the corners of the obstacle. Also, from the dynamics standpoint, there is an undesirable sharp change in the velocity vector at the corners of obstacles B and C. One can, of course, solve this specific example by including circular objects in the list of those allowed by the algorithm, but this will only shift this discussion to some other shapes.

For more detail on the Piano Mover's model, the reader is referred to the literature. The model's computational complexity for cases of rigid or hinged bodies has been studied extensively. The problem was shown to be computationally prohibitive [15–17]. A 2D case has been studied in Refs. 18–20. Cases where objects to be moved are polygons (polyhedra) or discs (spheres) moving amidst polygonal (polyhedral) obstacles are considered in [15, 18, 19, 21–23]. The first attempt to study the case of moving an object with a number of free-hinged links was initiated in 1968 by Pieper [24], in the context of motion planning for robot arm manipulators. Exact algorithms for this problem have been described [15, 16, 20], as have various heuristics (e.g., see Refs. 25 and 26).

The computational complexity of the problem was first reported by Reif [15], who showed that the general Piano Mover's problem is PSPACE-hard. He also sketched a possible solution for moving a solid object in polynomial time, by direct computation of the "forbidden" volumes in spaces of higher dimensions.[4] Reif also demonstrated that even the preliminary operation of approximating the robot workspace with a specified accuracy carries a high computational cost. Schwartz and Sharir [18] presented a polynomial-time algorithm for a 2D Piano Mover's problem with convex polygonal obstacles. It has been shown in a number of works (e.g., Lozano-Perez and Wesley [27]) that the process of moving in the task's configuration space carries additional computational costs. In general, even if the original obstacles are polyhedral, obstacles in the configuration space have nonplanar walls. In order to keep the problem manageable, various constraints are typically imposed.

Moravec [28] considered a path planning algorithm for a mobile robot moving in two dimensions, with the robot presented as a circle. In his treatment of a 2D path planning problem with a convex polygonal robot and convex polygonal obstacles, Brooks and Binford [29, 30] used a generalized cylinder presentation to reduce the planning problem to a graph search. A generalized cylinder is formed by a volume swept by a cross section (in general, of varying shape and size) moving along the cylinder axis, which in turn can be some spine curve.

The version of the Piano Movers problem where the robot can consist of a number of free-hinged links is more difficult. On the heuristic level this version was started by Pieper [24] and further investigated by Paul [31]. Both were attracted to the problem's obvious relation to control of robot arms with multiple degrees of freedom. Later, new approaches for this version have been considered in Refs. 16 and 20. The most general (although very expensive computationally) algorithm for moving a free-hinged body was given by Schwartz and Sharir [16]. The technique is based on the general method of cell decomposition; the robot and obstacles are assumed to be limited by algebraic surfaces. A more economical (but still prohibitive for many practical tasks) algorithm for the general case was reported by Canny [32]. A variety of special cases shown to lead to simpler algorithms were described by Hopcroft et al. [20].

2.9 MOTION PLANNING WITH INCOMPLETE INFORMATION

By the mid-1980s it became clear that the inherent uncertainty of a realistic robot environment and the subsequent need for real-time sensing called for a paradigm of motion planning that would fundamentally differ from the Piano Mover's paradigm. It was further realized that uncertainty and sensing were not some small irritating "engineering details" but major factors in the theoretical foundation of motion planning algorithms. As it turned out, uncertainty and sensing became the very center around which the new paradigm would be built. The result was the theory and practice of *robot motion planning with incomplete information*, or the *SIM (Sensing–Intelligence–Motion)* paradigm.

---

[4]Higher dimensions d appear when one takes into account the moving rigid object's orientation along its way; $d = 3$ for the 2D case, and $d = 6$ for the 3D case.

The information-theoretical (or uncertainty) aspect of the problem at hand points to connections with other fields. In general terms the problem of sensor-based motion planning can be seen as one of reaching a global goal using local means. Thus presented, it becomes a fundamental problem, various formulations of which have been studied in a number of areas. For example, in game theory (differential games and macroeconomics; see, e.g., Ref. 33) one is interested in conditions under which individualistic interests of many agents can result in predictable behavior of the whole group. In works on collective behavior, algorithms are designed whereby a group of individuals can organize a unified action at a specific moment based on local interaction only, without centralized control. In the Firing Squad Problem [34], soldiers are requested, using only pairwise communication, to agree on a moment when they fire all at once. In computer science, local operations are used to study database searches with uncertainty [35]. In geometry, attempts have been made to prove theorems of Euclidean geometry using local input information [36]. The difficult question of the relationship between uncertainty and algorithm complexity has been tackled in Ref. 37.

While some considerations, such as the importance of computational properties of their methods, still served as a bridge between the Piano Mover's and SIM paradigms, with time many divergent issues made them harder and harder to compare. One such issue is of course the SIM's favoring continuous computation over the Piano Mover's one-time computation. The other issue is the option of optimal solutions inherent in the Piano Mover's model but inherently impossible in the SIM model—not because of inferior algorithms, one should add hastily, but because of the inherent lack of relevant input information. Still another issue, as we shall see, is the difference in how both models deal with algorithm complexity (again, not because of algorithms' specifics but because of the nature of uncertainty). What counts in the Piano Mover's model is the complexity of the whole robot scene. In contrast, what counts in the SIM model is the amounts of robot's "wandering" in the scene and visits to some previously visited places in the scene. Let us consider these and other factors in more detail.

The SIM paradigm formulation includes an assumption that information about the robot's surroundings comes in real time, usually from its sensors. Except perhaps for some exotic sensors ("X-ray" vision and the like), sensory information is of local, rather than global, nature—sensors tell one something about their surroundings. In the SIM algorithms that will be developed in the following chapters, the only input information available to the robot at all times is its own coordinates and those of the target location. As the robot starts moving, new information appears from its sensors.

To exhaust the extreme case and demonstrate the algorithm completeness, we will start the algorithm development with the "ultra-local" tactile sensor. That is, the robot learns about an obstacle's presence only when it touches it physically. Later we will extend the resulting strategies to the case of proximal sensing, such as vision.

In motion planning with uncertainty, the guarantee of a solution—which is predicated on the algorithm convergence—should be distinguished from the guarantee of the solution optimality. As we will see, the former is feasible even in a very complex environment, whereas the latter is not feasible with the best of algorithms. No optimality of the solution can be promised even if only a small piece of information about the environment is missing. One's path that may look ill-conceived with hindsight may not have been planned any better with the information available at the time. A person visiting a big building for the first time should not be blamed for wandering around in search of a desired office. We are also familiar with similar patterns taking place in time, rather than in space. If a stock market investor had tomorrow's information, he would have become rich quickly. Given that he doesn't, his actual behavior may look less than optimal a few days or weeks later.

If the path optimality is not a good criterion, how does one judge the performance of a motion planning algorithm with uncertainty? Given the real-time nature of SIM algorithms, they are expected to allow reasonably fast processing and to produce "reasonable quality" paths. The first requirement is easier to grasp: It is clear that a robot cannot afford to spend a minute on calculation of a step of a continuous path that takes 20 ms to execute. For the algorithm performance, standard complexity theory performance estimates call for lower and upper bounds on the problem itself and on specific algorithms, as a function of the problem complexity.[5]

What is a "reasonable" path? In general, how do we assess algorithms' performance? The problem complexity is presented in complexity theory as a function of the number of elements in the problem at hand. In our case the scene complexity cannot be assessed in these terms because it may never be known, and even the notion itself of the scene complexity is very unclear. Unstructured environments typically include "shapeless" objects: Representing them with analytical entities is difficult because, again, objects are not known in advance.

Doing this would be also pointless because this representation is unrelated to what is easy or difficult for SIM algorithms. What we may think of as a complex shape may be as easy or difficult for a SIM algorithm as a simple shape. As another option, one might argue that because in any realistic system the robot moves in discrete steps, those steps might be used to build a real-time objects' approximation. This choice is also hard to defend because the size and number of those steps may differ from one robot control system to the other. What is left is to estimate the quality of the path itself that the algorithm generates.

A better measurement of algorithm efficiency in the case with uncertainty is a function tied to the length of paths that the algorithm generates. More precisely, the criterion measures the extent of the robot's "wandering" under a given algorithm: It assesses the maximum number of times, n, of the robot's retracing some segments of its path. When comparing algorithms, this upper bound will actually

---

[5]Less rigorous ways may include, for example, a direct comparison of an algorithm's results with those of other existing algorithms, or with the performance of an "average" human traveler.

be independent of the length of paths, in the following sense. We can say, for example, that the algorithm A is better than the algorithm B because their n numbers are 2 and 3.5, respectively. That is, under algorithm A the robot will visit the same piece of its path at most twice, whereas under algorithm B it can visit a piece of its path at most 3.5 times.

One inherent weakness in algorithms with incomplete information is that the problem dimensionality cannot be made arbitrarily high. This drawback, again, comes from the nature of dealing with uncertainty, not from the lack of good algorithms. Consider, for example, a point robot flying in the three-dimensional space, like a fly. If the robot meets an obstacle, it has an infinite number of possibilities for passing around it. Given the robot's limited knowledge about the scene, its difficulty is unsurmountable: While good sensing will often help, in the worst case the robot may have to search an infinite number of paths to find one acceptable path. Luckily, for the cases of practical interest—mobile robots moving on a 2D surface and 3D arm manipulators—the situation is inherently easier.

The lack of information about the robot environment dictates a shift of emphasis in the SIM paradigm from objects' geometry to their topology. Relying on geometric properties of objects, as in the Piano Mover's model, would make SIM algorithms too brittle. We will see that a more sensible approach is to rely on the scene's topological properties; this allows one to tolerate uncertainty in objects' geometry. Hence there is a corresponding shift in the SIM paradigm from computational geometry tools to those from topology.

There are two other factors that have not been mentioned yet and that are often neglected in both the Piano Mover's and SIM algorithms. One is the robot dynamics. If the robot is heavy and moves relatively fast, no strategy in the world will prevent it from collisions unless this strategy is capable of handling the relation between the robot dynamics, its speed, its sensing, and of course the robot's goal. A submarine cannot stop on a dime: Its motion control system has to process an impending collision in advance; how it avoids collision will also depend on what it plans to do next. In other words, a realistic motion planning system may well need to account for the *system dynamics*. This factor will be considered in Chapter 4.

The second factor relates to the robot's shape and dimensions. There is always a question of how an algorithm that assumes a point robot will work for a real robot with blood and flesh. A small robot can pass where a big robot will not. One can pass a narrow corridor with folded arms, but won't be able to do it with outstretched arms. Besides accounting for the robot dimensions, this also suggests the effect of robot kinematics on motion planning.

We will see in the next section that the first serious approaches to the motion planning problem started with an abstract problem of searching a graph. The major actors in the events had no idea that their work would be a contribution to robotics. Some formulated it as a maze-searching problem, in a rather narrow way, where a maze is defined as something that would be practically equivalent to a graph. We will see in the sequel that something is lost when replacing a scene by a graph: A graph may lose some information from the original problem.

Below we first review those first graph-searching approaches, switching then to the proper prior work on robot motion planning.

2.9.1 The Beginnings

Leonhard Euler, perhaps the most famous mathematician of all time, was born on April 15, 1707, in Basel, Switzerland, and spent most of his career, between 1727 and 1741 and then from 1766 until his death on September 18, 1783, in St. Petersburg, Russia, holding a prestigious position of academician—that is, a full member of Russian Academy of Sciences. As his fame grew, in 1733 he succeeded Daniel Bernoulli to the chair of mathematics in the Academy. It was at this first period of his St. Petersburg career, in 1736, that Euler proposed and solved a problem that was to become famous under the name *Köenigsberg Bridge Problem*. This work marked the beginning of two new mathematical disciplines, graph theory and topology. It also gives an important insight into the robot motion planning problem.

The city of Köenigsberg (called today Kaliningrad and being a part of Russia) was divided into two parts by the Pregel River, with the Island of Kneiphof in the middle. Seven bridges connected the island with the rest of the city (see Figure 2.18a). The question posed to Euler by the city's residents was this: Can a pedestrian, starting at some point, pass all seven bridges and return to the starting point so that he will traverse each bridge exactly once?

To find the solution for the puzzled residents of Köenigsberg, Euler decided to first reduce the problem to an equivalent abstract problem. In a leap of imagination, he said that the shapes and dimensions of the masses of lands that the bridges connect (A, B, C, D, Figure 2.18a) are immaterial for the problem. What matters are the *connectivity properties* of the scene, what today we would call the *topological properties* of space. This argument became the beginning of the discipline of topology. Euler denoted the land masses as vertices of a diagram, and he denoted the bridges as edges connecting the vertices (Figure 2.18b); hence the graph theory was born.

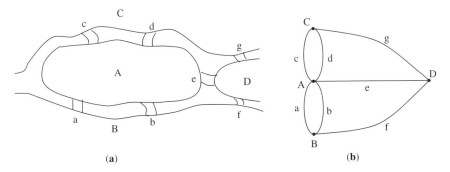

Figure 2.18 The Köenigsberg Bridge Problem.

The path that the Köenigsberg citizens wanted is called today a *Euler path*, and the graph it corresponds to is called a *Euler graph*. Define the number of *edges* incident to a graph *vertex* as the vertex *degree*. In today's formulation the related theorem sounds as follows:

Theorem 2.9.1 (Euler [38]). *A finite graph G is an Euler graph if and only if (a) G is connected and (b) every vertex of G is of an even degree.*

In the case of distinct starting and final points (which is the situation typical for the robot motion planning problems), exactly two vertices must be of an odd degree. For the Köenigsberg Bridge Problem the answer to the question posed to Euler is therefore "no" because all the vertices on the corresponding graph are of an odd degree.

As we saw above in the role of connectivity graphs in the Piano Mover's model, graph theory became an important tool in designing motion planning strategies with complete information. We will see in Chapters 3, 5, and 6 that topology became a no less important tool in sensor-based motion planning.

Neither Euler nor many of his followers asked explicitly what information about the scene was available to the traveler in the Köenigsberg Bridge Problem.[6] It was implicitly assumed that the traveler had complete information.[7] What if he didn't? What if at any moment of the trip the traveler's knowledge was limited by what he could see around him plus whatever he remembered from the path he had traversed already? What if this more realistic situation took place?

No live creature counts on knowing in advance all the objects on its journey, or calculates the precise path in advance. Algorithmically, the question about available input information puts the problem squarely into the domain of sensor-based motion planning, presenting it as a *maze-searching problem*. Clearly, even if the Köenigsberg bridges made a Euler graph, and if the traveler had no picture as in Figure 2.18, we can doubt that he would pass every bridge exactly once, except perhaps by sheer chance. If not, what would be the traveler's performance—say, with the best algorithm possible? Can a strategy be designed that will guarantee at least passing the whole graph when one starts with a zero knowledge about it? If so, how about doing it in some reasonably efficient manner? We will return to these questions later in this chapter.

This branch of motion planning—which can be formulated as moving in a graph without prior information about it—started long before Euler. Since the times of Theseus of Athens, people had great interest in labyrinths (mazes). After Theseus slew the Minotaur, he used the thread of glittering jewels given to him by Ariadne to find his way in the passages of the Labyrinth of Knossos. Many

---

[6]Eventually this question did appear in graph theory, though much later, as a question of existence of *local algorithms* [39].

[7]Interestingly, when in the 1960s and 1970s researchers turned to the problem of robot motion planning, the question of input information was not raised either. There seems to be something in human psychology that, unless told otherwise, between two choices—minimum and maximum of input information—we implicitly assume the latter.

medieval churches and castles in Europe had mazes in their gardens or as inside mosaics on the floor. Many mazes are built even today for public amusement and contemplation. Some labyrinth builders tried to emulate the famous labyrinth of the Chartres cathedral in France; one Chartres-type labyrinth appears in the Grace Cathedral in San Francisco. Even Washington, D.C., the United States capital, has its own tiny labyrinth in the small pretty St. Thomas Parish Park located in the heart of Dupont Circle.

Most collections of puzzles contain labyrinth problems. A "Bible" of labyrinths that is monumental in coverage, unique, and marvelously written is the book by Hermann Kern, *Through the Labyrinth: Designs and Meanings over 5000 Years* (originally published in Germany in 1983 [40] and translated into English in 2000 [41]).

A simple labyrinth is a set of corridors lying in the plane and connected in intricate ways. A labyrinth can be described by a graph whose edges represent corridors and vertices (vertices are the points where the corridors meet). Figure 2.19a shows the famous labyrinth in the garden at Hampton Court in London; the corresponding graph is shown in Figure 2.19b.

The problem is as follows: Given two points in a labyrinth, start S and target T, can a method be designed that would guarantee a path from S to T? It is usually assumed that the traveler has no beforehand knowledge about the labyrinth and has a way to mark the corridors and corridor intersections. In graph terms, this is a graph-searching problem: Given two graph vertices S and T, design a method to generate a path from S to T if one exists.

A general maze does not have to have explicit corridors and intersections. An arbitrary scene with obstacles presents a kind of maze. Any motion planning task for a robot in the plane can therefore be seen as a maze search. Such tasks can be naturally reduced to a graph search. Consider the scene in Figure 2.20a. Suppose a point robot plans to start at point S and reach point T. Suppose the robot knows its own location and that of T but has no beforehand knowledge about obstacles on its way. Then, it would be reasonable for the robot to take the shortest route to T, a straight line. Let us call this line the *Main line*, or *M-line*. If M-line happens to be free, the robot reckons, this would be the fastest route to T; if there are obstacles on the way, it will deal with them somehow. (The notion of Main line will appear often in the following chapters.)

Together with the scene, this strategy defines a graph that, although unknown to the robot, relates to the physical reality. Vertices of the graph are points S and T, as well as intersection points between M-line and obstacle boundaries; its edges are segments of M-line outside the obstacles and segments of obstacle boundaries (Figure 2.20b). The graph is called the *connectivity graph* of the maze. It has a simple structure: Vertices S and T are of degree one, and all other vertices are of degree three. Note that since each vertex has an odd number of edges incident to it, this is not a Euler graph, and so traversing the whole graph will result in at least some edges being traversed more than once. The graph can be easily transformed into a Euler graph—for example, by replacing with two segments each M-line segment that connects two obstacles.

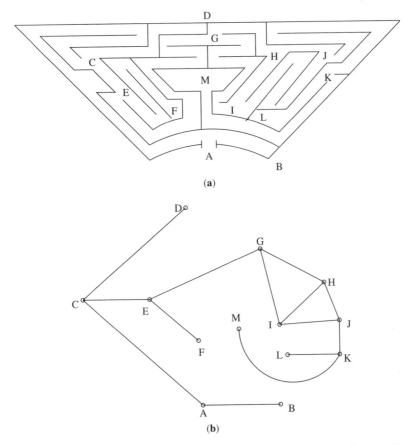

Figure 2.19 (a) The Hampton Court labyrinth; (b) the corresponding graph.

If all mazes can produce connectivity graphs of this particular structure, one may wonder if this property can warrant algorithms with better performance than algorithms for a *general graph*—that is, one with an arbitrary degrees of vertices. This raises still another question: Can mazes produce graphs with an arbitrary degrees of vertices? At first glance, the answer is yes: Many corridors in a maze can meet in the same spot, so connectivity graphs produced by realistic mazes must be general graphs.

Notice, however, that this argument ignores the fact that mazes appear in a continuous plane, not in some discrete domain where only one path exists in a limited space such as a corridor. Spaces between obstacles leave many options for moving in them. Even a corridor has a finite space between its walls: One can walk, for example, in one direction along one wall and walk back along the other wall; or one can move in a zigzag manner between walls. That is not what we represent by a graph; instead we want a "minimum" graph that describes the maze, and hence graph edges are maze walls and M-line segments,

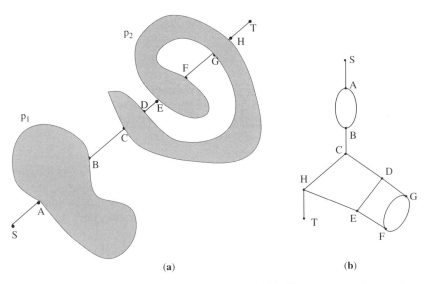

Figure 2.20   **(a)** A typical motion planning task. **(b)** The corresponding graph.

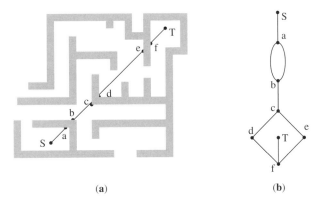

Figure 2.21   **(a)** In spite of the four-way corridor around points b, c, each vertex b, c in the corresponding connectivity graph **(b)** is of degree three.

and vertices are points where those segments meet. This means that all physical mazes can be reduced to a graph with the maximum vertex degree three (see Figure 2.21). Hence we are back to our first question: Does this special graph structure promise algorithms whose performance is better than that of algorithms for general graphs?

Denote the length of M-line as D, denote its segments outside the M-line as d_i, and denote the segments of obstacle boundaries cut by the M-line as p_i (Figure 2.20a). Those lengths can signify weights on the graph edges. Then the total "length" of the graph is no more than $D + \sum_i p_i$.

The first systematic procedure for solving maze problems—in fact, for covering a general graph—seems to be the one suggested by Wiener in 1873. His algorithm is as follows:

Wiener's Algorithm: Starting at point S, proceed along the graph edges as far as possible, selecting at each vertex an edge that has not been traversed before. At a vertex where such motion is no longer possible, retrace the sequence of edges until you arrive at a vertex with some unused edges.

Clearly, Wiener's method presents a version of what is called today an exhaustive search. One will also recognize in this algorithm a geometric version of the "width-first" procedure that appears in many textbooks in computer science. The procedure does guarantee covering all edges of the graph, but it will produce many repetitive visits to the same vertices.

Does an algorithm have to be that inefficient? For example, how about the following simple algorithm found in children stories? "Keep your left hand on the wall as you walk, and you will eventually get out of the labyrinth." This naive procedure will only work if all the walls in the labyrinth are connected. Clearly, if one keeps the hand on the wall that is a part of an island inside the labyrinth, one will walk in the labyrinth forever.

To understand the limits of achievable performance of algorithms with incomplete information, let us see first what that limit is for algorithms with complete information, for a general graph.

Theorem 2.9.2 (Ore [42]). *In a finite connected graph, it is always possible to construct a cyclic directed path passing through each edge once and only once in each direction.*

This promises a much better performance than in Wiener's algorithm, but is such performance feasible for an algorithm with incomplete information? The positive answer was given in the method by Tremaux (reported by Lucas in 1892 [43]). It was shown in 1895 by Tarry [42, 44] that the same result can be obtained in a more economical way. In Tarry's algorithm, called *Tarry's rule*, for a given graph, when the point robot arrives at a vertex v, the following input information is assumed: (i) the subset of those edges incident to v that the robot traversed before when leaving v—that is, those edges that were traversed in the direction pointing away from v; (ii) the *entrance edge* via which the robot first arrived at v. The procedure is very simple:

Tarry's Rule:

1. Upon arrival at v, continue via an edge (v, v') that was not yet traversed in the direction of v to v'.
2. Choose the entrance edge as a last resort.

Under Tarry's algorithm, assuming that the start and target vertices are distinct and there exists a path between them, the target vertex is guaranteed to be reached, and every edge will be traversed exactly twice, once in each direction, with the exception of the two edges incident to the start and target vertices, which will be traversed once each. Using our terminology, the upper bound on the length of paths generated by the algorithm is as follows:

Theorem 2.9.3. *For any finite maze, Tarry's algorithm will generate a path of length P such that*

$$P \leq 2D + 2\sum_i p_i \qquad (2.23)$$

where D is the length of M-line, and p_i are perimeters of obstacles in the maze.

Finally this gives us an answer to our question above: What performance can be expected for an unknown Euler graph.

It took a long time, over 70 years, for the next improvement to come. In 1970, Fraenkel proposed a more economical algorithm [45, 46]. Though it has the same performance as Tarry's in the worst case, it performs better if the robot is lucky with the maze; then, some or even all graph edges may be traversed just once. Fraenkel's algorithm is more complex that Tarry's. It makes use of a counter, which is set to zero at the start vertex. The algorithm operates as follows:

Fraenkel's Algorithm:

1. Whenever one arrives at a vertex not visited before, increase the counter by 1.
2. When arriving at a vertex v such that before entering it there was at least one edge incident to it that was not traversed before, and upon arrival at v there remains at most one such edge, decrease the counter by 1.
3. As long as the counter is positive, the tour is conducted according to the Tarry's algorithm, except, whenever possible, an edge not traversed before is preferred to an edge already traversed.
4. As soon as the counter contains zero, leave all edges via their entrance edges.

The accompanying theorem, whose proof is relatively involved, is derived for the case when the start and target vertices coincide. The theorem states that under Fraenkel's algorithm the target vertex is guaranteed to be reached if reachable, and every edge will be traversed at least once but never more than twice, once in each direction. Using our terminology and the M-line concept, the upper bound on the length of paths generated by the Fraenkel's algorithm is as follows:

Theorem 2.9.4. *For any finite maze, Fraenkel's algorithm generates a path of length P such that*

$$P \leq 2D + 2\sum_i p_i \qquad (2.24)$$

where D is the length of M-line, and p_i are perimeters of obstacles in the maze.

In other words, the worst-case estimates of the length of generated paths for Trumaux's, Tarry's, and Fraenkel's algorithms are identical. The performance of Fraenkel's algorithm can be better, and never worse, than that of the two other algorithms. As an example, if the graph presents a Euler graph, Fraenkel's robot will traverse each edge only once.

2.9.2 Maze-to-Graph Transition

It is interesting to note that until the advent of robotics, all work on labyrinth search methods was limited to graphs. Each of the strategies above is based solely on graph-theoretical considerations, irrespective of the geometry and topology of mazes that produce those connectivity graphs. That is why constructs like the M-line are foreign to those methods. (M-line was not of course a part of the works above; it was introduced here to make this material consistent with the algorithmic work that will follow.) One can only speculate with regard to the reasons: Perhaps it might be the power of Euler's ideas and the appeal of models of graph theory.

Whatever the reason, the universal substitution of mazes by graphs made the researchers overlook some additional information and some rich problems and formulations that are relevant to physical mazes but are easily lost in the transition to general graphs. These are, for example: (a) the fact that any physical obstacle boundary must present a closed curve, and this fact can be used for motion planning; (b) the fact that the continuous space between obstacles present an infinite number of options for moving in free space between obstacles; and (c) the fact that in space there is a sense of direction (one can use, for example, a compass) which disappears in a graph. (See more on this later in this and next chapter.)

Strategies that take into account such considerations stay somewhat separate from the algorithms cited above that deal directly with graph processing. As input information is assumed in these algorithms to come from on-line sensing, we will call them sensor-based algorithms and consider them in the next section, before embarking on development and analysis of such algorithms in the following chapters.

2.9.3 Sensor-Based Motion Planning

The problem of robot path planning in an uncertain environment has been first considered in the context of heuristic approaches and as applied to autonomous vehicle navigation. Although robot arm manipulators are very important for theory and practice, little has been done for them until later, when the underlying issues became clearer. An incomplete list of path planning heuristics includes Refs. 28 and 47–52.

Not rarely, attempts for planning with incomplete information have their starting point in the Piano Mover's model and in planning with complete information. For example, in heuristic algorithms considered in Refs. 47, 48 and 50, a piece of the path is formed from the edges of a connectivity graph resulting from modeling the robot's surrounding area for which information is available at the moment (for example, from the robot's vision sensor). As the robot moves to the next area, the process repeats. This means that little can be said about the procedures' chances for reaching the goal. Obstacles are usually approximated with polygons; the corresponding connectivity graph is formed by straightline segments that connect obstacle vertices, the robot starting point, and its target point, with a constraint on nonintersection of graph edges with obstacles.

In these works, path planning is limited to the robot's immediate surroundings, the area for which sensing information on the scene is available from robot sensors. Within this limited area, the problem is actually treated as one with complete information. Sometimes the navigation problem is treated as a hierarchical problem [48, 53], where the upper level is concerned with global navigation for which the information is assumed available, while the lower level is doing local navigation based on sensory feedback. A heuristic procedure for moving a robot arm manipulator among unknown obstacles is described in Ref. 54.

Because the above heuristic algorithms have no theoretical assurance of convergence, it is hard to judge how complete they are. Their explicit or implicit reliance on the so-called common sense is founded on the assumption that humans are good at orienting and navigation in space and at solving geometrical search problems. This assumption is questionable, however, especially in the case of arm manipulators. As we will see in Chapter 7, when lacking global input information and directional clues, human operators are confused, lose their sense of orientation, and exhibit inferior performance. Nevertheless, in relatively simple scenes, such heuristic procedures have been shown to produce an acceptable performance.

More recently, algorithms have been reported that do not have the above limitations—they treat obstacles as they come, have a proof of convergence, and so on—and are closer to the SIM model. All these works deal with motion planning for mobile robots; the strategies they propose are in many ways close to the algorithms studied further in Chapter 3. These works will be reviewed later, in Section 3.8, once we are ready to discuss the underlying issues.

With time the SIM paradigm acquired popularity and found a way to applications. Algorithms with guaranteed convergence appeared, along with a plethora of heuristic schemes. Since knowing the robot location is important for motion planning, some approaches attempted to address robot localization and motion planning within the same framework.[8] Other approaches assume that, similar to human and animals' motion planning, the robot's location in space should come from sensors or from some separate sensor processing software, and so they concentrate on motion planning and collision-avoidance strategies.

Consider the scene shown in Figure 2.22. A point robot starts at point S and attempts to reach the target point T. Since the robot knows at all times where point T is, a simple strategy would be to walk toward T whenever possible. Once the robot's sensor informs it about the obstacle O_1 on its way, it will start passing around it, for only as long as it takes to clear the direction toward T, and then continue toward T. Note that the efficiency of this strategy is independent of the complexity of obstacles in the scene: No matter how complex (say, fiord-like) an obstacle boundary is, the robot will simply walk along this boundary.

One can easily build examples where this simple idea will not work, but we shall see in the sequel that slightly more complex ideas of this kind can work and even guarantee a solution in an arbitrary scene, in spite of the high uncertainty and scant knowledge about the scene. Even more interesting, despite the fact that arm manipulators present a much more complex case for navigation than do mobile robots, such strategies are feasible for robot arm manipulators as well. To repeat, in these strategies, (a) the robot can start with zero information about the scene,

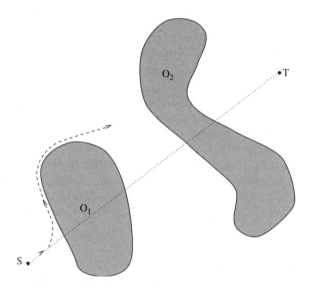

Figure 2.22   A point robot starts at point S and attempts to reach the target location T. No knowledge about the scene is available beforehand, and no computations are done prior to the motion. As the robot encounters an obstacle, it passes it around and then continues toward T. If feasible, such a strategy would allow real-time motion planning, and its complexity would be a constant function of the scene complexity.

---

[8]One name for procedures that combine localization and motion planning is *SLAM*, which stands for Simultaneous Localization and Motion Planning (see, e.g., Ref. 55).

(b) the robot uses only a small amount of local information about obstacles delivered by its sensors, and (c) the complexity of motion planning is a constant function of the complexity of obstacles (interpreted as above, as the maximum number of times the robot visits some pieces of its path). We will build these algorithms in the following chapters. For now, it is clear that, if feasible, such procedures will likely save the robot a tremendous amount of data processing compared to models with complete information.

The only complete (nonheuristic) algorithm for path planning in an uncertain environment that was produced in this earlier period seems to be the Pledge algorithm described by Abelson and diSessa [36]. The algorithm is shown to converge; no performance bounds are given (its performance was assessed later in Ref. 56). However, the algorithm addresses a problem different from ours: The robot's task is to escape from an arbitrary maze. It can be shown that the Pledge algorithm cannot be used for the common mobile robot task of reaching a specific point inside or outside the maze.

That the convergence of motion planning algorithms with uncertainty cannot be left to one's intuition is underscored by the following example, where a seemingly reasonable strategy can produce disappointing results. Consider this algorithm; let us call it Optimist[9]:

1. Walk directly toward the target until one of these occurs:
   (a) The target is reached. The procedure stops.
   (b) An obstacle is encountered. Go to Step 2.
2. Turn left and follow the obstacle boundary until one of these occurs:
   (a) The target is reached. The procedure stops.
   (b) The direction toward the target clears. Go to Step 1.

Common sense suggests that this procedure should behave reasonably well, at least in simpler scenes. Indeed, even complex-looking examples can be readily designed where the algorithm Optimist will successfully bring the robot to the target location. Unfortunately, it is equally easy to produce simple scenes in which the algorithm will fail. In the scene shown in Figure 2.23a, for example, the algorithm would take the robot to infinity instead of the target, and in the scene of Figure 2.23b the algorithm forces the robot into infinite looping. (Depending on the scheme's details, it may produce the loop 1 or the loop 2.) Attempts to fix this scheme with other common-sense modifications—for example, by alternating the left and right direction of turns in Step 2 of the algorithm—will likely only shift the problem: the algorithm will perhaps succeed in the scenes in Figure 2.23 but fail in some other scenes.

This example suggests that unless convergence of an algorithm is proven formally, the danger of the robot going astray under its guidance is real. As we will see later, the problem becomes even more unintuitive in the case of

---

[9]The procedure has been frequently suggested to me at various meetings.

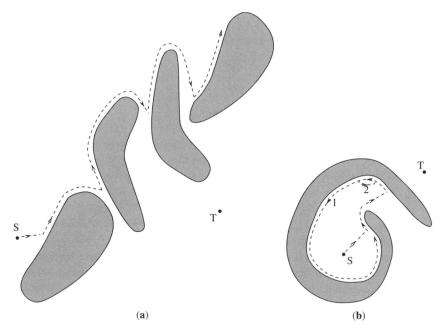

(a)   (b)

Figure 2.23 In scene **(a)** algorithm Optimist will take the robot arbitrarily far from the target T. In scene **(b)** depending on its small details, it will go into one of infinite loops shown.

arm manipulators. Hence, from now on, we will concentrate on the SIM (sensing–intelligence–motion) paradigm, and in particular on provable sensor-based motion planning algorithms.

As said above, instead of focusing on geometry of space, as in the Piano Mover's model, SIM procedures exploit topological properties of space. Limiting ourselves for now to the 2D plane, notice that an obstacle in a 2D scene is a simple closed curve. If one starts at some point outside the obstacle and walks around it—say, clockwise—eventually one will arrive at the starting point. This is true, independent of the direction of motion: If one walks instead counterclockwise, one will still arrive at the same starting point. This property does not depend on whether the obstacle is a square or a triangle or a circle or an arbitrary object of complex shape.

However complex the robot workspace is—and it will become even more complex in the case of 3D arm manipulators—the said property still holds. If we manage to design algorithms that can exploit this property, they will likely be very stable to the uncertainties of a real-world scenes. We can then turn to other complications that a real-world algorithm has to respect: finite dimensions of the robot itself, improving the algorithm performance with sensors like vision, the effect of robot dynamics on motion planning, and so on. We are now ready to tackle those issues in the following chapters.

2.10 EXERCISES

1. Develop direct and inverse kinematics equations, for both position and velocity, for a two-link planar arm manipulator, the so-called RP arm, where R means "revolute joint" and P means "prismatic" (or sliding) joint (see Figure 2.E.1). The sliding link l_2 is perpendicular to the revolute link l_1, and has the front and rear ends; the front end holds the arm's end effector (the hand). Draw a sketch. Analyze degeneracies, if any. Notation: $\theta_1 = [0, 2\pi]$, $l_2 = [l_{2\,min}, l_{2\,max}]$; ranges of both joints, respectively: $l_2 = (l_{2\,max} - l_{2\,min})$; $l_1 = \text{const} > 0$ – lengths of links.

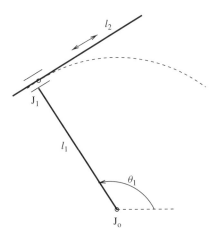

Figure 2.E.1

2. Design a straight-line path of bounded accuracy for a planar RR (revolute–revolute) arm manipulator, given the starting S and target T positions, $(\theta_{1S}, \theta_{2S})$ and $(\theta_{1T}, \theta_{2T})$:

$$\theta_{1S} = \pi/4, \qquad \theta_{2S} = \pi/2, \qquad \theta_{1T} = 0, \qquad \theta_{2T} = \pi/6$$

3. The lengths of arm links are $l_1 = 50$ and $l_2 = 70$. Angles θ_1 and θ_2 are measured counterclockwise, as shown in Figure 2.E.2.
   Find the minimum number of knot points for the path that will guarantee that the deviation of the actual path from the straight line (S, T) will be within the error $\delta = 2$. The knot points are not constrained to lie on the line (S, T) or to be spread uniformly between points S and T. Discuss significance of these conditions. Draw a sketch. Explain why your knot number is minimum.

4. Consider the best- and worst-case performance of Tarry's algorithm in a planar graph. The algorithm's objective is to traverse the whole graph and return to the starting vertex. Design a planar graph that would provide to Tarry algorithm different options for motion, and such that the algorithm would

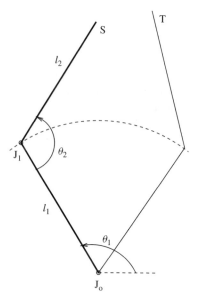

Figure 2.E.2

achieve in it its best-case performance if it were "lucky" with its choices of directions of motion, and its worst-case performance if it were "unlucky." Explain your reasoning.

5. Assuming two C-shaped obstacles in the plane, along with an M-line that connects two distinct points S and T and intersects both obstacles, design two examples that would result in the best-case and worst-case performance, respectively, of Tarry's algorithm. An obstacle can be mirror image reversed if desired. Obstacles can touch each other, in which case the point robot would not be able to pass between them at the contact point(s). Evaluate the algorithm's performance in each case.

CHAPTER 3

Motion Planning for a Mobile Robot

Thou mayst not wander in that labyrinth; There Minotaurs and ugly treasons lurk.

— *William Shakespeare, King Henry the Sixth*

What is the difference between exploring and being lost?

— *Dan Eldon, photojournalist*

As discussed in Chapter 1, to plan a path for a mobile robot means to find a continuous trajectory leading from its initial position to its target position. In this chapter we consider a case where the robot is a point and where the scene in which the robot travels is the two-dimensional plane. The scene is populated with unknown obstacles of arbitrary shapes and dimensions. The robot knows its own position at all times, and it also knows the position of the target that it attempts to reach. Other than that, the only source of robot's information about the surroundings is its sensor. This means that the input information is of a local character and that it is always partial and incomplete. In fact, the sensor is a simple tactile sensor: It will detect an obstacle only when the robot touches it. "Finding a trajectory" is therefore a process that goes on in parallel with the journey: The robot will finish finding the path only when it arrives at the target location.

We will need this model simplicity and the assumption of a point robot only at the beginning, to develop the basic concepts and algorithms and to produce the upper and lower bound estimates on the robot performance. Later we will extend our algorithmic machinery to more complex and more practical cases, such as nonpoint (physical) mobile robots and robot arm manipulators, as well as to more complex sensing, such as vision or proximity sensing. To reflect the abstract nature of a point robot, we will interchangeably use for it the term *moving automaton* (*MA*, for brevity), following some literature cited in this chapter.

Other than those above, no further simplifications will be necessary. We will not need, for example, the simplifying assumptions typical of approaches that deal with complete input information such as approximation of obstacles with algebraic and semialgebraic sets; representation of the scene with intermediate structures such as connectivity graphs; reduction of the scene to a discrete space; and so on. Our robot will treat obstacles as they are, as they are sensed by its sensor. It will deal with the real continuous space—which means that all points of the scene are available to the robot for the purpose of motion planning.

The approach based on this model (which will be more carefully formalized later) forms the *sensor-based motion planning paradigm*, or, as we called it above, SIM (Sensing–Intelligence–Motion). Using algorithms that come out of this paradigm, the robot is continuously analyzing the incoming sensing information about its current surroundings and is continuously planning its path. The emphasis on strictly local input information is somewhat similar to the approach used by Abelson and diSessa [36] for treating geometric phenomena based on local information: They ask, for example, if a turtle walking along the sides of a triangle and seeing only a small area around it at every instant would have enough information to prove triangle-related theorems of Euclidean geometry. In general terms, the question being posed is, Can one make global inferences based solely on local information? Our question is very similar: Can one guarantee a global solution—that is, a path between the start and target locations of the robot—based solely on local sensing?

Algorithms that we will develop here are *deterministic*. That is, by running the same algorithm a few times in the same scene and with the same start and target points, the robot should produce identical paths. This point is crucial: One confusion in some works on robot motion planning comes from a view that the uncertainty that is inherent in the problem of motion planning with incomplete information necessarily calls for probabilistic approaches. This is not so.

As discussed in Chapter 1, the sensor-based motion planning paradigm is distinct from the paradigm where complete information about the scene is known to the robot beforehand—the so-called *Piano Mover's model* [16] or *motion planning with complete information*. The main question we ask in this and the following chapters is whether, under our model of sensor-based motion planning, *provable* (*complete* and *convergent* are equivalent terms) path planning algorithms can be designed. If the answer is yes, this will mean that no matter how complex the scene is, under our algorithms the robot will find a path from start to target, or else will conclude in a finite time that no such path exists if that is the case.

Sometimes, approaches that can be classified as sensor-based planning are referred to in literature as *reactive planning*. This term is somewhat unfortunate: While it acknowledges the local nature of robot sensing and control, it implicitly suggests that a sensor-based algorithm has no way of inferring any global characteristics of space from local sensing data ("the robot just reacts"), and hence cannot guarantee anything in global terms. As we will see, the sensor-based planning paradigm can very well account for space global properties and can guarantee algorithms' global *convergence*.

Recall that by judiciously using the limited information they managed to get about their surroundings, our ancestors were able to reach faraway lands while avoiding many obstacles, literally and figuratively, on their way. They had no maps. Sometimes along the way they created maps, and sometimes maps were created by those who followed them. This suggests that one does not have to know everything about the scene in order to solve the go-from-A-to-B motion planning problem. By always knowing one's position in space (recall the careful triangulation of stars the seaman have done), by keeping in mind where the target position is relative to one's position, and by remembering two or three key locations along the way, one should be able to infer some important properties of the space in which one travels, which will be sufficient for getting there. Our goal is to develop strategies that make this possible.

Note that the task we pose to the robot does not include producing a map of the scene in which it travels. All we ask the robot to do is go from point A to point B, from its current position to some target position. This is an important distinction. If all I need to do is find a specific room in an unfamiliar building, I have no reason to go into an expensive effort of creating a map of the building. If I start visiting the same room in that same building often enough, eventually I will likely work out a more or less optimal route to the room—though even then I will likely not know of many nooks and crannies of the building (which would have to appear in the map). In other words, map making is a different task that arises from a different objective. A map may perhaps appear as a by-product of some path planning algorithm; this would be a rather expensive way to do path planning, but this may happen. We thus emphasize that one should distinguish between path planning and map making.

Assuming for now that sensor-based planning algorithms are viable and computationally simple enough for real-time operation and also assuming that they can be extended to more complex cases—nonpoint (physical) robots, arm manipulators, and complex nontactile sensing—the SIM paradigm is clearly very attractive. It is attractive, first of all, from the practical standpoint:

1. Sensors are a standard fare in engineering and robot technology.
2. The SIM paradigm captures much of what we observe in nature. Humans and animals solve complex motion planning tasks all the time, day in and day out, while operating with local sensing information. It would be wonderful to teach robots to do the same.
3. The paradigm does away with complex gathering of information about the robot's surroundings, replacing it with a continuous processing of incoming sensor information. This, in turn, allows one not to worry about the shapes and locations of obstacles in the scene, and perhaps even handle scenes with moving or shape-changing obstacles.
4. From the control standpoint, sensor-based motion planning introduces the powerful notion of *sensor feedback control*, thus transforming path planning into a continuous on-line control process. The fact that local sensing information is sufficient to solve the global task (which we still need to prove) is good news: Local information is likely to be simple and easy to process.

These attractive points of sensor-based planning stands out when comparing it with the paradigm of motion planning with complete information (the Piano Mover's model). The latter requires the complete information about the scene, and it requires it up front. Except in very simple cases, it also requires formidable calculations; this rules out a real-time operation and, of course, handling moving or shape-changing obstacles.

From the standpoint of theory, the main attraction of sensor-based planning is the surprising fact that in spite of the local character of robot sensing and the high level of uncertainly—after all, practically nothing may be known about the environment at any given moment—SIM algorithms can guarantee reaching a global goal, even in the most complex environment.

As mentioned before, those positive sides of the SIM paradigm come at a price. Because of the dynamic character of incoming sensor information—namely, at any given moment of the planning process the future is not known, and every new step brings in new information—the path cannot be preplanned, and so its *global optimality is ruled out*. In contrast, the Piano Mover's approach can in principle produce an optimal solution, simply because it knows everything there is to know.[1] In sensor-based planning, one looks for a "reasonable path," a path that looks acceptable compared to what a human or other algorithms would produce under similar conditions. For a more formal assessment of performance of sensor-based algorithms, we will develop some bounds on the length of paths generated by the algorithms. In Chapter 7 we will try to assess human performance in motion planning.

Given our continuous model, we will not be able to use the discrete criteria typically used for evaluating algorithms of computational geometry—for example, assessing a task complexity as a function of the number of vertices of (polygonal or otherwise algebraically defined) obstacles. Instead, a new *path-length performance criterion* based on the length of generated paths as a function of obstacle perimeters will be developed.

To generalize performance assessment of our path planning algorithms, we will develop the lower bound on paths generated by any sensor-based planning algorithm, expressed as the length of path that the best algorithm would produce in the worst case. As known in complexity theory, the difficulty of this task lies in "fighting an unknown enemy"—we do not know how that best algorithm may look like.

This lower bound will give us a yardstick for assessing individual path planning algorithms. For each of those we will be interested in the upper bound on the algorithm performance—the worst-case scenario for a specific algorithm. Such results will allow us to compare different algorithms and to see how far are they from an "ideal" algorithm.

All sensor-based planning algorithms can be divided into these two nonoverlapping intuitively transparent classes:

---

[1]In practice, while obtaining the optimal solution is often too computationally expensive, the ever-increasing computer speeds make this feasible for more and more problems.

Class 1. Algorithms in which the robot explores each obstacle that it encounters completely before it goes to the next obstacle or to the target.

Class 2. Algorithms where the robot can leave an obstacle that it encounters without exploring it completely.

The distinction is important. Algorithms of Class 1 are quite "thorough"—one may say, quite conservative. Often this irritating thoroughness carries the price: From the human standpoint, paths generated by a Class 1 algorithm may seem unnecessarily long and perhaps a bit silly. We will see, however, that this same thoroughness brings big benefits in more difficult cases. Class 2 algorithms, on the other hand, are more adventurous—they are "more human", they "take risks." When meeting an obstacle, the robot operating under a Class 2 algorithm will have no way of knowing if it has met it before. More often than not, a Class 2 algorithm will win in real-life scenes, though it may lose badly in an unlucky scene.

As we will see, the sensor-based motion planning paradigm exploits two essential topological properties of space and objects in it—the *orientability* and *continuity* of manifolds. These are expressed in topology by the Jordan Curve Theorem [57], which states:

> Any closed curve homeomorphic to a circle drawn around and in the vicinity of a given point on an orientable surface divides the surface into two separate domains, for which the curve is their common boundary.

The threateningly sounding "orientable surface" clause is not a real constraint. For our two-dimensional case, the *Moebius strip* and *Klein bottle* are the only examples of nonorientable surfaces. Sensor-based planning algorithms would not work on these surfaces. Luckily, the world of real-life robotics never deals with such objects.

In physical terms, the Jordan Curve Theorem means the following: (a) If our mobile robot starts walking around an obstacle, it can safely assume that at some moment it will come back to the point where it started. (b) There is no way for the robot, while walking around an obstacle, to find itself "inside" the obstacle. (c) If a straight line—for example, the robot's intended path from start to target—crosses an obstacle, there is a point where the straight line enters the obstacle and a point where it comes out of it. If, because of the obstacle's complex shape, the line crosses it a number of times, there will be an equal number of entering and leaving points. (The special case where the straight line touches the obstacle without crossing it is easy to handle separately—the robot can simply ignore the obstacle.)

These are corollaries of the Jordan Curve Theorem. They will be very explicitly used in the sensor-based algorithms, and they are the basis of the algorithms' convergence. One positive side effect of our reliance on topology is that geometry of space is of little importance. An obstacle can be polygonal or circular, or of a shape that for all practical purposes is impossible to define in mathematical terms; for our algorithm it is only a closed curve, and so handling one is as easy as the other. In practice, reliance on space topology helps us tremendously in computational savings: There is no need to know objects' shapes and dimensions in advance, and there is no need to describe and store object descriptions once they have been visited.

In Section 3.1 below, the formal model for the sensor-based motion planning paradigm is introduced. The universal lower bound on paths generated by any algorithm operating under this model is then produced in Section 3.2. One can see the bound as the length of a path that the best algorithm in the world will generate in the most "uncooperating" scene. In Sections 3.3.1 and 3.3.2, two provably correct path planning algorithms are described, called Bug1 and Bug2, one from Class 1 and the other from Class 2, and their convergence properties and performance upper bounds are derived. Together the two are called *basic algorithms*, to indicate that they are the base for later strategies in more complex cases. They also seem to be the first and simplest provable sensor-based planning algorithms known. We will formulate tests for target reachability for both algorithms and will establish the (worst-case) upper bounds on the length of paths they generate.

Analysis of the two algorithms will demonstrate that a better upper bound on an algorithm's path length does not guarantee shorter paths. Depending on the scene, one algorithm can produce a shorter path than the other. In fact, though Bug2's upper bound is much worse than that of Bug1, Bug2 will be likely preferred in real-life tasks.

In Sections 3.4 and 3.5 we will look at further ways to obtain better algorithms and, importantly, to obtain tighter performance bounds. In Section 3.6 we will expand the basic algorithms—which, remember, deal with tactile sensing—to richer sensing, such as vision. Sections 3.7 to 3.10 deal with further extensions to real-world (nonpoint) robots, and compare different algorithms. Exercises for this chapter appear in Section 3.11.

3.1 THE MODEL

The model includes two parts: One is related to geometry of the robot (automaton) environment, and the other is related to characteristics and capabilities of the automaton. To save on multiple uses of words "robot" and "automaton," we will call it MA, for "moving automaton."

Environment. The scene in which MA operates is a plane. The scene may be populated with obstacles, and it has two given points in it: the MA starting location, S, and the target location, T. Each obstacle's boundary is a simple closed curve of finite length, such that a straight line will cross it in only finitely many points. The case when the straight line is tangential to an obstacle at a point or coincides with a finite segment of the obstacle is not a "crossing." Obstacles do not touch each other; that is, a point on an obstacle belongs to one and only one obstacle (if two obstacles do touch, they will be considered one obstacle). The scene can contain only a locally finite number of obstacles. This means that any disc of finite radius intersects a finite set of obstacles. Note that the model does not require that the scene or the overall set of obstacles be finite.

Robot. MA is a point. This means that an opening of any size between two distinct obstacles can be passed by MA. MA's motion skills include three actions: It knows how to move toward point T on a straight line, how to move along the obstacle boundary, and how to start moving and how to stop. The only input information that MA is provided with is (1) coordinates of points S and T as well as MA's current locations and (2) the fact of contacting an obstacle. The latter means that MA has a tactile sensor. With this information, MA can thus calculate, for example, its direction toward point T and its distance from it. MA's memory for storing data or intermediate results is limited to a few computer words.

Definition 3.1.1. *A local direction is a once-and-for-all decided direction for passing around an obstacle. For the two-dimensional problem, it can be either left or right.*

That is, if the robot encounters an obstacle and intends to pass it around, it will walk around the obstacle clockwise if the chosen local direction is "left," and walk around it counterclockwise if the local direction is "right." Because of the inherent uncertainty involved, every time MA meets an obstacle, there is no information or criteria it can use to decide whether it should turn left or right to go around the obstacle. For the sake of consistency and without losing generality, unless stated otherwise, let us assume that the local direction is always *left*, as in Figure 3.5.

Definition 3.1.2. *MA is said to define a hit point on the obstacle, denoted H, when, while moving along a straight line toward point T, it contacts the obstacle at the point H. It defines a leave point, L, on the obstacle when it leaves the obstacle at point L in order to continue its walk toward point T. (See Figure 3.5.)*

In case MA moves along a straight line toward point T and the line touches some obstacle tangentially, there is no need to invoke the procedure for walking around the obstacle—MA will simply continue its straight-line walk toward point T. This means that no H or L points will be defined in this case. Consequently, no point of an obstacle can be defined as both an H and an L point. In order to define an H or an L point, the corresponding straight line has to produce a "real" crossing of the obstacle; that is, in the vicinity of the crossing, a finite segment of the line will lie inside the obstacle and a finite segment of the line will lie outside the obstacle.

Below we will need the following notation:

D is Euclidean distance between points S and T.

$d(A, B)$ is Euclidean distance between points A and B in the scene; $d(S, T) = D$.

$d(A)$ is used as a shorthand notation for $d(A, T)$.

$d(A_i)$ signifies the fact that point A is located on the boundary of the ith obstacle met by MA on its way to point T.

P is the total length of the path generated by MA on its way from S to T.

p_i is the perimeter of ith obstacle encountered by MA.

$\sum_i p_i$ is the sum of perimeters of obstacles met by MA on its way to T, or of obstacles contained in a specific area of the scene; this quantity will be used to assess performance of a path planning algorithm or to compare path planning algorithms.

3.2 UNIVERSAL LOWER BOUND FOR THE PATH PLANNING PROBLEM

This lower bound, formulated in Theorem 3.2.1 below, informs us what performance can be expected in the worst case from any path planning algorithm operating within our model. The bound is formulated in terms of the length of paths generated by MA on its way from point S to point T. We will see later that the bound is a powerful means for measuring performance of different path planning procedures.

Theorem 3.2.1 ([58]). *For any path planning algorithm satisfying the assumptions of our model, any (however large) $P > 0$, any (however small) $D > 0$, and any (however small) $\delta > 0$, there exists a scene in which the algorithm will generate a path of length no less than P,*

$$P \geq D + \sum_i p_i - \delta \tag{3.1}$$

where D is the distance between points S and T, and p_i are perimeters of obstacles intersecting the disk of radius D centered at point T.

Proof: We want to prove that for any known or unknown algorithm X a scene can be designed such that the length of the path generated by X in it will satisfy (3.1).[2] Algorithm X can be of any type: It can be deterministic or random; its intermediate steps may or may not depend on intermediate results; and so on. The only thing we know about X is that it operates within the framework of our model above. The proof consists of designing a scene with a special set of obstacles and then proving that this scene will force X to generate a path not shorter than P in (3.1).

---

[2]In Section 3.5 we will learn of a lower bound that is better and tighter: $P \geq D + 1.5 \sum_i p_i - \delta$. The proof of that bound is somewhat involved, so in order to demonstrate the underlying ideas we prefer to consider in detail the easier proof of the bound (3.1).

We will use the following scheme to design the required scene (called the *resultant scene*). The scene is built in two stages. At the first stage, a *virtual obstacle* is introduced. Parts of this obstacle or the whole of it, but not more, will eventually become, when the second stage is completed, the *actual obstacle(s)* of the resultant scene.

Consider a virtual obstacle shown in Figure 3.1a. It presents a corridor of finite width $2W > \delta$ and of finite length L. The top end of the corridor is closed. The corridor is positioned such that the point S is located at the middle point of its closed end; the corridor opens in the direction opposite to the line (S, T). The thickness of the corridor walls is negligible compared to its other dimensions. Still in the first stage, MA is allowed to walk from S to T along the path prescribed by the algorithm X. Depending on the X's procedure, MA may or may not touch the virtual obstacle.

When the path is complete, the second stage starts. A segment of the virtual obstacle is said to be *actualized* if all points of the inside wall of the segment have been touched by MA. If MA has contiguously touched the inside wall of the virtual obstacle at some length l, then the actualized segment is exactly of length l. If MA touched the virtual obstacle at a point and then bounced back, the corresponding actualized area is considered to be a wall segment of length δ around the point of contact. If two segments of the MA's path along the virtual obstacle are separated by an area of the virtual obstacle that MA has not touched, then MA is said to have actualized two separate segments of the virtual obstacle.

We produce the resultant scene by designating as actual obstacles only those areas of the virtual obstacle that have been actualized. Thus, if an actualized

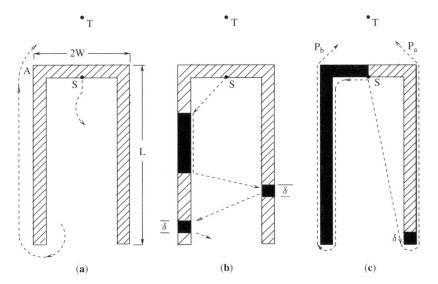

Figure 3.1 Illustration for Theorem 3.2.1. Actualized segments of the virtual obstacle are shown in solid black. S, start point; T, target point.

segment is of length l, then the perimeter of the corresponding actual obstacle is equal to $2l$; this takes into account the inside and outside walls of the segment and also the fact that the thickness of the wall is negligible (see Figure 3.1).

This method of producing the resultant scene is justified by the fact that, under the accepted model, the behavior of MA is affected only by those obstacles that it touches along its way. Indeed, under algorithm X the very same path would have been produced in two different scenes: in the scene with the virtual obstacle and in the resultant scene. One can therefore argue that the areas of the virtual obstacle that MA has not touched along its way might have never existed, and that algorithm X produced its path not in the scene with the virtual obstacle but in the resultant scene. This means the performance of MA in the resultant scene can be judged against (3.1). This completes the design of the scene. Note that depending on the MA's behavior under algorithm X, zero, one, or more actualized obstacles can appear in the scene (Figure 3.1b).

We now have to prove that the MA's path in the resultant scene satisfies inequality (3.1). Since MA starts at a distance $D = d(S, T)$ from point T, it obviously cannot avoid the term D in (3.1). Hence we concentrate on the second term in (3.1). One can see by now that the main idea behind the described process of designing the resultant scene is to force MA to generate, for each actual obstacle, a segment of the path at least as long as the total length of that obstacle's boundary. Note that this characteristic of the path is independent of the algorithm X.

The MA's path in the scene can be divided into two parts, $P1$ and $P2$; $P1$ corresponds to the MA's traveling inside the corridor, and $P2$ corresponds to its traveling outside the corridor. We use the same notation to indicate the length of the corresponding part. Both parts can become intermixed since, after having left the corridor, MA can temporarily return into it. Since part $P2$ starts at the exit point of the corridor, then

$$P2 \geq L + C \qquad (3.2)$$

where $C = \sqrt{D^2 + W^2}$ is the hypotenuse AT of the triangle ATS (Figure 3.1a). As for part $P1$ of the path inside the corridor, it will be, depending on the algorithm X, some curve. Observe that in order to defeat the bound—that is, produce a path shorter than the bound (3.1)—algorithm X has to decrease the "path per obstacle" ratio as much as possible. What is important for the proof is that, from the "path per obstacle" standpoint, every segment of $P1$ that does not result in creating an equivalent segment of the actualized obstacle makes the path worse. All possible alternatives for $P1$ can be clustered into three groups. We now consider these groups separately.

1. Part $P1$ of the path never touches walls of the virtual obstacle (Figure 3.1a). As a result, no actual obstacles will be created in this case, $\sum_i p_i = 0$. Then the resulting path is $P > D$, and so for an algorithm X that produces this kind of path the theorem holds. Moreover, at the final evaluation, where

only actual obstacles count, the algorithm X will not be judged as efficient: It creates an additional path component at least equal to $(2 \cdot L + (C - D))$, in a scene with no obstacles!

2. MA touches more than once one or both inside walls of the virtual obstacle (Figure 3.1b). That is, between consecutive touches of walls, MA is temporarily "out of touch" with the virtual obstacle. As a result, part $P1$ of the path will produce a number of disconnected actual obstacles. The smallest of these, of length δ, corresponds to point touches. Observe that in terms of the "path per obstacle" assessment, this kind of strategy is not very wise either. First, for each actual obstacle, a segment of the path at least as long as the obstacle perimeter is created. Second, additional segments of $P1$, those due to traveling between the actual obstacles, are produced. Each of these additional segments is at least not smaller than $2W$, if the two consecutive touches correspond to the opposite walls of the virtual obstacle, or at least not smaller than the distance between two sequentially visited disconnected actual obstacles on the same wall. Therefore, the length P of the path exceeds the right side in (3.1), and so the theorem holds.

3. MA touches the inside walls of the virtual obstacle at most once. This case includes various possibilities, from a point touching, which creates a single actual obstacle of length δ, to the case when MA closely follows the inside wall of the virtual obstacle. As one can see in Figure 3.1c, this case contains interesting paths. The shortest possible path would be created if MA goes directly from point S to the furthest point of the virtual obstacle and then directly to point T (path P_a, Figure 3.1c). (Given the fact that MA knows nothing about the obstacles, a path that good can be generated only by an accident.) The total perimeter of the obstacle(s) here is 2δ, and the theorem clearly holds.

Finally, the most efficient path, from the "path per obstacle" standpoint, is produced if MA closely follows the inside wall of the virtual obstacle and then goes directly to point T (path P_b, Figure 3.1c). Here MA is doing its best in trying to compensate each segment of the path with an equivalent segment of the actual obstacle. In this case, the generated path P is equal to

$$P = \sum_i p_i + \sqrt{D^2 + W^2} - W \tag{3.3}$$

(In the path P_b in Figure 3.1c, there is only one term in $\sum_i p_i$.) Since no constraints have been imposed on the choice of lengths D and W, take them such that

$$\delta \geq D + W - \sqrt{D^2 + W^2} \tag{3.4}$$

which is always possible because the right side in (3.4) is nonnegative for any D and W. Reverse both the sign and the inequality in (3.4), and add

$(D + \sum_i p_i)$ to its both sides. With a little manipulation, we obtain

$$\sum_i p_i + \sqrt{D^2 + W^2} - W \geq D + \sum_i p_i - \delta \qquad (3.5)$$

Comparing (3.3) and (3.5), observe that (3.1) is satisfied.

This exhausts all possible cases of path generation by the algorithm X. **Q.E.D.**

We conclude this section with two remarks. First, by appropriately select-ing multiple virtual obstacles, Theorem 3.2.1 can be extended to an arbitrary number of obstacles. Second, for the lower bound (3.1) to hold, the constraints on the information available to MA can be relaxed significantly. Namely, the only required constraint is that at any time moment MA does not have complete information about the scene.

We are now ready to consider specific sensor-based path planning algorithms. In the following sections we will introduce three algorithms, analyze their per-formance, and derive the upper bounds on the length of the paths they generate.

3.3 BASIC ALGORITHMS

3.3.1 First Basic Algorithm: Bug1

This procedure is executed at every point of the MA's (continuous) path [17, 58]. Before describing it formally, consider the behavior of MA when operating under this procedure (Figure 3.2). According to the definitions above, when on its way from point S (Start) to point T (Target), MA encounters an ith obstacle, it defines on it a *hit point* $H_i, i = 1, 2, \ldots$. When leaving the ith obstacle in order to continue toward T, MA defines a *leave point* L_i. Initially $i = 1$, $L_0 = S$. The procedure will use three registers—R_1, R_2, and R_3—to store intermediate information. All three are reset to zero when a new hit point is defined. The use of the registers is as follows:

- R_1 is used to store coordinates of the latest point, Q_m, of the minimum distance between the obstacle boundary and point T; this takes one com-parison at each path point. (In case of many choices for Q_m, any one of them can be taken.)
- R_2 integrates the length of the ith obstacle boundary starting at H_i.
- R_3 integrates the length of the ith obstacle boundary starting at Q_m.

We are now ready to describe the algorithm's procedure. The test for target reachability mentioned in Step 3 of the procedure will be explained further in this section.

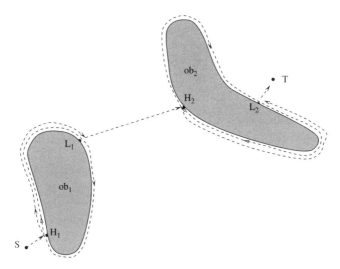

Figure 3.2 The path of the robot (dashed lines) under algorithm Bug1. ob_1 and ob_2 are obstacles, H_1 and H_2 are hit points, L_1 and L_2 are leave points.

Bug1 Procedure

1. From point L_{i-1}, move toward point T (Target) along the straight line until one of these occurs:
   (a) Point T is reached. The procedure stops.
   (b) An obstacle is encountered and a hit point, H_i, is defined. Go to Step 2.
2. Using the local direction, follow the obstacle boundary. If point T is reached, stop. Otherwise, after having traversed the whole boundary and having returned to H_i, define a new leave point $L_i = Q_m$. Go to Step 3.
3. Based on the contents of registers R_2 and R_3, determine the shorter way along the boundary to point L_i, and use it to reach L_i. Apply the test for target reachability. If point T is not reachable, the procedure stops. Otherwise, set $i = i + 1$ and go to Step 1.

Analysis of Algorithm Bug1

Lemma 3.3.1. *Under Bug1 algorithm, when MA leaves a leave point of an obstacle in order to continue toward point T, it will never return to this obstacle again.*

Proof: Assume that on its way from point S to point T, MA does meet some obstacles. We number those obstacles in the order in which MA encounters them. Then the following sequence of distances appears:

$$D, \ d(H_1), \ d(L_1), \ d(H_2), \ d(L_2), \ d(H_3), \ d(L_3), \ldots$$

If point S happens to be on an obstacle boundary and the line (S, T) crosses that obstacle, then $D = d(H_1)$.

According to our model, if MA's path touches an obstacle tangentially, then MA needs not walk around it; it will simply continue its straight-line walk toward point T. In all other cases of meeting an ith obstacle, unless point T lies on an obstacle boundary, a relation $d(H_i) > d(L_i)$ holds. This is because, on the one hand, according to the model, any straight line (except a line that touches the obstacle tangentially) crosses the obstacle at least in two distinct points. This is simply a reflection of the finite "thickness" of obstacles. On the other hand, according to algorithm Bug1, point L_i is the closest point from obstacle i to point T. Starting from L_i, MA walks straight to point T until (if ever) it meets the $(i + 1)$th obstacle. Since, by the model, obstacles do not touch one another, then $d(L_i) > d(H_{i+1})$. Our sequence of distances, therefore, satisfies the relation

$$d(H_1) > d(L_1) > d(H_2) > d(L_2) > d(H_3) > d(L_3) \ldots \qquad (3.6)$$

where $d(H_1)$ is or is not equal to D. Since $d(L_i)$ is the shortest distance from the ith obstacle to point T, and since (3.6) guarantees that algorithm Bug1 monotonically decreases the distances $d(H_i)$ and $d(L_i)$ to point T, Lemma 3.3.1 follows. **Q.E.D.**

The important conclusion from Lemma 3.3.1 is that algorithm Bug1 guarantees to never create cycles.

Corollary 3.3.1. *Under Bug1, independent of the geometry of an obstacle, MA defines on it no more than one hit and no more than one leave point.*

To assess the algorithm's performance—in particular, we will be interested in the upper bound on the length of paths that it generates—an assurance is needed that on its way to point T, MA can encounter only a finite number of obstacles. This is not obvious: While following the algorithm, MA may be "looking" at the target not only from different distances but also from different directions. That is, besides moving toward point T, it may also rotate around it (see Figure 3.3). Depending on the scene, this rotation may go first, say, clockwise, then counterclockwise, then again clockwise, and so on. Hence we have the following lemma.

Lemma 3.3.2. *Under Bug1, on its way to the Target, MA can meet only a finite number of obstacles.*

Proof: Although, while walking around an obstacle, MA may sometimes be at distances much larger than D from point T (see Figure 3.3), the straight-line segments of its path toward the point T are always within the same circle of radius D centered at point T. This is guaranteed by inequality (3.6). Since,

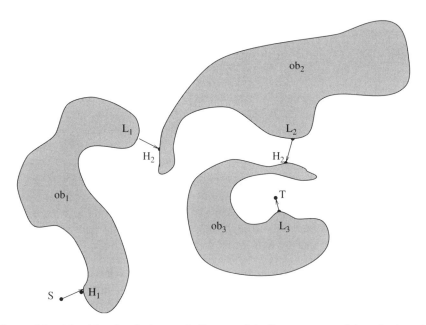

Figure 3.3   Algorithm Bug1. Arrows indicate straight-line segments of the robot's path. Path segments around obstacles are not shown; they are similar to those in Figure 3.2.

according to our model, any disc of finite radius can intersect only a finite number of obstacles, the lemma follows. **Q.E.D.**

Corollary 3.3.2.   *The only obstacles that MA can be meet under algorithm Bug1 are those that intersect the disk of radius D centered at target T.*

Together, Lemma 3.3.1, Lemma 3.3.2, and Corollary 3.3.2 guarantee convergence of the algorithm Bug1.

Theorem 3.3.1.   *Algorithm Bug1 is convergent.*

We are now ready to tackle the performance of algorithm Bug1. As discussed, it will be established in terms of the length of paths that the algorithm generates. The following theorem gives an upper bound on the path lengths produced by Bug1.

Theorem 3.3.2.   *The length of paths produced by algorithm Bug1 obeys the limit,*

$$P \leq D + 1.5 \cdot \sum_i p_i \qquad (3.7)$$

where D is the distance (Start, Target), and $\sum_i p_i$ includes perimeters of obstacles intersecting the disk of radius D centered at the Target.

Proof: Any path generated by algorithm Bug1 can be looked at as consisting of two parts: (a) straight-line segments of the path while walking in free space between obstacles and (b) path segments when walking around obstacles. Due to inequality (3.6), the sum of the straight-line segments will never exceed D. As to path segments around obstacles, algorithm Bug1 requires that in order to define a leave point on the ith obstacle, MA has to first make a "full circle" along its boundary. This produces a path segment equal to one perimeter, p_i, of the ith obstacle, with its end at the hit point. By the time MA has completed this circle and is ready to walk again around the ith obstacles from the hit to the leave point, in order to then depart for point T, the procedure prescribes it to go along the shortest path. By then, MA knows the direction (going left or going right) of the shorter path to the leave point. Therefore, its path segment between the hit and leave points along the ith obstacle boundary will not exceed $0.5 \cdot p_i$. Summing up the estimates for straight-line segments of the path and segments around the obstacles met by MA on its way to point T, obtain (3.7). **Q.E.D.**

Further analysis of algorithm Bug1 shows that our model's requirement that MA knows its own coordinates at all times can be eased. It suffices if MA knows only its distance to and direction toward the target T. This information would allow it to position itself at the circle of a given radius centered at T. Assume that instead of coordinates of the current point Q_m of minimum distance between the obstacle and T, we store in register R_1 the minimum distance itself. Then in Step 3 of the algorithm, MA can reach point Q_m by comparing its current distance to the target with the content of register R_1. If more than one point of the current obstacle lie at the minimum distance from point T, any one of them can be used as the leave point, without affecting the algorithm's convergence.

In practice, this reformulated requirement may widen the variety of sensors the robot can use. For example, if the target sends out, equally in all directions, a low-frequency radio signal, a radio detector on the robot can (a) determine the direction on the target as one from which the signal is maximum and (b) determine the distance to it from the signal amplitude.

Test for Target Reachability. The test for target reachability used in algorithm Big1 is designed as follows. Every time MA completes its exploration of a new obstacle i, it defines on it a leave point L_i. Then MA leaves the ith obstacle at L_i and starts toward the target T along the straight line (L_i, T). According to Lemma 3.3.1, MA will never return again to the ith obstacle. Since point L_i is by definition the closest point of obstacle i to point T, there will be no parts of the obstacle i between points L_i and T. Because, by the model, obstacles do not touch each other, point L_i cannot belong to any other obstacle but i. Therefore, if, after having arrived at L_i in Step 3 of the algorithm, MA discovers that the straight line (L_i, T) crosses some obstacle at the leave point L_i, this can only mean that this is the ith obstacle and hence target T is not reachable—either point S or point T is *trapped* inside the ith obstacle.

To show that this is true, let O be a simple closed curve; let X be some point in the scene that does not belong to O; let L be the point on O closest to X;

and let (L, X) be the straight-line segment connecting L and X. All these are in the plane. Segment (L, X) is said to be *directed outward* if a finite part of it in the vicinity of point L is located outside of curve O. Otherwise, if segment (L, X) penetrates inside the curve O in the vicinity of L, it is said to be *directed inward*.

The following statement holds: If segment (L, X) is directed inward, then X is inside curve O. This condition is necessary because if X were outside curve O, then some other point of O that would be closer to X than to L would appear in the intersection of (L, X) and O. By definition of the point L, this is impossible. The condition is also sufficient because if segment (L, X) is directed inward and L is the point on curve O that is the closest to X, then segment (L, X) cannot cross any other point of O, and therefore X must lie inside O. This fact is used in the following test that appears as a part in Step 3 of algorithm Bug1:

Test for Target Reachability.   If, while using algorithm Bug1, after having defined a point L on an obstacle, MA discovers that the straight line segment (L, Target) crosses the obstacle at point L, then the target is not reachable.

One can check the test on the example shown in Figure 3.4. Starting at point T, the robot encounters an obstacle and establishes on it a hit point H. Using the local direction "left," it then does a full exploration of the (accessible) boundary of the obstacle. Once it arrives back at point H, its register R_1 will contain the location of the point on the boundary that is the closest to T. This happens to be

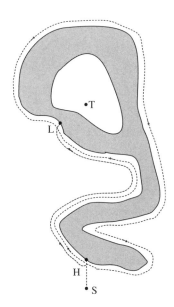

Figure 3.4   Algorithm Bug1. An example with an unreachable target (a trap).

point L. The robot then walks to L by the shortest route (which it knows from the information it now has) and establishes on it the leave point L. At this point, algorithm Bug1 prescribes it to move toward T. While performing the test for target reachability, however, the robot will note that the line (L, T) enters the obstacle at L and hence will conclude that the target is not reachable.

3.3.2   Second Basic Algorithm: Bug2

Similar to the algorithm Bug1, the procedure Bug2 is executed at every point of the robot's (continuous) path. As before, the goal is to generate a path from the start to the target position. As will be evident later, three important properties distinguish algorithm Bug2 from Bug1: Under Bug2, (a) MA can encounter the same obstacle more than once, (b) algorithm Bug2 has no way of distinguishing between different obstacles, and (c) the straight line (S, T) that connects the starting and target points plays a crucial role in the algorithm's workings. The latter line is called *M-line* (for *Main line*). In imprecise words, the reason M-line is so important is that the procedure uses it to index its progress toward the target and to ensure that the robot does not get lost.

Because of these differences, we need to change the notation slightly: Subscript i will be used only when referring to more than one obstacle, and superscript j will be used to indicate the jth occurrence of a hit or leave points on the same or on a different obstacle. Initially, $j = 1$; $L^0 =$ Start. Similar to Bug1, the Bug2 procedure includes a test for target reachability, which is built into Steps 2b and 2c of the procedure. The test is explained later in this section. The reader may find it helpful to follow the procedure using an example in Figure 3.5.

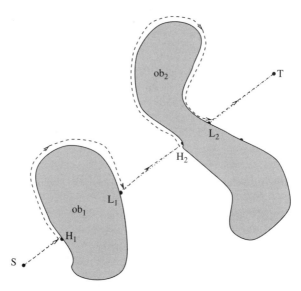

Figure 3.5   Robot's path (dashed line) under Algorithm Bug2.

Bug2 Procedure

1. From point L^{j-1}, move along the M-line (straight line (S, T)) until one of these occurs:
   (a) Target T is reached. The procedure stops.
   (b) An obstacle is encountered and a hit point, H^j, is defined. Go to Step 2.
2. Using the accepted local direction, follow the obstacle boundary until one of these occurs:
   (a) Target T is reached. The procedure stops.
   (b) M-line is met at a point Q such that distance $d(Q) < d(H^j)$, and straight line (Q, T) does not cross the current obstacle at point Q. Define the leave point $L^j = Q$. Set $j = j + 1$. Go to Step 1.
   (c) MA returns to H^j and thus completes a closed curve along the obstacle boundary, without having defined the next hit point, H^{j+1}. Then, the target point T is trapped and cannot be reached. The procedure stops.

Unlike with algorithm Bug1, more than one hit and more than one leave point can be generated on a single obstacle under algorithm Bug2 (see the example in Figure 3.6). Note also that the relationship between perimeters of the obstacles and the length of paths generated by Bug2 is not as clear as in the case of algorithm Bug1. In Bug1, the perimeter of an obstacle met by MA is traversed at least once and never more than 1.5 times. In Bug2, more options appear. A path segment around an obstacle generated by MA is sometimes shorter than the obstacle perimeter (Figure 3.5), which is good news: We finally see something "intelligent." Or, when a straight-line path segment of the path meets an obstacle almost tangentially and MA happened to be walking around the obstacle in an "unfortunate" direction, the path can become equal to the obstacle's full perimeter (Figure 3.7). Finally, as Figure 3.6a demonstrates, the situation can get even worse: MA may have to pass along some segments of a maze-like obstacle more than once and more than twice. (We will return to this case later in this section.)

Analysis of Algorithm Bug2

Lemma 3.3.3. *Under Bug2, on its way to the target, MA can meet only a finite number of obstacles.*

Proof: Although, while walking around an obstacle, MA may at times find itself at distances much larger than D from point T (Target), its straight-line path segments toward T are always within the same circle of radius D centered at T. This is guaranteed by the algorithm's condition that $d(L^j, T) > d(H^j, T)$ (see Step 2 of Bug2 procedure). Since, by the model, any disc of finite radius can intersect with only a finite number of obstacles, the lemma follows. **Q.E.D.**

Corollary 3.3.3. *The only obstacles that MA can meet while operating under algorithm Bug2 are those that intersect the disc of radius D centered at the target.*

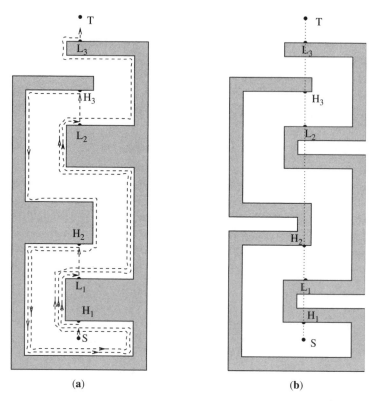

(a)   (b)

Figure 3.6 (a, b) Robot's path around a maze-like obstacle under Algorithm Bug2 (in-position case). Both obstacles (a) and (b) are similar, except in (a) the M-line (straight line (S, T)) crosses the obstacle 10 times, and in (b) it crosses 14 times. MA passes through the same path segment(s) at most three times (here, through segment (H_1, L_1)). Thus, at most two local cycles are created in this examples.

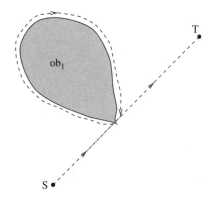

Figure 3.7 In this example, under Algorithm Bug2 the robot will make almost a full circle around this convex obstacle.

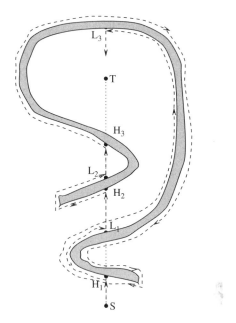

Figure 3.8 Robot's path in an in-position case; here point S is outside of the obstacle, and T is inside.

Moreover, the only obstacles that can be met by MA are those that intersect the M-line (straight line (Start, Target)).

Definition 3.3.1. *For the given local direction, a* local cycle *is created when MA has to pass some point of its path more than once.*

In the example in Figure 3.5, no local cycles are created; in Figures 3.6 and 3.8 there are local cycles.

Definition 3.3.2. *The term* in-position *refers to a mutual position of points (Start, Target) and a given obstacle, such that (1) the M-line crosses the obstacle boundary at least once, and (2) either Start or Target lie inside the convex hull of the obstacle. The term* out-position *refers to a mutual position of points (Start, Target) and a given obstacle, such that both points Start and Target lie outside the convex hull of the obstacle. A given scene is referred to as an* in-position *case if at least one obstacle in the scene creates an in-position situation; otherwise, the scene presents an* out-position *case.*

For example, the scene in Figure 3.3 is an in-position case. Without obstacle ob_3, the scene would have been an out-position case.

We denote n_i to be the number of intersections between the M-line (straight line (S, T)) and the ith obstacle; n_i is thus a characteristic of the set (scene, Start, Target) and not of the algorithm. Obviously, for any convex obstacle, $n_i = 2$.

If an obstacle is not convex but still $n_i = 2$, the path generated by Bug2 can be as simple as that for a convex obstacle (see, e.g., Figure 3.5, obstacle ob_2). Even with more complex obstacles, such as that in Figure 3.6, the situation and the resulting path can be quite simple. For example, in this same scene, if the M-line happened to be horizontal, we would have $n_i = 2$ and a very simple path.

The path can become more complicated if $n_i > 2$ and we are dealing with an in-position case. In Figure 3.6a, the segment of the boundary from point H_1 to point $L1$, $(H1, L1)$, will be traversed three times: segments $(L1, L2)$ and $(H2, H1)$, twice each; and segments $(L2, L3)$ and $(H3, H2)$, once each. On the other hand, if in this example the M-line line extends below, so that point S is under the whole obstacle (that is, this becomes an out-position case), the path will again become very simple, with no local cycles (in spite of high n_i number).

Lemma 3.3.4. *Under Bug2, MA will pass any point of the ith obstacle boundary at most $n_i/2$ times.*

Proof: As one can see, procedure Bug2 does not distinguish whether two consecutive obstacle crossings by the M-line (straight line (S, T)) correspond to the same or to different obstacles. Without loss of generality, assume that only one obstacle is present; then we can drop the index i. For each hit point H^j, the procedure will make MA walk around the obstacle until it reaches the corresponding leave point, L^j. Therefore, all H and L points appear in pairs, (H^j, L^j). Because, by the model, all obstacles are of finite "thickness," for each pair (H^j, L^j) an inequality holds, $d(H^j) > d(L^j)$. After leaving L^j, MA walks along a straight line to the next hit point, H^{j+1}. Since, according to the model, the distance between two crossings of the obstacle by a straight line is finite, we have $d(L^j) > d(H^{j+1})$. This produces a chain of inequalities for all H and L points,

$$d(H^1) > d(L^1) > d(H^2) > d(L^2) > d(H^3) > d(L^3) > \cdots \qquad (3.8)$$

Therefore, although any H or L point may be passed more than once, it will be defined as an H (correspondingly, L) point only once. That point can hence generate only one new passing of the same segment of the obstacle perimeter. In other words, each pair (H^j, L^j) can give rise to only one passing of a segment of the obstacle boundary. This means that n_i crossings will produce at most $n_i/2$ passings of the same path segment. **Q.E.D.**

The lemma guarantees that the procedure terminates, and it gives a limit on the number of generated local cycles. Using the lemma, we can now produce an upper bound on the length of paths generated by algorithm Bug2.

Theorem 3.3.3. *The length of a path generated by algorithm Bug2 will never exceed the limit*

$$P = D + \sum_i \frac{n_i p_i}{2} \qquad (3.9)$$

where D is the distance (Start, Target), and p_i refers to perimeters of obstacles that intersect the M-line (straight line segment (Start, Target)). This means Bug2 is convergent.

Proof: Any path can be looked at as consisting of two parts: (a) straight-line segments of the M-line between the obstacles that intersect it and (b) path segments that relate to walking around obstacle boundaries. Because of inequality (3.8), the sum of the straight line segments will never exceed D. As to path segments around obstacles, there is an upper bound guaranteed by Lemma 3.3.4 for each obstacle met by MA on its path: No more than $n_i/2$ passings along the same segment of the obstacle boundary will take place. Because of Lemma 3.3.3 (see its proof), only those obstacles that intersect the M-line should be counted. Summing up the straight-line segments and segments that correspond to walking around obstacles, obtain (3.9). **Q.E.D.**

Theorem 3.3.3 suggests that in some "bad" scenes, under algorithm Bug2, MA may be forced to go around obstacles any large, albeit finite, number of times. An important question, therefore, is how typical such "bad" scenes are.

In particular, other things being equal, what characteristics of the scene influence the length of the path? Theorem 3.3.4 and its corollary below address this question. They suggest that the mutual position of point S, point T, and obstacles in the scene can affect the path length rather dramatically. Together, they significantly improve the upper bound on the length of paths generated by Bug2—in out-position scenes in general and in scenes with convex obstacles in particular.

Theorem 3.3.4. *Under algorithm Bug2, in the case of an out-position scene, MA will pass any point of an obstacle boundary at most once.*

In other words, if the mutual position of the obstacle and of points S and T satisfies the out-position definition, the estimate on the length of paths generated by Bug2 reaches the universal lower bound (3.1). That is a very good news indeed. Out-position situations are rather common for mobile robots.[3] We know already that in some situations, algorithm Bug2 is extremely efficient and traverses only a fraction of obstacle boundaries. Now the theorem tells us that as long as the robot deals with an out-position situation, even in the most unlucky case it will not traverse more than 1.5 times the obstacle boundaries involved.

Proof: Figure 3.9 is used to illustrate the proof. Shaded areas in the figure correspond to one or many obstacles. Dashed boundaries indicate that obstacle boundaries in these areas can be of any shape.

Consider an obstacle met by MA on its way to the Target, and consider an arbitrary point Q on the obstacle boundary (not shown in the figure). Assume that Q is not a hit point. Because the obstacle boundary is a simple closed curve, the only way that MA can reach point Q is to come to it from a previously

---

[3]We will see later that out-position situations are a rarity for arm manipulators.

Figure 3.9   Illustration for Theorem 3.3.4.

defined hit point. Now, move from Q along the already generated path segment in the direction opposite to the accepted local direction, until the closest hit point on the path is encountered; say, that point is H^j. We are interested only in those cases where Q is involved in at least one local cycle—that is, when MA passes point Q more than once. For this event to occur, MA has to pass point H^j at least as many times. In other words, if MA does not pass H^j more than once, it cannot pass Q more than once.

According to the Bug2 procedure, the first time MA reaches point H^j it approaches it along the M-line (straight line (Start, Target))—or, more precisely, along the straight line segment (L^{j-1}, T). MA then turns left and starts walking around the obstacle. To form a local cycle on this path segment, MA has to return to point H^j again. Since a point can become a hit point only once (see the proof for Lemma 3.3.4), the next time MA returns to point H^j it must approach it from the right (see Figure 3.9), along the obstacle boundary. Therefore, after having defined H^j, in order to reach it again, this time from the right, MA must somehow cross the M-line and enter its right semiplane. This can take place in one of only two ways: outside or inside the interval (S, T). Consider both cases.

1. The crossing occurs outside the interval (S, T). This case can correspond only to an in-position configuration (see Definition 3.3.2). Theorem 3.3.4, therefore, does not apply.

2. The crossing occurs inside the interval (S, T). We want to prove now that such a crossing of the path with the interval (S, T) cannot produce local cycles. Notice that the crossing cannot occur anywhere within the interval (S, H^j) because otherwise at least a part of the straight-line segment (L^{j-1}, H^j) would be included inside the obstacle. This is impossible

because MA is known to have walked along the whole segment (L^{j-1}, H^j). If the crossing occurs within the interval (H^j, T), then at the crossing point MA would define the corresponding leave point, L^j, and start moving along the line (S, T) toward the target T until it defined the next hit point, H^{j+1}, or reached the target. Therefore, between points H^j and L^j, MA could not have reached into the right semiplane of the M-line (see Figure 3.9).

Since the above argument holds for any Q and the corresponding H^j, we conclude that in an out-position case MA will never cross the interval (Start, Target) into the right semiplane, which prevents it from producing local cycles. **Q.E.D.**

So far, no constraints on the shape of the obstacles have been imposed. In a special case when all the obstacles in the scene are convex, no in-position configurations can appear, and the upper bound on the length of paths generated by Bug2 can be improved:

Corollary 3.3.4. *If all obstacles in the scene are convex, then in the worst case the length of the path produced by algorithm Bug2 is*

$$P = D + \sum_i p_i \qquad (3.10)$$

and, on the average,

$$P = D + 0.5 \cdot \sum_i p_i \qquad (3.11)$$

where D is distance (Start, Target), and p_i refer to perimeters of the obstacles that intersect the straight line segment (Start, Target).

Consider a statistically representative number of scenes with a random distribution of convex obstacles in each scene, a random distribution of points Start and Target over the set of scenes, and a fixed local direction as defined above. The M-line will cross obstacles that it intersects in many different ways. Then, for some obstacles, MA will be forced to cover the bigger part of their perimeters (as in the case of obstacle ob_1, Figure 3.5); for some other obstacles, MA will cover only a smaller part of their perimeters (as with obstacle ob_2, Figure 3.5).

On the average, one would expect a path that satisfies (3.11). As for (3.10), Figure 3.7 presents an example of such a "noncooperating" obstacle. Corollary 3.3.4 thus ensures that for a wide range of scenes the length of paths generated by algorithm Bug2 will not exceed the universal lower bound (3.1).

Test for Target Reachability. As suggested by Lemma 3.3.4, under Bug2 MA may pass the same point H^j of a given obstacle more than once, producing a finite number p of local cycles, $p = 0, 1, 2, \ldots$. The proof of the lemma indicates that after having defined a point H^j, MA will never define this point again as an H or an L point. Therefore, on each of the subsequent local cycles (if any), point H^j will be passed not along the M-line but along the obstacle boundary. After having left point H^j, MA can expect one of the following to occur:

MA will never return again to H^j; this happens, for example, if it leaves the current obstacle altogether or reaches the Target T.

MA will define at least the first pair of points $(L^j, H^{j+1}), \ldots,$ and will then return to point H^j, to start a new local cycle.

MA will come back to point H^j without having defined a point L^j on the previous cycle. This means that MA could find no other intersection point Q of the line (H^j, T) with the current obstacle such that Q would be closer to the point T than H^j, and the line (Q, T) would not cross the current obstacle at Q. This can happen only if either MA or point T are trapped inside the current obstacle (see Figure 3.10). The condition is both necessary and sufficient, which can be shown similar to the proof in the target reachability test for algorithm Bug1 (Section 3.3.1).

Based on this observation, we now formulate the test for target reachability for algorithm Bug2.

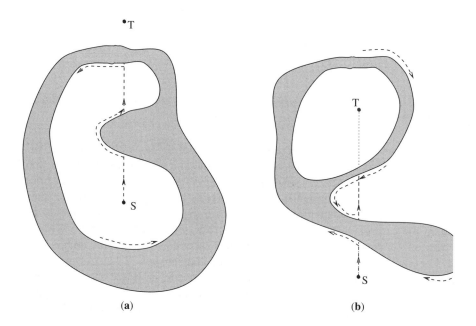

(a)   (b)

Figure 3.10   Examples where no path between points S and T is possible (traps), algorithm Bug2. The path is the dashed line. After having defined the hit point H_2, the robot returns to it before it defines any new leave point. Therefore, the target is not reachable.

Test for Target Reachability. If, on the pth local cycle, $p = 0, 1, \ldots$, after having defined a hit point H^j, MA returns to this point before it defines at least the first two out of the possible set of points $L^j, H^{j+1}, \ldots, H^k$, this means that MA has been *trapped* and hence the target is not reachable.

We have learned that in in-position situations algorithm Bug2 may become inefficient and create local cycles, visiting some areas of its path more than once. How can we characterize those situations? Does starting or ending "inside" the obstacle—that is, having an in-position situation—necessarily lead to such inefficiency? This is clearly not so, as one can see from the following example of Bug2 operating in a maze (labyrinth). Consider a version of the labyrinth problem where the robot, starting at one point inside the labyrinth, must reach some other point inside the labyrinth. The well-known mice-in-the-labyrinth problem is sometimes formulated this way. Consider an example[4] shown in Figure 3.11.

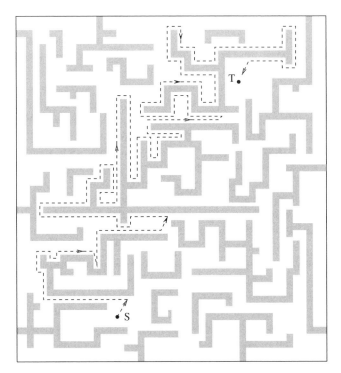

Figure 3.11 Example of a walk (dashed line) in a maze under algorithm Bug2. S, Start; T, Target.

---

[4]To fit the common convention of maze search literature, we present a discrete version of the continuous path planning problem: The maze is a rectangular cell structure, with each cell being a little square; any cell crossed by the M-line (straight line (S, T)) is considered to be lying on the line. This same discussion can be carried out using an arbitrary curvilinear maze.

Given the fact that no bird's-eye view of the maze is available to MA (at each moment it can see only the small cell that it is passing), the MA's path looks remarkably efficient and purposeful. (It would look even better if MA's sensing was something better than simple tactile sensing; see Figure 3.20 and more on this topic in Section 3.6.) One reason for this is, of course, that no local cycles are produced here. In spite of its seeming complexity, this maze is actually an easy scene for the Bug2 algorithm.

Let's return to our question, How can we classify in-position situations, so as to recognize which one would cause troubles to the algorithm Bug2? This question is not clear at the present time. The answer, likely tied to the topological properties of the combination (scene, Start, Target), is still awaiting a probing researcher.

3.4 COMBINING GOOD FEATURES OF BASIC ALGORITHMS

Each of the algorithms Bug1 and Bug2 has a clear and simple, and quite distinct, underlying idea: Bug1 "sticks" to every obstacle it meets until it explores it fully; Bug2 sticks to the M-line (line (Start, Target)). Each has its pluses and minuses. Algorithm Bug1 never creates local cycles; its worse-case performance looks remarkably good, but it tends to be "overcautious" and will never cover less than the full perimeter of an obstacle on its way. Algorithm Bug2, on the other hand, is more "human" in that it can "take a risk." It takes advantage of simpler situations; it can do quite well even in complex scenes in spite of its frighteningly high worst-case performance—but it may become quite inefficient, much more so than Bug1, in some "unlucky" cases.

The difficulties that algorithm Bug2 may face are tied to local cycles— situations when the robot must make circles, visiting the same points of the obstacle boundaries more than once. The source of these difficulties lies in what we called *in-position* situations (see the Bug2 analysis above). The problem is of topological nature. As the above estimates of Bug2 "average" behavior show, its performance in *out-positions* situations may be remarkably good; these are situations that mobile robots will likely encounter in real-life scenes.

On the other hand, fixing the procedure so as to handle in-position situations well would be an important improvement. One simple idea for doing this is to attempt a procedure that combines the better features of both basic algorithms. (As always, when attempting to combine very distinct ideas, the punishment will be the loss of simplicity and elegance of both algorithms.) We will call this procedure BugM1 (for "modified") [59]. The procedure combines the efficiency of algorithm Bug2 in simpler scenes (where MA will pass only portions, instead of full perimeters, of obstacles, as in Figure 3.5) with the more conservative, but in the limit the more economical, strategy of algorithm Bug1 (see the bound (3.7)). The idea is simple: Since Bug2 is quite good except in cases with local cycles, let us try to switch to Bug1 whenever MA concludes that it is in a local cycle. As a result, for a given point on a BugM1 path, the number of local cycles containing this point will never be larger than two; in other words, MA will never pass the same point of the obstacle boundary more than three times, producing the upper bound

$$P \geq D + 3 \cdot \sum_i p_i \tag{3.12}$$

Algorithm BugM1 is executed at every point of the continuous path. Instead of using the fixed M-line (straight line (S, T)), as in Bug2, BugM1 uses a straight-line segment (L_i^j, T) with a changing point L_i^j; here, L_i^j indicates the jth leave point on obstacle i. The procedure uses three registers, R_1, R_2, and R_3, to store intermediate information. All three are reset to zero when a new hit point H_i^j is defined:

- Register R_1 stores coordinates of the current point, Q_m, of minimum distance between the obstacle boundary and the Target.
- R_2 integrates the length of the obstacle boundary starting at H_i^j.
- R_3 integrates the length of the obstacle boundary starting at Q_m. (In case of many choices for Q_m, any one of them can be taken.)

The test for target reachability that appears in Step 2d of the procedure is explained lower in this section. Initially, $i = 1$, $j = 1$; L_o^o = Start. The BugM1 procedure includes these steps:

1. From point L_{i-1}^{j-1}, move along the line $(L_o^{j-1}$, Target) toward Target until one of these occurs:
   (a) Target is reached. The procedure stops.
   (b) An ith obstacle is encountered and a hit point, H_i^j, is defined. Go to Step 2.
2. Using the accepted local direction, follow the obstacle boundary until one of these occurs:
   (a) Target is reached. The procedure stops.
   (b) Line $(L_o^{j-1}$, Target) is met inside the interval $(L_o^{j-1}$, Target), at a point Q such that distance $d(Q) < d(H^j)$, and the line $(Q$, Target) does not cross the current obstacle at point Q. Define the leave point $L_i^j = Q$. Set $j = j + 1$. Go to Step 1.
   (c) Line $(L_o^{j-1}$, Target) is met outside the interval $(L_o^{j-1}$, Target). Go to Step 3.
   (d) The robot returns to H_i^j and thus completes a closed curve (of the obstacle boundary) without having defined the next hit point. The target cannot be reached. The procedure stops.

3. Continue following the obstacle boundary. If the target is reached, stop. Otherwise, after having traversed the whole boundary and having returned to point H_i^j, define a new leave point $L_i^j = Q_m$. Go to Step 4.

4. Using the contents of registers R_2 and R_3, determine the shorter way along the obstacle boundary to point L_i^j, and use it to get to L_i^j. Apply the test for Target reachability (see below). If the target is not reachable, the procedure stops. Otherwise, designate $L_i^o = L_i^j$, set $i = i + 1$, $j = 1$, and go to Step 1.

As mentioned above, the procedure itself BugM1 is obviously longer and "messier" compared to the elegantly simple procedures Bug1 and Bug2. That is the price for combining two algorithms governed by very different principles. Note also that since at times BugM1 may leave an obstacle before it fully explores it, according to our classification above it falls into the Class 2.

What is the mechanism of algorithm BugM1 convergence? Depending on the scene, the algorithm's flow fits one of the following two cases.

Case 1. If the condition in Step 2c of the procedure is never satisfied, then the algorithm flow follows that of Bug2—for which convergence has been already established. In this case, the straight lines $(L_i^j$, Target) always coincide with the M-line (straight line (Start, Target)), and no local cycles appear.

Case 2. If, on the other hand, the scene presents an in-position case, then the condition in Step 2c is satisfied at least once; that is, MA crosses the straight line $(L_o^{j-1}$, Target) outside the interval $(L_o^{j-1}$, Target). This indicates that there is a danger of multiple local cycles, and so MA switches to a more conservative procedure Bug1, instead of risking an uncertain number of local cycles it might now expect from the procedure Bug2 (see Lemma 3.3.4). MA does this by executing Steps 3 and 4 of BugM1, which are identical to Steps 2 and 3 of Bug1.

After one execution of Steps 3 and 4 of the BugM1 procedure, the last leave point on the ith obstacle is defined, L_i^j, which is guaranteed to be closer to point T than the corresponding hit point, H_i^j [see inequality (3.7), Lemma 3.3.1]. Then MA leaves the ith obstacle, never to return to it again (Lemma 3.3.1). From now on, the algorithm (in its Steps 1 and 2) will be using the straight line $(L_i^o$, Target) as the "leading thread." [Note that, in general, the line $(L_i^o$, Target) does not coincide with the straight lines (L_{i-1}^o, T) or (S, T)]. One execution of the sequence of Steps 3 and 4 of BugM1 is equivalent to one execution of Steps 2 and 3 of Bug1, which guarantees the reduction by one of the number of obstacles that MA will meet on its way. Therefore, as in Bug1, the convergence of this case is guaranteed by Lemma 3.3.1, Lemma 3.3.2, and Corollary 3.3.2. Since Case 1 and Case 2 above are independent and together exhaust all possible cases, the procedure BugM1 converges.

3.5 GOING AFTER TIGHTER BOUNDS

The above analysis raises two questions:

1. There is a gap between the bound given by (3.1), $P \geq D + \sum_i p_i - \delta$ (the universal lower bound for the planning problem), and the bound given by (3.7), $P \leq D + 1.5 \cdot \sum_i p_i$ (the upper bound for Bug1 algorithm).
   What is there in the gap? Can the lower bound (3.1) be tightened upwards—or, inversely, are there algorithms that can reach it?
2. How big and diverse are Classes 1 and 2?

To remind the reader, Class 1 combines algorithms in which the robot never leaves an obstacle unless and until it explores it completely. Class 2 combines algorithms that are complementary to those in Class 1: In them the robot can leave an obstacle and walk further, and even return to this obstacle again at some future time, without exploring it in full.

A decisive step toward answering the above questions was made in 1991 by A. Sankaranarayanan and M. Vidyasagar [60]. They proposed to (a) analyze the complexity of Classes 1 and 2 of sensor-based planning algorithms separately and (b) obtain the lower bounds on the lengths of generated paths for each of them. This promised tighter bounds compared to (3.1). Then, since together both classes cover all possible algorithms, the lower of the obtained bounds would become the universal lower bound. Proceeding in this direction, Sankaranarayanan and Vidyasagar obtained the lower bound for Class 1 algorithms as

$$P \geq D + 1.5 \sum_i p_i \qquad (3.13)$$

and the lower bound for Class 2 algorithms as

$$P \geq D + 2 \cdot \sum_i p_i \qquad (3.14)$$

As before, P is the length of a generated path, D is the distance (Start, Target), and p_i refers to perimeters of obstacles met by the robot on its way to the target. There are three important conclusions from these results:

- It is the bound (3.13), and not (3.1), that is today the universal lower bound: in the worst case no sensor-based motion planning algorithm can produce a path shorter than P in (3.13).
- According to the bound (3.13), algorithm Bug1 reaches the universal lower bound. That is, no algorithm in Class 1 will be able to do better than Bug1 in the worst case.
- According to bounds (3.13) and (3.14), in the worst case no algorithm from either of the two classes can do better than Bug1.

How much variety and how many algorithms are there in Classes 1 and 2? For Class 1, the answer is simple: At this time, algorithm Bug1 is the only representative of Class 1. The future will tell whether this represents just the lack of interest in the research community to such algorithms or something else. One can surmise that it is both: The underlying mechanism of this class of algorithms does not promise much richness or unusual algorithms, and this gives little incentive for active research.

In contrast, a lively innovation and variety has characterized the development in Class 2 algorithms. At least a dozen or so algorithms have appeared in literature since the problem was first formulated and the basic algorithms were reported. Since some such algorithms make use of the types of sensing that are more elaborate than basic tactile sensing used in this section, we defer a survey in this area until Section 3.8, after we discuss in the next section the effect of more complex sensing on sensor-based motion planning.

3.6 VISION AND MOTION PLANNING

In the previous section we developed the framework for designing sensor-based path planning algorithms with proven convergence. We designed some algorithms and studied their properties and performance. For clarity, we limited the sensing that the robot possesses to (the most simple) tactile sensing. While tactile sensing plays an important role in real-world robotics—in particular in short-range motion planning for object manipulation and for escaping from tight places—for general collision avoidance, richer remote sensing such as computer vision or range sensing present more promising options.

The term "range" here refers to devices that directly provide distance information, such as a laser ranger. A stereo vision device would be another option. In order to successfully negotiate a scene with obstacles, a mobile robot can make a good use of distance information to objects it is passing.

Here we are interested in exploring how path planning algorithms would be affected by the sensing input that is richer and more complex than tactile sensing. In particular, can algorithms that operate with richer sensory data take advantage of additional sensor information and deliver better *path length performance*—to put it simply, shorter paths—than when using tactile sensing? Does proximal or distant sensing really help in motion planning compared to tactile sensing, and, if so, in what way and under what conditions? Although this question is far from trivial and is important for both theory and practice (this is manifested by a recent continuous flow of experimental works with "seeing" robots), there have been little attempts to address this question on the algorithmic level.

We are thus interested in algorithms that can make use of a range finder or stereo vision and that, on the one hand, are provably correct and, on the other hand, would let, say, a mobile robot deliver a reasonable performance in nontrivial scenes. It turns out that the answers to the above question are not trivial as well. First, yes, algorithms can be modified so as to take advantage of better sensing. Second, extensive modifications of "tactile" motion planning algorithms are needed in order to fully utilize additional sensing capabilities. We will consider in detail two principles for provably correct motion planning with vision. As we will see, the resulting algorithms exhibit different "styles" of behavior and are not, in general, superior to each other. Third and very interestingly, while one can expect great improvements in real-world tasks, in general richer sensing has no effect on algorithm path length performance bounds.

Algorithms that we are about to consider will demonstrate an ability that is often referred to in the literature as *active vision* [61, 62]. This ability goes deeply into the nature of interaction between sensing and control. As experimentalists well know, scanning the scene and making sense of acquired information is a time-consuming operation. As a rule, the robot's "eye" sees a bewildering amount of details, almost all of which are irrelevant for the robot's goal of finding its way around. One needs a powerful mechanism that would reject what is irrelevant and immediately use what is relevant so that one can continue the motion and continue gathering more visual data. We humans, and of course all other species in nature that use vision, have such mechanisms.

As one will see in this section, motion planning algorithms with vision that we will develop will provide the robot with such mechanisms. As a rule, the robot will not scan the whole scene; it will behave much as a human when walking along the street, looking for relevant information and making decisions when the right information is gathered. While the process is continuous, for the sake of this discussion it helps to consider it as a quasi-discrete.

Consider a moment when the robot is about to pass some location. A moment earlier, the robot was at some prior location. It knows the direction toward the target location of its journey (or, sometimes, some intermediate target in the visible part of the scene). The first thing it does is look in that direction, to see if this brings new information about the scene that was not available at the prior position. Perhaps it will look in the direction of its target location. If it sees an obstacle in that direction, it may widen its "scan," to see how it can pass around this obstacle. There may be some point on the obstacle that the robot will decide to head to, with the idea that more information may appear along the way and the plan may be modified accordingly.

Similar to how any of us behaves when walking, it makes no sense for the robot to do a 360° scan at every step—or ever. Based on what the robot sees ahead at any moment, it decides on the next step, executes it, and looks again for more information. In other words, *robot's sensing dictates the next step motion, and the next step dictates where to look for new relevant information.* It is this sensing-planning control loop that guides the robot's active vision, and it is executed continuously.

The first algorithm that we will consider, called *VisBug-21*, is a rather simple-minded and conservative procedure. (The number "2" in its name refers to the Bug2 algorithm that is used as its base, and "1" refers to the first vision algorithm.) It uses range data to "cut corners" that would have been produced by a "tactile" algorithm Bug2 operating in the same scene. The advantage of this modification is clear. Envision the behavior of two people, one with sight and the other blindfolded. Envision each of them walking in the same direction around the perimeter of a complex-shaped building. The path of the person with sight will be (at least, often enough) a shorter approximation of the path of the blind-folded person.

The second algorithm, called *VisBug-22*, is more opportunistic in nature: it tries to use every chance to get closer to the target. (The number in its name signifies that it is the vision algorithm 2 based on the Bug2 procedure.)

Section 3.6.1 is devoted to the algorithms' underlying model and basic ideas. The algorithms themselves, related analysis, and examples demonstrating the algorithms' performance appear in Sections 3.6.2 and 3.6.3.

3.6.1 The Model

Our assumptions about the scene in which the robot travels and about the robot itself are very much the same as for the basic algorithms (Section 3.1). The available input information includes knowing at all times the robot's current location, C, and the locations of starting and target points, S and T. We also assume that a very limited memory does not allow the robot more than remembering a few "interesting" points.

The difference in the two models relates to the robot sensing ability. In the case at hand the robot has a capability, referred to as *vision*, to detect an obstacle, and the distance to any visible point of it, along any direction from point C, within the sensor's *field of vision*. The field of vision presents a disc of radius r_v, called *radius of vision*, centered at C. A point Q in the scene is *visible* if it is located within the field of vision and if the straight-line segment CQ does not cross any obstacles.

The robot is capable of using its vision to *scan* its surroundings during which it *identifies* obstacles, or the lack of thereof, that intersect its field of vision. We will see that the robot will use this capability rather sparingly; the particular use of scanning will depend on the algorithm. Ideally the robot will scan a part of the scene only in those specific directions that make sense from the standpoint of motion planning. The robot may, for example, identify some intermediate target point within its field of vision and walk straight toward that point. Or, in an "unfortunate" (for its vision) case when the robot walks along the boundary of a convex obstacle, its effective radius of vision in the direction of intended motion (that is, around the obstacle) will shrink to zero.

As before, the straight-line segment (S, T) between the start S and target T points—it is called the *Main line* or *M-line*—is the desirable path. Given its current position C_i, at moment i the robot will execute an elementary operation that includes scanning some minimum sector of its current field of vision in the direction it is following, enough to define its *next intermediate target*, point T_i. Then the robot makes a little step in the direction of T_i, and the process repeats. T_i is thus a moving target; its choice will somehow relate to the robot's global goal. In the algorithms, every T_i will lie on the M-line or on an obstacle boundary.

For a path segment whose point T_i moves along the M-line, the firstly defined T_i that lies at the intersection of M-line and an obstacle is a special point called the *hit point*, H. Recall that in algorithms Bug1 or Bug2 a hit point would be reached physically. In algorithms with vision a hit point may be defined from a distance, thanks to the robot's vision, and the robot will not necessarily pass through this location. For a path segment whose point T_i moves along an obstacle boundary, the firstly defined T_i that lies on the M-line is a special point called the *leave point*, L. Again, the robot may or may not pass physically through that point. As we will see, the main difference between the two algorithms VisBug-21 and VisBug-22 is in how they define intermediate targets T_i. Their resulting paths will likely be quite different. Naturally, the current T_i is always at a distance from the robot no more than r_v.

While scanning its field of vision, the robot may be detecting some contiguous sets of visible points—for example, a segment of the obstacle boundary. A point Q is *contiguous* to another point S *over the set* $\{P\}$, if three conditions are met: (i) $S \in \{P\}$, (ii) Q and $\{P\}$ are visible, and (iii) Q can be continuously connected with S using only points of $\{P\}$. A *set is contiguous* if any pair of its points are contiguous to each other over the set. We will see that no memorization of contiguous sets will be needed; that is, while "watching" a contiguous set, the robot's only concern will be whether two points that it is currently interested in are contiguous to each other.

A *local direction* is a once-and-for-all determined direction for passing around an obstacle; facing the obstacle, it can be either left or right. Because of incomplete information, neither local direction can be judged better than the other. For the sake of clarity, assume the local direction is always *left*.

The M-line divides the environment into two half-planes. The half-plane that lies to the local direction's side of M-line is called the *main semiplane*. The other half-plane is called the *secondary semiplane*. Thus, with the local direction "left," the left half-plane when looking from S toward T is the main semiplane.

Figure 3.12 exemplifies the defined terms. Shaded areas represent obstacles; the straight-line segment ST is the M-line; the robot's current location, C, is in the secondary (right) semiplane; its field of vision is of radius r_v. If, while standing at C, the robot were to perform a complete scan, it would identify three contiguous segments of obstacle boundaries, $a_1a_2a_3$, $a_4a_5a_6a_7a_8$, and $a_9a_{10}a_{11}$, and two contiguous segments of M-line, b_1b_2 and b_3b_4.

A Sketch of Algorithmic Ideas. To understand how vision sensing can be incorporated in the algorithms, consider first how the "pure" basic algorithm Bug2 would behave in the scene shown in Figure 3.12. Assuming a local direction "left," Bug2 would generate the path shown in Figure 3.13. Intuitively, replacing tactile sensing with vision should smooth sharp corners in the path and perhaps allow the robot to cut corners in appropriate places.

However, because of concern for algorithms' convergence, we cannot introduce vision in a direct way. One intuitively appealing idea is, for example, to make the robot always walk toward the farthest visible "corner" of an obstacle in the robot's preferred direction. An example can be easily constructed showing that this idea cannot work—it will ruin the algorithm convergence. (We have already seen examples of treachery of intuitively appealing ideas; see Figure 2.23—it applies to the use of vision as well.)

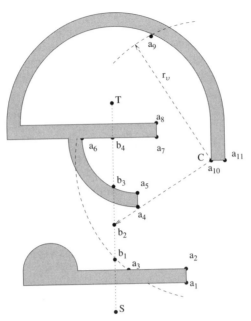

Figure 3.12 Shaded areas are obstacles. At its current location C, the robot will see within its radius of vision r_v segments of obstacle boundaries $a_1a_2a_3$, $a_4a_5a_6a_7a_8$, and $a_9a_{10}a_{11}$. It will also conclude that segments b_1b_2 and b_3b_4 of the M-line are visible.

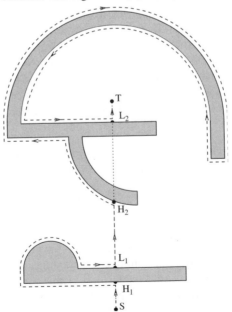

Figure 3.13 Scene 1: Path generated by the algorithm Bug2.

Since algorithm Bug2 is known to converge, one way to incorporate vision is to instruct the robot at each step of its path to "mentally" reconstruct in its current field of vision the path segment that would have been produced by Bug2 (let us call it the *Bug2 path*). The farthest point of that segment can then be made the current intermediate target point, and the robot would make a step toward that point. And then the process repeats. To be meaningful, this would require an assurance of continuity of the considered Bug2 path segment; that is, unless we know for sure that every point of the segment is on the Bug2 path, we cannot take a risk of using this segment. Just knowing the fact of segment continuity is sufficient; there is no need to remember the segment itself. As it turns out, deciding whether a given point lies on the Bug2 path—in which case we will call it a *Bug2 point*—is not a trivial task. The resulting algorithm is called *VisBug-21*, and the path it generates is referred to as the *VisBug-21 path*.

The other algorithm, called VisBug-22, is also tied to the mechanism of Bug2 procedure, but more loosely. The algorithm behaves more opportunistically compared to VisBug-21. Instead of the VisBug-21 process of replacing some "mentally" reconstructed Bug2 path segments with straight-line shortcuts afforded by vision, under VisBug-22 the robot can deviate from Bug2 path segments if this looks more promising and if this is not in conflict with the convergence conditions. As we will see, this makes VisBug-22 a rather radical departure from the Bug2 procedure—with one result being that Bug2 cannot serve any longer as a source of convergence. Hence convergence conditions in VisBug-22 will have to be established independently.

In case one wonders why we are not interested here in producing a vision-laden algorithm extension for the Bug1 algorithm, it is because savings in path length similar to the VisBug-21 and VisBug-22 algorithms are less likely in this direction. Also, as mentioned above, exploring every obstacle completely does not present an attractive algorithm for mobile robot navigation.

Combining Bug1 with vision can be a viable idea in other motion planning tasks, though. One problem in computer vision is recognizing an object or finding a specific item on the object's surface. One may want, for example, to automatically detect a bar code on an item in a supermarket, by rotating the object to view it completely. Alternatively, depending on the object's dimensions, it may be the viewer who moves around the object. How do we plan this rotating motion? Holding the camera at some distance from the object gives the viewer some advantages. For example, since from a distance the camera will see a bigger part of the object, a smaller number of images will be needed to obtain the complete description of the object [63].

Given the same initial conditions, algorithms VisBug-21 and VisBug-22 will likely produce different paths in the same scene. Depending on the scene, one of them will produce a shorter path than the other, and this may reverse in the next scene. Both algorithms hence present viable options. Each algorithm includes a *test for target reachability* that can be traced to the Bug2 algorithm and is based on the following necessary and sufficient condition:

Test for Target Reachability. If, after having defined the last hit point as its intermediate target, the robot returns to it before it defines the next hit point, then either the robot or the target point is *trapped* and hence the target is not reachable. (For more detail, see the corresponding text for algorithm Bug2.)

The following notation is used in the rest of this section:

- C_i and T_i are the robot's position and intermediate target at step i.
- $|AB|$ is the straight-line segment whose endpoints are A and B; it may also designate the length of this segment.
- (AB) is the obstacle boundary segment whose endpoints are A and B, or the length of this segment.
- $[AB]$ is the path segment between points A and B that would be generated by algorithm Bug2, or the length of this path segment.
- $\{AB\}$ is the path segment between points A and B that would be generated by algorithm VisBug-21 or VisBug-22, or the length of this path segment.

It will be evident from the context whether a given notation is referring to a segment or its length. When more than one segment appears between points A and B, the context will resolve the ambiguity.

3.6.2   Algorithm VisBug-21

The algorithm consists of the *Main body*, which does the motion planning proper, and a procedure called *Compute T_i-21*, which produces the next intermediate target T_i and includes the test for target reachability. As we will see, typically the flow of action in the main body is confined to its step S1. Step S2 is executed only in the special case when the robot is moving along a (locally) convex obstacle boundary and so it cannot use its vision to define the next intermediate target T_i. For reasons that will become clear later, the algorithm distinguishes between the case when point T_i lies in the main semiplane and the case when T_i lies in the secondary semiplane. Initially, $C = T_i = S$.

Main Body. The procedure is executed at each point of the continuous path. It includes the following steps:

- S1: Move toward point T_i while executing *Compute T_i-21* and performing the following test:

  If $C = T$ the procedure stops.

  Else if Target is unreachable the procedure stops.

  Else if $C = T_i$ go to step S2.
- S2: Move along the obstacle boundary while executing *Compute T_i-21* and performing the following test:

  If $C = T$ the procedure stops.

Else if Target is unreachable the procedure stops.

Else if $C \neq T_i$ go to step S1.

Procedure *Compute T_i-21* includes the following steps:

- Step 1: If T is visible, define $T_i = T$; procedure stops.

  Else if T_i is on an obstacle boundary go to Step 3.

  Else go to Step 2.

- Step 2: Define point Q as the endpoint of the maximum length contiguous segment $|T_i Q|$ of M-line that extends from point T_i in the direction of point T.

  If an obstacle has been identified crossing the M-line at point Q, then define a hit point, $H = Q$; assign $X = Q$, define $T_i = Q$; go to Step 3.

  Else define $T_i = Q$; go to Step 4.

- Step 3: Define point Q as the endpoint of the maximum length contiguous segment of obstacle boundary, $(T_i Q)$, extending from T_i in the local direction.

  If an obstacle has been identified crossing M-line at a point $P \in (T_i Q)$, $|PT| < |HT|$, assign $X = P$; if, in addition, $|PT|$ does not cross the obstacle at P, define a leave point, $L = P$, define $T_i = P$, and go to Step 2.

  If the lastly defined hit point, H, is identified again and $H \in (T_i Q)$, then Target is not reachable; procedure stops.

  Else define $T_i = Q$; go to Step 4.

- Step 4: If T_i is on the M-line define $Q = T_i$, otherwise define $Q = X$.

  If some points $\{P\}$ on the M-line are identified such that $|S'T| < |QT|$, $S' \in \{P\}$, and C is in the main semiplane, then find the point $S' \in \{P\}$ that produces the shortest distance $|S'T|$; define $T_i = S'$; go to Step 2.

  Else procedure stops.

In brief, procedure *Compute T_i-21* operates as follows. Step 1 is executed at the last stage, when target T becomes visible (as at point A, Figure 3.15). A special case in which points of the M-line noncontiguous to the previously considered sets of points are tested as candidates for the next intermediate Target T_i is handled in Step 4. All the remaining situations relate to choosing the next point T_i among the Bug2 path points contiguous to the previously defined point T_i; these are treated in Steps 2 and 3. Specifically, in Step 2 candidate points along the M-line are processed, and hit points are defined. In Step 3, candidate points along obstacle boundaries are processed, and leave points are defined. The test for target reachability is also performed in Step 3. It is conceivable that at some locations C of the robot the procedure will execute, perhaps even more than once, some combinations of Steps 2, 3, and 4. While doing this, contiguous and noncontiguous segments of the Bug2 path along the M-line and along obstacle boundaries may be considered before the next intermediate target T_i is defined. Then the robot makes a physical step toward T_i.

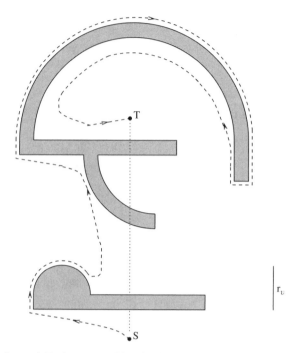

Figure 3.14 Scene 1: Path generated by algorithms VisBug-21 or VisBug-22. The radius of vision is r_v.

Analysis of VisBug-21. Examples shown in Figures 3.14 and 3.15 demonstrate the effect of radius of vision r_v on performance of algorithm VisBug-21. (Compare this with the Bug2 performance in the same environment, Figure 3.13). In the analysis that follows, we first look at the algorithm performance and then address the issue of convergence. Since the path generated by VisBug-21 can diverge significantly from the path that would be produced under the same conditions by algorithm Bug2, it is to be shown that the path-length performance of VisBug-21 is never worse than that of Bug2. One would expect this to be true, and it is ensured by the following lemma.

Lemma 3.6.1. *For a given scene and a given set of Start and Target points, the path produced by algorithm VisBug-21 is never longer than that produced by algorithm Bug2.*

Proof: Assume the scene and start S and target T points are fixed. Consider the robot's position, C_i, and its corresponding intermediate target, T_i, at step i of the path, $i = 0, 1, \ldots$. We wish to show that the lemma holds not only for the whole path from S to T, but also for an arbitrary step i of the path. This amounts to showing that the inequality

$$\{SC_i\} + |C_i T_i| \leq [ST_i] \tag{3.15}$$

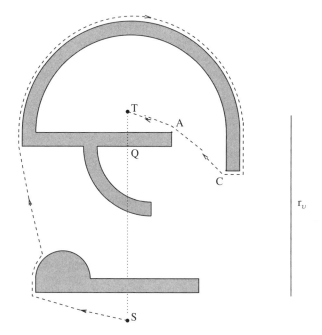

Figure 3.15 Scene 1: Path generated by VisBug-21. The radius of vision r_v is larger than that in Figure 3.14.

holds for any i. The proof is by induction. Consider the initial stage, $i = 0$; it corresponds to $C_0 = S$. Clearly, $|ST_0| \leq [ST_0]$. This can be written as $\{SC_0\} + |C_0T_0| \leq [ST_0]$, which corresponds to the inequality (3.15) when $i = 0$. To proceed by induction, assume that inequality (3.15) holds for step $(i - 1)$ of the path, $i > 1$:

$$\{SC_{i-1}\} + |C_{i-1}T_{i-1}| \leq [ST_{i-1}] \tag{3.16}$$

Each step of the robot's path takes place in one of two ways: either $C_{i-1} \neq T_{i-1}$ or $C_{i-1} = T_{i-1}$. The latter case takes place when the robot moves along the locally convex part of an obstacle boundary; the former case comprises all the remaining situations. Consider the first case, $C_{i-1} \neq T_{i-1}$. Here the robot will take a step of length $|C_{i-1}C_i|$ along a straight line toward T_{i-1}; Eq. (3.16) can thus be rewritten as

$$\{SC_{i-1}\} + |C_{i-1}C_i| + |C_iT_{i-1}| \leq [ST_{i-1}] \tag{3.17}$$

In (3.17), the first two terms form $\{SC_i\}$, and so

$$\{SC_i\} + |C_iT_{i-1}| \leq [ST_{i-1}] \tag{3.18}$$

At point C_i the robot will define the next intermediate target, T_i. Now add to (3.18) the obvious inequality $|T_{i-1}T_i| \leq [T_{i-1}T_i]$:

$$\{SC_i\} + |C_iT_{i-1}| + |T_{i-1}T_i| \leq [ST_{i-1}] + [T_{i-1}T_i] = [ST_i] \qquad (3.19)$$

By the Triangle Inequality, we have

$$|C_iT_i| \leq |C_iT_{i-1}| + |T_{i-1}T_i| \qquad (3.20)$$

Therefore, it follows from (3.19) and (3.20) that

$$\{SC_i\} + |C_iT_i| \leq [ST_i] \qquad (3.21)$$

which proves (3.15).

Consider now the second case, $C_{i-1} = T_{i-1}$. Here the robot takes a step of length $(C_{i-1}C_i)$ along the obstacle boundary (the Bug2 path, $[C_{i-1}C_i]$). Equation (3.16) becomes

$$\{SC_{i-1}\} + [C_{i-1}C_i] \leq [SC_{i-1}] + [C_{i-1}C_i] \qquad (3.22)$$

where the left-hand side amounts to $\{SC_i\}$ and the right-hand side to $[SC_i]$. At point C_i, the robot will define the next intermediate target, T_i. Since $|C_iT_i| \leq [C_iT_i]$, inequality (3.22) can be written as

$$\{SC_i\} + |C_iT_i| \leq [SC_i] + [C_iT_i] = [ST_i] \qquad (3.23)$$

which, again, produces (3.15). Since, by the algorithm's design, at some finite i, $C_i = T$, then

$$\{ST\} \leq [ST] \qquad (3.24)$$

which completes the proof. **Q.E.D.**

One can also see from Figure 3.15 that when r_v goes to infinity, algorithm VisBug-21 will generate locally optimal paths, in the following sense. Take two obstacles or two parts of the same obstacle, k and $k+1$, that are visited by the robot, in this order. During the robot's passing around obstacle k, once a point on obstacle $k+1$ is identified as the next intermediate target, the gap between k and $k+1$ will be traversed along the straight line, which presents the locally shortest path.

When defining its intermediate targets, algorithm VisBug-21 could sometimes use points on the M-line that are not necessarily contiguous to the prior intermediate targets. This would result in a more efficient use of robot's vision: By "cutting corners," the robot would be able to skip some obstacles that intersect the M-line and that it would otherwise have to traverse. However, from the standpoint of algorithm convergence, this is not an innocent operation: It is important to make sure that in such cases the candidate points on the M-line that are "approved" by the algorithm do indeed lie on the Bug2 path.

This is assured by Step 4 of the procedure *Compute T_i-21*, where a noncontiguous point Q on the M-line is considered a possible candidate for an intermediate target only if the robot's current location C is in the main semiplane. The purpose of this condition is to preserve convergence. We will now show that this arrangement always produces intermediate targets that lie on the Bug2 path.

Consider a current location C of the robot along the VisBug-21 path, a current intermediate target T_i, and some visible point Q on the M-line that is being considered as a candidate for the next intermediate target, T_{i+1}. Apparently, Q can be accepted as the intermediate target T_{i+1} only if it lies further along the Bug2 path than T_i.

Recall that in order to ensure convergence, algorithm Bug2 organizes the set of hit and leave points, H_j and L_j, along the M-line so as to form a sequence of segments

$$|ST| > |H_1T| > |L_1T| > |H_2T| > |L_2T| > \cdots \qquad (3.25)$$

that shrinks to T. This inequality dictates two conditions that candidate points for Q must satisfy in order for VisBug-21 to converge: (i) When the current intermediate target T_i lies on the M-line, then only those points Q should be considered for which $|QT| < |T_iT|$. (ii) When T_i is not on the M-line, it lies on the obstacle boundary, in which case there must be the latest crossing point X between M-line and the obstacle boundary, such that the boundary segment (XT_i) is a part of the Bug2 path. In this case, only those points Q should be considered for which $|QT| < |XT|$. Since points Q, T_i, and X are already known, both of these conditions can be easily checked. Let us assume that these conditions are satisfied. Note that the crossing point X does not necessarily correspond to a hit point for either Bug2 or VisBug-21 algorithms. The following statement holds.

Lemma 3.6.2. *For point Q to be further along the Bug2 path than the intermediate target T_i, it is sufficient that the current robot position C lies in the main semiplane.*

Proof: Assume that C lies in the main semiplane; this includes a special case when C lies on the M-line. Then, all possible situations can be classified into three cases:

(1) Both T_i and C lie on the M-line.
(2) T_i lies on the M-line, whereas C does not.
(3) T_i does not lie on the M-line. Let us consider each of these cases.

1. Here the robot is moving along the M-line toward T; thus, T_i is between C and T (Figure 3.16a). Since T_i is by definition on the Bug2 path, and both T_i and Q are visible from point C, then point Q must be on the Bug2 path. And, because of condition (i) above, Q must be further along the Bug2 path than T_i.

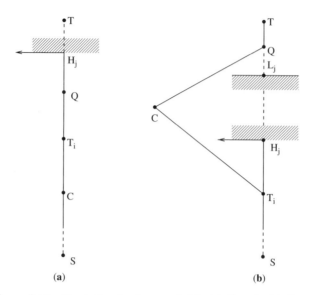

Figure 3.16   Illustration for Lemma 3.6.2: **(a)** Case 1. **(b)** Case 2.

2. This case is shown in Figure 3.16b. If no obstacles are crossing the M-line between points T_i and Q, then the lemma obviously holds. If, however, there is at least one such obstacle, then a hit point, H_j, would be defined. By design of the Bug2 algorithm, the line segment $T_i H_j$ lies on the Bug2 path. At H_j the Bug2 path would turn left and proceed along the obstacle boundary as shown. For each hit point, there must be a matching leave point. Where does the corresponding leave point, L_j, lie?

Consider the triangle $T_i C Q$. Because of the visibility condition, the obstacle cannot cross line segments CT_i or CQ. Also, the obstacle cannot cross the line segment $T_i H_j$, because otherwise some other hit point would have been defined between T_i and H_j. Therefore, the obstacle boundary and the corresponding segment of the Bug2 path must cross the M-line somewhere between H_j and Q. This produces the leave point L_j. Thereafter, because of condition (i) above, the Bug2 path either goes directly to Q, or meets another obstacle, in which case the same argument applies. Therefore, Q is on the Bug2 path and it is further along this path than T_i.

3. Before considering this case in detail, we make two observations.

Observation 1. Within the assumptions of the lemma, if T_i is not on the M-line, then the current position C of the robot is not on the M-line either. Indeed, if T_i is not on the M-line, then there must exist an obstacle that is responsible for the latest hit point, H_j, and thereafter the intermediate target T_i. This obstacle prevents the robot from seeing any point Q on the M-line that would satisfy the requirement (ii) above.

Observation 2. If point C is not on the M-line, then the line segment $|CT_i|$ will never cross the open line segment $|H_j T|$ ("open" here means that the

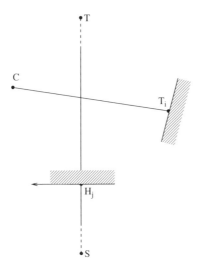

Figure 3.17   Illustration for Lemma 3.6.2: Observation 2.

segment's endpoints are not included). Here H_j is the lastly defined hit point. Indeed, for such a crossing to take place, T_i must lie in the secondary semiplane (Figure 3.17). For this to happen, the Bug2 path would have to proceed from H_j first into the main semiplane and then enter the secondary semiplane somewhere outside of the line segment $|H_j T|$; otherwise, the leave point, L_j, would be established and the Bug2 path would stay in the main semiplane at least until the next hit point, H_{j+1}, is defined. Note, however, that any such way of entering the secondary semiplane would produce segments of the Bug2 path that are not contiguous (because of the visibility condition) to the rest of the Bug2 path. By the algorithm, no points on such segments can be chosen as intermediate targets T_i—which means that if the point C is in the main semiplane, then the line segments $|CT_i|$ and $|H_j T|$ never intersect.

Situations that fall into the case in question can in turn be divided into three groups:

3a. Point C is located on the obstacle boundary, and $C = T_i$. This happens when the robot walks along a locally convex obstacle boundary (point C', Figure 3.18). Consider the curvilinear triangle $X_j C' Q$. Continuing the boundary segment $(X_j C')$ after the point C', the obstacle (and the corresponding segment of the Bug2 path) will either curve inside the triangle, with $|QT|$ lying outside the triangle (Figure 3.18a), or curve outside the triangle, leaving $|QT|$ inside (Figure 3.18b). Since the obstacle can cross neither the line $|C'Q|$ nor the boundary segment $(X_j C')$, it (and the corresponding segment of the Bug2 path) must eventually intersect the M-line somewhere between X_j and Q before intersecting $|QT|$. The rest of the argument is identical to case 2 above.

3b. Point C is on the obstacle boundary, and $C \neq T_i$ (Figure 3.18). Consider the curvilinear triangle $X_j CQ$. Again, the obstacle can cross neither the line of

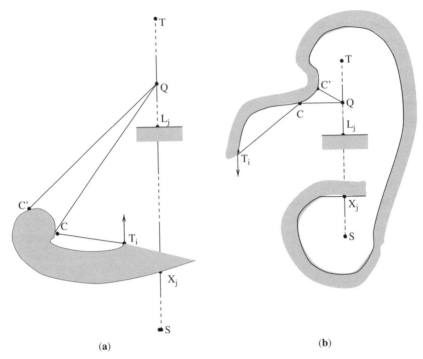

(a)   (b)

Figure 3.18   Illustration for Lemma 3.6.2: Cases 3a and 3b.

visibility $|CQ|$ nor the boundary segment (X_jC), and so the obstacle (and the corresponding segment of the Bug2 path) will either curve inside the triangle, with $|QT|$ left outside of it, or curve outside the triangle, with $|QT|$ lying inside. The rest is identical to case 2.

3c. Point C is not on the obstacle boundary. Then, a curvilinear quadrangle is formed, X_jT_iCQ (Figure 3.19). Again, the obstacle will either curve inside the quadrangle, with $|QT|$ outside of it, or curve outside the quadrangle, with $|QT|$ lying inside. Since neither the lines of visibility $|CT_i|$ and $|CQ|$ nor the boundary segment (X_jT_i) can be crossed, the obstacle (and the corresponding segment of the Bug2 path) will eventually cross $|X_jQ|$ before intersecting $|QT|$ and form the leave point L_j. The rest of the argument is identical to case 2. **Q.E.D.**

If the robot is currently located in the secondary semiplane, then it is indeed possible that a point that lies on the M-line and seems otherwise a good candidate for the next intermediate target T_i does not lie on the Bug2 path. This means that point should not even be considered as a candidate for T_i. Such an example is shown in Figure 3.15: While at location C, the robot will reject the seemingly attractive point Q (Step 2 of the algorithm) because it does not lie on the Bug2 path. We are now ready to establish convergence of algorithm VisBug-21.

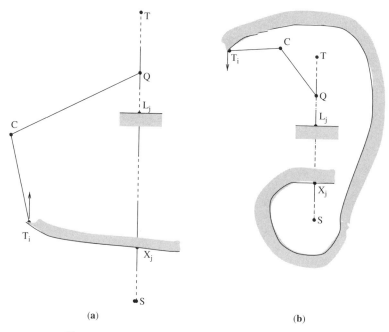

Figure 3.19   Illustration for Lemma 3.6.2: Case 3c.

Theorem 3.6.1. *Algorithm VisBug-21 is convergent.*

Proof: By the definition of intermediate target point T_i, for any T_i that corresponds to the robot's location C, T_i is reachable from C. According to the algorithm, the robot's next step is always in the direction of T_i. This means that if the locus of points T_i is converging to T, so will the locus of points C. In turn, we know that if the locus of points T_i lies on the Bug2 path, then it indeed converges to T. The question now is whether all the points T_i generated by VisBug-21 lie on the Bug2 path.

All steps of the procedure *Compute T_i-21*, except Step 4, explicitly test each candidate for the next T_i for being contiguous to the previous T_i and belonging to the Bug2 path. The only questionable points are points T_i on the M-line that are produced in Step 4 of the procedure: They are not required to be contiguous to the previous T_i. In such cases, points in question are chosen as T_i points only if the robot's location C lies in the main semiplane, in which case the conditions of Lemma 3.6.2 apply. This means that all the intermediate targets T_i generated by algorithm VisBug-21 path lie on the Bug2 path, and therefore converge to T. **Q.E.D.**

Compared to a tactile sensing-based algorithm, the advantage of using vision is of course in the extra information due to the scene visibility. If the robot is thrown

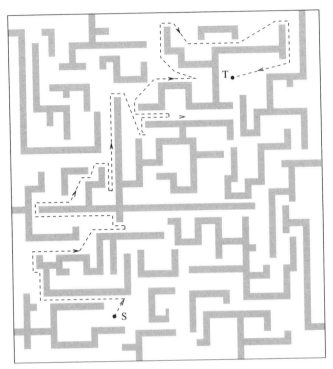

Figure 3.20 Example of a walk (dashed line) in a maze under Algorithm VisBug-21 (compare with Figure 3.11). S, Start; T, Target.

in a crowded scene where at any given moment it can see only a small part of the scene, the efficacy of vision will be obviously limited. Nevertheless, unless the scene is artificially made impossible for vision, one can expect gains from it. This can be seen in performance of VisBug-21 algorithm in the maze borrowed from Section 3.3.2 (see Figure 3.20). For simplicity, assume that the robot's radius of vision goes to infinity. While this ability would be mostly defeated here, the path still looks significantly better than it does under the "tactile" algorithm Bug2 (compare with Figure 3.11).

3.6.3 Algorithm VisBug-22

The structure of this algorithm is somewhat similar to VisBug-21. The difference is that here the robot makes no attempt to ensure that intermediate targets T_i lie on the Bug2 path. Instead, it tries "to shoot as far as possible"; that is, it chooses as intermediate targets those points that lie on the M-line and are as close to the target T as possible. The resulting behavior is different from the algorithm VisBug-21, and so is the mechanism of convergence.

Consider a scene with given Start S and Target T points, and consider a third point, S', that lies on the M-line somewhere between S and T. Following the term *Bug2 path* that we used before, we define a *quasi-Bug2 path segment* as a contiguous segment that starts at S' and produces a part of the path that algorithm Bug2 *would have generated* if points S' and T were its starting and target points, respectively. Because point S' does not need to be on the Bug2 path, a quasi-Bug2 path segment may or may not be a part of the Bug2 path.

The algorithm VisBug-22 will check points along the Bug2 path or a quasi-Bug2 path segment until the best point on the M-line—that is, one that is the closest to T—is identified. This point S' then becomes the starting point of another quasi-Bug2 path segment. Then the process repeats. As a result, unlike algorithms Bug2 and VisBug-21, where each hit point has its matching leave point, in VisBug-22 no such matching needs to occur. To be chosen as the starting point S' of the next quasi-Bug2 path segment, a point must satisfy certain requirements that ensure convergence. These will be considered later, after we describe the whole procedure.

The algorithm includes the *Main body*, which is identical to that of algorithm VisBug-21 (refer to Section 3.6.2), and the procedure *Compute T_i-22*. The purpose of the latter is to produce the next intermediate target T_i for a given current position C; it is also responsible for the test for target reachability. Initially, $C = S = T_i$.

Procedure Compute T_i-22:

- Step 1: If Target T is visible, define $T_i = T$; procedure stops.
  Else if T_i is on an obstacle boundary, go to Step 3.
  Else go to Step 2.
- Step 2: Define point Q as the endpoint of the maximum-length contiguous segment of the M-line, $|T_i Q|$, extending from T_i in the direction of T.
  If an obstacle has been identified crossing the M-line at point Q, then define a hit point, $H = Q$; define $T_i = Q$; go to Step 3.
  Else define $T_i = Q$; go to Step 4.
- Step 3: Define point Q as the endpoint of the maximum length contiguous segment of the obstacle boundary, $(T_i Q)$, extending from T_i in the local direction.
  If an obstacle has been identified crossing the M-line at a point $P \in (T_i Q)$, $|P, T| < |HT|$, and line $|PT|$ does not cross the obstacle at P, then define a leave point, $L = P$, define $T_i = P$, and go to Step 2.
  If the lastly defined hit point H is identified again and $H \in (T_i Q)$, then the target is not reachable; procedure stops.
  Else define $T_i = Q$; go to Step 4.
- Step 4: If T_i is on the M-line, define $Q = T_i$; otherwise define $Q = H$.

If points $\{P\}$ on the M-line are identified such that $|S'T| < |QT|$, $S' \in \{P\}$, then find the point $S' \in \{P\}$ that produces the shortest distance $|S'T|$; define $T_i = S'$; go to Step 2.

Else the procedure stops.

Performance and Convergence of VisBug-22. The algorithm's performance is demonstrated in Figures 3.14 and 3.21; the values of radius of vision r_v used are the same as in examples for VisBug-21 (Figures 3.14 and 3.15). Compare these with the performance of algorithm Bug2 in the same scene (Figure 3.13). As one can see from Figure 3.14, algorithms VisBug-21 and VisBug-22 can sometimes generate identical paths. This is not so in general. Also, neither algorithm can be said to be superior to the other in all scenes. For example, in one scene algorithm VisBug-21 performs better (Figure 3.15) than algorithm VisBug-22 (Figure 3.21), but luck switches to VisBug-22 in another scene shown in Figure 3.23.

Convergence of algorithm VisBug-22 follows simply from the fact that all the starting points, S', of the successive quasi-Bug2 path segments lie on the M-line, and they are organized in such a way as to produce a finite sequence of distances shrinking to T:

$$|S'_1T| > |S'_2T| > |S'_3T| > \cdots \tag{3.26}$$

where points S' are numbered in the order of their appearance.

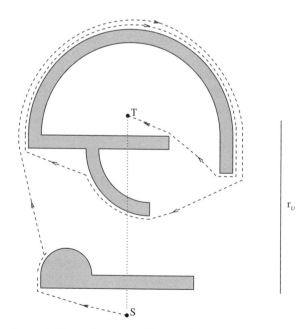

Figure 3.21   Scene 1. The path generated by algorithm VisBug-22. The radius of vision r_v here is larger than that in Figure 3.14, and is equal to that in Figure 3.15.

3.7 FROM A POINT ROBOT TO A PHYSICAL ROBOT

So far our mobile robot has been a point. What changes if we try to extend the motion planning algorithms considered in this chapter to real mobile robots with flesh and finite dimensions?[5] A theoretically correct way to address this question is to replace the original problem of guiding the robot in the two-dimensional workspace (*W*-space) with its reflection in the corresponding *configuration space (C-space)*. (This will be done systematically in Chapter 5 when considering the motion planning problem for robot arm manipulators.) C-space is the space of the robot's control variables, its *degrees of freedom*. In *C*-space the robot becomes a point. Since our robot has two degrees of freedom, which correspond to its coordinates *X* and *Y* in the planar *W*-space, its *C*-space is also a plane.[6] Obstacles will change accordingly.

If the robot is of a simple convex shape—for example, circular or rectangular, as most mobile robots are, or can be satisfactorily approximated by such shapes—the corresponding *C*-space can be obtained simply by "growing" the obstacles to compensate for the robot's "shrinking" to a point. This well-known approach has been used already in the earlier works on motion planning (see Section 2.7). For simple robot shapes the *C*-space is "almost the same" as the *W*-space, and motion planning can be done in *W*-space, keeping in mind this transformation. One can see, for example, that asking whether the circular robot *R* of diameter *D* shown in Figure 3.22 can pass between two obstacles, O_1 and O_2, is equivalent to asking if the minimum distance between the grown obstacles, each grown by $D/2$, is more than zero.

Recall that explicitly building the C-space is possible only in the paradigm of motion planning with complete information (the Piano Mover's model). Since

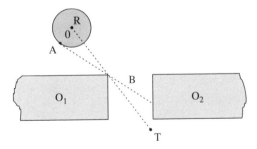

Figure 3.22 Effect of robot shape and geometry on motion planning.

---

[5]A related question is, What kind of sensing does such a robot need in order to protect its whole body from potential collisions? This will be considered in more detail in Chapters 5 and 8.

[6]Including other control variables—for example, the robot orientation—would make *C*-space three- or even higher-dimensional and will complicate the problem accordingly. In practice, the effect of orientation can be often considered independent from the translation controls in *X* and *Y* directions. Then the said complication can be avoided. These more special questions are not pursued in this text. Some of these are discussed in Ref. 64.

in the SIM paradigm input information is never complete and appears as the robot moves, there is no information to calculate the C-space. One can, however, design algorithms based on C-space properties, and this is what will happen in the following chapters.

For the practical side of our question, What changes for the algorithms considered in this chapter when applying them to real mobile robots with finite dimensions? the answer is, nothing changes. Recall that the algorithms VisBug make decisions "on the fly," in real time. They make the robot either (a) move in free space by following the M-line or (b) follow obstacle boundaries. For example, when following an obstacle boundary, if the robot arrives at a gap between two obstacles, it may or may not be able to pass it. If the gap is too narrow for the robot to pass through, it will perceive both obstacles as one. When following the obstacle boundaries, the robot will switch from one obstacle to the other without even noticing this fact.

Additional heuristics can be added to improve the algorithm efficiency, as long as care is taken not to imperil the algorithm convergence. For example, if the robot sees its target T through a gap between two obstacles, it may attempt to measure the width of the gap to make sure that it will be able to pass it, before it actually moves to the gap. Or, if the robot's shape is more complex than a circle, it may try to move through the gap by varying its orientation.

An interesting question appears when studying the effect of location of sensor(s) on the robot body on motion planning. Assume the robot R shown in Figure 3.22 has a range sensor. If the sensor is located at the robot's center then, as the dotted line of vision OT shows, the robot will see the gap between two obstacles and will act accordingly. But, if the sensor happened to be attached to the point A on the robot periphery, then the dotted line AB shows that the robot will not be able to see if the gap is real. The situation can be even more complex: For example, it is not uncommon for real-world mobile robots to have a battery of sonar sensors attached along the circumference of the robot body. Then different sensors may see different objects, the robot's intelligence needs to reconcile those different readings, and a more careful scheme is needed to model the C-space sensing. Little work has been done in this area; some such schemes have been explored by Skewis [64].

3.8 OTHER APPROACHES

Recall the division of all sensor-based motion planning algorithms into two classes (Section 3.5). Class 1 combines algorithms in which the robot never leaves an obstacle unless and until it explores it completely. Class 2 combines algorithms that are complementary to those in Class 1: In them the robot can leave an obstacle and walk further, and even return to this obstacle again at some future time, without exploring it in full.

As mentioned above, today Class 1 includes only one algorithm, Bug1. The reason for this paucity likely lies in the inherent conservatism of algorithms in this Class. Their dogmatically reaching the upper performance bound in the most simple scenes, where a more agile strategy would likely do much better, does not leave much room for creativity. Besides, Bug1 does already reach the universal lower bound (3.13) of the sensor-based motion planning problem, so there is not much one can expect in advancing theory.

Performance aside, one may wonder how much variety there is in Class 1. Most likely not much, but we do not know for sure.

This makes Bug1 an unrivaled champion among "tactile" algorithms—a somewhat ironic title, given that Bug1 is not a likely candidate for real-world robots, except perhaps in special applications. One such application is generation of the outline of a scene in question, which is producing contours of obstacles present in the scene. You supply the robot with a few pairs of S and T points to ensure that it will visit all obstacles of interest, make sure that it stores the files of coordinates while walking around obstacles, and let it go. As a side product of those trips, the robot will bring back a map of the scene. The technique is especially good for obtaining such outlines and contours in a database, with an imaginary point robot. This method of obtaining the map, while not theoretically sound, is quite competitive compared to other techniques.[7] With this technique the robot cannot, of course, uncover pieces of free space trapped inside obstacles, unless the robot explicitly starts there.

The situation is more interesting with Class 2 algorithms, including extensions to vision and range sensing. If the *Bug family* (as some researchers started calling these procedures) is to grow, this will likely be happening in Class 2. Between 1984 and now (1984 being the year of first publications on sensor-based motion planning), over a dozen provable algorithms from this class have been reported in the literature. Besides, many heuristic procedures relying on the "engineering approach" have been described; their convergence and performance properties in arbitrary scenes is anybody's guess.

The following brief review of significant work on provable Class 2 sensor-based algorithms is admittedly incomplete. Ideas similar to those explored in this book—that is, with a focus on topological rather than geometrical properties of space—has been considered by a number of researchers. Algorithms Alg1 and Alg2 by Sankaranarayanan and Vidyasagar [65] successfully fight the unpleasant tendency of the Bug2 algorithm to produce multiple local cycles in some special scenes (see Section 3.5). Whereas local cycles can be stopped via a straightforward combination of Bug1 and Bug2 procedures (see the BugM1 algorithm, Section 3.5), Alg1 and Alg2 do it better and they do it more economically. Importantly, they reach the path length lower bound (3.14) for the Class 2 algorithms, and by doing so they "close" the Class 2 of sensor-based planning algorithms, similar to how Bug1 closes Class 1.

Also in this group are elegant provable algorithms TangentBug [66] by Kamon, Rivlin, and Rimon, and DistBug [67, 68] by Kamon and Rivlin. Algorithm

---

[7]This is not to suggest that Bug1 is an algorithm for map-making. Map-making (*terrain acquisition* and *terrain coverage* are other terms one finds in literature) is a different problem. See, for example, Ref. 1.

TangentBug, in turn, has inspired procedures WedgeBug and RoverBug [69, 70] by Laubach, Burdick, and Matthies, which try to take into account issues specific for NASA planet rover exploration. A number of schemes with and without proven convergence have been reported by Noborio [71].

Given the practical needs, it is not surprising that many attempts in sensor-based planning strategies focus on distance sensing—stereo vision, laser range sensing, and the like. Some earlier attempts in this area tend to stick to more familiar graph-theoretical approaches of computer science, and consequently treat space in a discrete rather than continuous manner. A good example of this approach is the visibility-graph based approach by Rao et al. [72].

Standing apart is the approach described by Choset et al. [73, 74], which can be seen as an attempt to fill the gap between the two paradigms, motion planning with complete information (Piano Mover's model) and motion planning with incomplete information [other names are sensor-based planning, or Sensing–Intelligence–Motion (SIM)]. The idea is to use sensor-based planning to first build the map and then the Voronoi diagram of the scene, so that the future robot trips in this same area could be along shorter paths—for example, along links of the acquired Voronoi diagram. These ideas, and applications that inspire them, are different from the go-from-A-to-B problem considered in this book and thus beyond our scope. They are closer to the systematic space exploration and map-making. The latter, called in the literature *terrain acquisition* or *terrain coverage*, might be of use in tasks like robot-assisted map-making, floor vacuuming, lawn mowing, and so on (see, e.g., Refs. 1 and 75).

While most of the above works provide careful analysis of performance and convergence, the "engineering approach" heuristics to sensor-based motion planning procedures usually discuss their performance in terms of "consistently better than" or "better in our experiments," and so on. Since idiosyncracies of these algorithms are rarely analyzed, their utility is hard to assess. There have been examples when an algorithm published as provable turned out to be ruefully divergent even in simple scenes.[8]

Related to the area of two-dimensional motion planning are also works directed toward motion planning for a "point robot" moving in three-dimensional space. Note that the increase in dimensionality changes rather dramatically the formal foundation of the sensor-based paradigm. When moving in the (two-dimensional) plane, if the point robot encounters an obstacle, it has a choice of only two ways to pass around it: from the left or from the right, clockwise or counterclockwise. When a point robot encounters an object in the three-dimensional space, it is faced with an infinite number of directions for passing around the object. This means that unlike in the two-dimensional case, the topological properties of three-dimensional space cannot be used directly anymore when seeking guarantees of algorithm completeness.

---

[8]As the principles of design of motion planning algorithms have become clearer, in the last 10–15 years the level of sophistication has gone up significantly. Today the homework in a graduate course on motion planning can include an assignment to design a new provable sensor-based algorithm, or to decide if some published algorithm is or is not convergent.

Accordingly, objectives of works in this area are usually toward complete exploration of objects. One such application is visual exploration of objects (see, e.g., Refs. 63 and 76): One attempts, for example, to come up with an economical way of automatically manipulating an object on the supermarket counter in order to locate on it the bar code.

Extending our go-from-A-to-B problem to the mobile robot navigation in three-dimensional space will likely necessitate "artificial" constraints on the robot environment (which we were lucky not to need in the two-dimensional case), such as constraints on the shapes of objects, the robot's shape, some recognizable properties of objects' surfaces, and so on. One area where constraints appear naturally, as part of the system kinematic design, is motion planning for three-dimensional arm manipulators. The very fact that the arm links are tied into some kinematic structure and that the arm's base is bolted to its base provide additional constraints that can be exploited in three-dimensional sensor-based motion planning algorithms. This is an exciting area, with much theoretical insight and much importance to practice. We will consider such schemes in Chapter 6.

3.9 WHICH ALGORITHM TO CHOOSE?

With the variety of existing sensor-based approaches and algorithms, one is entitled to ask a question: How do I choose the right sensor-based planning algorithm for my job? When addressing this question, we can safely exclude the Class 1 algorithms: For the reasons mentioned above, except in very special cases, they are of little use in practice.

As to Class 2, while usually different algorithms from this group produce different paths, one would be hard-pressed to recommend one of them over the others. As we have seen above, if in a given scene algorithm A performs better than algorithm B, their luck may reverse in the next scene. For example, in the scene shown in Figures 3.15 and 3.21, algorithm VisBug-21 outperforms algorithm VisBug-22, and then the opposite happens in the scene shown in Figure 3.23. One is left with an impression that when used with more advanced sensing, like vision and range finders, in terms of their motion planning skills just about any algorithm will do, as long as it guarantees convergence.

Some people like the concept of a benchmark example for comparing different algorithms. In our case this would be, say, a fixed benchmark scene with a fixed pair of start and target points. Today there is no such benchmark scene, and it is doubtful that a meaningful benchmark could be established. For example, the elaborate labyrinth in Figure 3.11 turns out to be very easy for the Bug2 algorithm, whereas the seemingly simpler scene in Figure 3.6 makes the same algorithm produce a torturous path. It is conceivable that some other algorithm would have demonstrated an exemplary performance in the scene of Figure 3.6, only to look less brave in another scene. Adding vision tends to smooth algorithms' idiosyncrasies and to make different algorithms behave more similarly, especially in real-life scenes with relatively simple obstacles, but the said relationship stays.

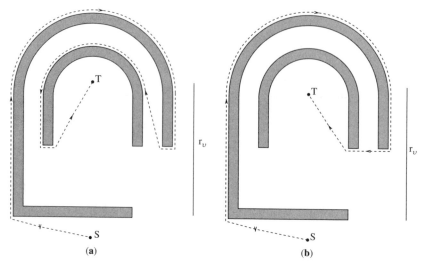

Figure 3.23 Scene 2. Paths generated (**a**) by algorithm VisBug-21 and (**b**) by algorithm VisBug-22.

Furthermore, even seemingly simple questions—(1) Does using vision sensing guarantee a shorter path compared to using tactile sensing? or (2) Does a better (that is, farther) vision buy us better performance compared to an inferior (that is, more myopic) vision?—have no simple answers. Let us consider these questions in more detail.

1. *Does using vision sensing guarantee a shorter path compared to using tactile sensing?* The answer is no. Consider the simple example in Figure 3.24. The robot's start S and target T points are very close to and on the opposite sides of the convex obstacle that lies between them. By far the main part of the robot path will involve walking around the obstacle. During this time the robot will have little opportunity to use its vision because at every step it will see only a tiny piece of the obstacle boundary; the rest of it will be curving "around the corner."

So, in this example, robot vision will behave much like tactile sensing. As a result, the path generated by algorithm VisBug-21 or VisBug-22 or by some other "seeing" algorithm will be roughly no shorter than a path generated by a "tactile" algorithm, no matter what the robot's radius of vision r_v is. If points S and T are further away from the obstacle, the value of r_v will matter more in the initial and final phases of the path but still not when walking along the obstacle boundary.

When comparing "tactile" and "seeing" algorithms, the comparative performance is easier to analyze for less opportunistic algorithms, such as VisBug-21: Since the latter emulates a specific "tactile" algorithm by continuously short-cutting toward the furthest visible point on that algorithm's path, the resulting path will usually be shorter, and never longer, than that of the emulated "tactile" algorithm (see, e.g., Figure 3.14).

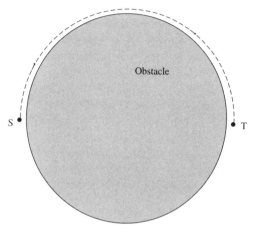

Figure 3.24 In this scene, the path generated by an algorithm with vision would be almost identical to the path generated by a "tactile" planning algorithm.

With more opportunistic algorithms, like VisBig-22, even this property breaks down: While paths that algorithm VisBig-22 generates are often significantly shorter than paths produced by algorithm Bug2, this cannot be guaranteed (compare Figures 3.13 and 3.21).

2. *Does better vision (a larger radius of vision, r_v) guarantee better performance compared to an inferior vision (a smaller radius of vision)?* We know already that for VisBig-22 this is definitely not so—a larger radius of vision does not guarantee shorter paths (compare Figures 3.21 and 3.14). Interestingly, even for a more stable VisBug-21, it is not so. The example in Figure 3.25 shows that, while VisBug-21 always does better with vision than with tactile sensing, more vision—that is, a larger r_v—does not necessarily buy better performance. In this scene the robot will produce a shorter path when equipped with a smaller radius of vision (Figures 3.25a) than when equipped with a larger radius of vision (Figures 3.25b).

The problem lies, of course, in the fundamental properties of uncertainty. As long as some, even a small piece, of relevant information is missing, anything may happen. A more experienced hiker will often find a shorter path, but once in a while a beginner hiker will outperform an experienced hiker. In the stock market, an experienced stock broker will usually outperform an amateur investor, but once in a while their luck will reverse.[9] In situations with uncertainty, more experience certainly helps, but it helps only on the average, not in every single case.

---

[9]On a quick glance, the same principle seems to apply to the game of chess, but it does not. Unlike in other examples above, in chess the uncertainty comes not from the lack of information—complete information is right there on the table, available to both players—but from the limited amount of information that one can process in limited time. In a given time an experienced player will check more candidate moves than will a novice.

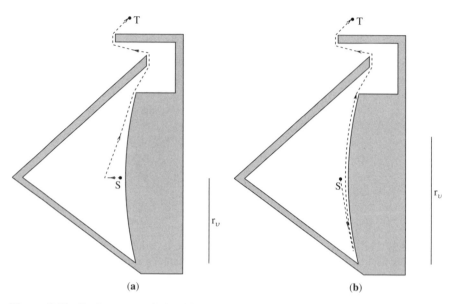

Figure 3.25 Performance of algorithm VisBug-21 in the same scene **(a)** with a smaller radius of vision and **(b)** with a larger radius of vision. The smaller (worse) vision results in a shorter path!

These examples demonstrate the variety of types of uncertainty. Notice another interesting fact: While the experienced hiker and experienced stock broker can make use of a probabilistic analysis, it is of no use in the problem of motion planning with incomplete information. A direction to pass around an obstacle that seems to promise a shorter path to the target may offer unpleasant surprises around the corner, compared to a direction that seemed less attractive before but is objectively the winner. It is far from clear how (and whether) one can impose probabilities on this process in any meaningful way. That is one reason why, in spite of high uncertainty, sensor-based motion planning is essentially a deterministic process.

3.10 DISCUSSION

The somewhat surprising examples above (see the last few figures in the previous section) suggest that further theoretical analysis of general properties of Class 2 algorithms may be of more benefit to science and engineering than proliferation of algorithms that make little difference in real-world tasks. One interesting possibility would be to attempt a meaningful classification of scenes, with a predictive power over the performance of various algorithmic schemes. Our conclusions from the worst-case bounds on algorithm performance also beg for a similar analysis in terms of some other, perhaps richer than the worst-case, criteria.

This said, the material in this chapter demonstrates a remarkable success in the last 10–15 years in the state of the art in sensor-based robot motion planning. In spite of the formidable uncertainty and an immense diversity of possible obstacles and scenes, a good number of algorithms discussed above guarantee convergence: That is, a mobile robot equipped with one of these procedures is guaranteed to reach the target position if the target can in principle be reached; if the target is not reachable, the robot will make this conclusion in finite time. The algorithms guarantee that the paths they produce will not circle in one area an indefinite number of times, or even a large number of times (say, no more than two or three).

Twenty years ago, most specialists would doubt that such results were even possible. On the theoretical level, today's results mean, to much surprise from the standpoint of earlier views on the subject, that purely local input information is not an obstacle to obtaining global solutions, even in cases of formidable complexity.

Interesting results raise our appetite for more results. Answers bring more questions, and this is certainly true for the area at hand. Below we discuss a number of issues and questions for which today we do not have answers.

Bounds on Performance of Algorithms with Vision. Unlike with "tactile" algorithms, today there are no upper bounds on performance of motion planning algorithms with vision, such as VisBug-21 or VisBug-22 (Section 3.6). While from the standpoint of theory it would be of interest to obtain bounds similar to the bound (3.13) for "tactile" algorithms, they would likely be of limited generality, for the following reasons.

First, to make such bounds informative, we would likely want to incorporate into them characteristics of the robot's vision—at least the radius of vision r_v, and perhaps the resolution, accuracy, and so on. After all, the reason for developing these bounds would be to know how vision affects robot performance compared to the primitive tactile sensing. One would expect, in particular, that vision improves performance. As explained above, this cannot be expected in general. Vision does improve performance, but only "on the average," where the meaning of "average" is not clear. Recall some examples in the previous section: In some scenes a robot with a larger radius of vision r_v will perform worse than a robot with a smaller r_v. Making the upper bound reflect such idiosyncrasies would be desirable but also difficult.

Second, how far the robot can see depends not only on its vision but also on the scene it operates in. As the example in Figure 3.24 demonstrates, some scenes can bring the efficiency of vision to almost that of tactile sensing. This suggests that characteristics of the scene, or of classes of scenes, should be part of the upper bounds as well. But, as geometry does not like probabilities, the latter is not a likely tool: It is very hard to generalize on distributions of locations and shapes of obstacles in the scene.

Third, given a scene and a radius of vision r_v, a vastly different path performance will be produced for different pairs of start and target points in that same scene.

Moving Obstacles. The model of motion planning considered in this chapter (Section 3.1) assumes that obstacles in the robot's environment are all static—that is, do not move. But obstacles in the real world may move. Let us call an environment where obstacles may be moving the *dynamic (changing, time-sensitive) environment.* Can sensor-based planning strategies be developed capable of handling a dynamic environment? Even more specifically, can strategies that we developed in this chapter be used in, or modified to account for, a dynamic environment?

The answer is a qualified yes. Since our model and algorithms do not include any assumptions about specifics of the geometry and dimensions of obstacles (or the robot itself), they are in principle ideally suited for handling a dynamic environment. In fact, one can use the Bug and VisBug family algorithms in a dynamic environment without any changes. Will they always work? The answer is, "it depends," and the reason for the qualified answer is easy to understand.

Assume that our robot moves with its maximum speed. Imagine that while operating under one of our algorithms—it does not matter which one—the robot starts passing around an obstacle that happens to be of more or less complex shape. Imagine also that the obstacle itself moves. Clearly, if the obstacle's speed is higher than the speed of the robot, the robot's chance to pass around the obstacle and ever reach the target is in doubt. If on top of that the obstacle happens to also be rotating, so that it basically cancels the robot's attempts to pass around it, the answer is not even in doubt: The robot's situation is hopeless.

In other words, the motion parameters of obstacles matter a great deal. We now have two options to choose from. One is to use algorithms as they are, but drop the promise of convergence. If the obstacles' speeds are low enough compared to the robot, or if obstacles move more or less in one place, like a tree in the wind, then the robot will likely get where it intends. Even if obstacles move faster than the robot, but their shapes or directions of motion do not create situations as in the example above, the algorithms will still work well. But, if the situation is like the one above, there will be no convergence.

Or we can choose another option. We can guarantee convergence of an algorithm, but impose some additional constraints on the motion of objects in the robot workspace. If a specific environment satisfies our constraints, convergence is guaranteed. The softer those constraints, the more universal the resulting algorithms. There has been very little research in this area.

For those who need a real-world incentive for such work, here is an example. Today there are hundreds of human-made dead satellites in the space around Earth. One can bet that all of them have been designed, built, and launched at high cost. Some of them are beyond repair and should be hauled to a satellite cemetery. Some others could be revived after a relatively simple repair—for example, by replacing their batteries. For long time, NASA (National Aeronautics and Space Administration) and other agencies have been thinking of designing a robot space vehicle capable of doing such jobs.

Imagine we designed such a system: It is agile and compact; it is capable of docking, repair, and hauling of space objects; and, to allow maneuvering around space objects, it is equipped with a provable sensor-based motion planning algorithm. Our robot—call it R-SAT—arrives to some old satellite "in a coma"—call it X. The satellite X is not only moving along its orbit around the Earth, it is also tumbling in space in some arbitrary ways. Before R-SAT starts on its repair job, it will have to fly around X, to review its condition and its useability. It may need to attach itself to the satellite for a more involved analysis. To do this—fly around or attach to the satellite surface—the robot needs to be capable of speeds that would allow these operations.

If the robot arrives at the site without any prior analysis of the satellite X condition, this amounts to our choosing the first option above: No convergence of R-SAT motion planning around X is guaranteed. On the other hand, a decision to send R-SAT to satellite X might have been made after some serious remote analysis of the X's rate of tumbling. The analysis might have concluded that the rate of tumbling of satellite X was well within the abilities of the R-SAT robot. In our terms, this corresponds to adhering to the second option and to satisfying the right constraints—and then the R-SAT's motion planning will have a guaranteed convergence.

Multirobot Groups. One area where the said constraints on obstacles' motion come naturally is multirobot systems. Imagine a group of mobile robots operating in a planar scene. In line with our usual assumption of a high level of uncertainty, assume that the robots are of different shapes and the system is highly decentralized. That is, each robot makes its own motion planning decisions without informing other robots, and so each robot knows nothing about the motion planning intentions of other robots. When feasible, this type of control is very reliable and well protected against communication and other errors.

A decentralized control in multirobot groups is desirable in many settings. For example, it would be of much value in a "robotic" battlefield, where a continuous centralized control from a single commander would amount to sacrificing the system reliability and fault tolerance. The commander may give general commands from time to time—for instance, on changing goals for the whole group or for specific robots (which is an equivalent of prescribing each robot's next target position)—but most of the time the robots will be making their own motion planning decisions.

Each robot presents a moving obstacle to other robots. (Then there may also be static obstacles in the workspace.) There is, however, an important difference between this situation and the situation above with arbitrary moving obstacles. You cannot have any beforehand agreement with an arbitrary obstacle, but you can have one with other robots. What kind of agreement would be unconstraining enough and would not depend on shapes and dimensions and locations? The system designers may prescribe, for example, that if two robots meet, each robot will attempt to pass around the other only clockwise. This effectively eliminates the above difficulty with the algorithm convergence in the situation with moving obstacles.[10] (More details on this model can be found in Ref. 77.)

Needs for More Complex Algorithms. One area where good analysis of algorithms is extremely important for theory and practice is sensor-based motion planning for robot arm manipulators. Robot manipulators operate sometimes in a two-dimensional space, but more often they operate in the three-dimensional space. They have complex kinematics, and they have parts that change their relative positions in complex ways during the motion. Not rarely, their workspace is filled with obstacles and with other machinery (which is also obstacles).

Careful motion planning is essential. Unlike with mobile robots, which usually have simple shapes and can be controlled in an intuitively clear fashion, intuition helps little in designing new algorithms or even predicting the behavior of existing algorithms for robot arm manipulators.

As mentioned above, performance of Bug2 algorithm deteriorates when dealing with situations that we called *in-position*. In fact, this will be likely so for all Class 2 motion planning algorithms. Paths tend to become longer, and the robot may produce local cycles that keep "circling" in some segments of the path. The chance of in-position situations becomes very persistent, almost guaranteed, with arm manipulators. This puts a premium on good planning algorithms. This area is very interesting and very unintuitive. Recall that today about 1,000,000 industrial arms manipulators are busy fueling the world economy. Two chapters of this book, Chapters 5 and 6, are devoted to the topic of sensor-based motion planning for arm manipulators.

The importance of motion planning algorithms for robot arm manipulators is also reinforced by its connection to teleoperation systems. Space-operator-guided robots (such as arm manipulators on the Space Shuttle and International Space Station), robot systems for cleaning nuclear reactors, robot systems for detonating mines, and robot systems for helping in safety operations are all examples of teleoperation systems. Human operators are known to make mistakes in such tasks. They have difficulty learning necessary skills, and they tend to compensate difficulties by slowing the operation down to crawling. (Some such problems will be discussed in Chapter 7.) This rules out tasks where at least a "normal" human speed is a necessity.

One potential way out of this difficulty is to divide responsibilities between the operator and the robot's own intelligence, whereby the operator is responsible for higher-level tasks—planning the overall task, changing the plan on the fly if needed, or calling the task off if needed—whereas the lower-level tasks like obstacle collision avoidance would be the robot's responsibility. The two types of intelligence, human and robot intelligence, would then be combined in one control system in a synergistic manner. Designing the robot's part of the system would require (a) the type of algorithms that will be considered in Chapters 5 and 6 and (b) sensing hardware of the kind that we will explore in Chapter 8.

---

[10]Note that this is the spirit of the automobile traffic rules.

Turning back to motion planning algorithms for mobile robots, note that nowhere until now have we talked about the effect of robot dynamics on motion planning. This implicitly assumed, for example, that any sharp turn in the robot's path dictated by the planning algorithm was deemed feasible. For a robot with flesh and reasonable mass and speed, this is of course not so. In the next chapter we will turn to the connection between robot dynamics and motion planning.

3.11 EXERCISES

1. Recall that in the so-called *out-position* situations (Section 3.3.2) the algorithm Bug2 has a very favorable performance: The robot is guaranteed to have no cycles in the path (i.e., to never pass a path segment more than once). On the other hand, the *in-position* situations can sometimes produce long paths with local cycles. For a given scene, the in-position was defined in Section 3.3.2 as a situation when either Start or Target points, or both, lie inside the convex hull of obstacles that the line (Start, Target) intersects. Note that the in-position situation is only a sufficient condition for trouble: Simple examples can be designed where no cycles are produced in spite of the in-position condition being satisfied.

    Try to come up with a *necessary and sufficient* condition—call it GOODCON—that would guarantee a no-cycle performance by Bug2 algorithm. Your statement would say: "Algorithm Bug2 will produce no cycles in the path if and only if condition GOODCON is satisfied."

2. The following sensor-based motion planning algorithm, called AlgX (see the procedure below), has been suggested for moving a mobile point automaton (MA) in a planar environment with unknown arbitrarily shaped obstacles. MA knows its own position and that of the target location T, and it has tactile sensing; that is, it learns about an obstacle only when it touches it. AlgX makes use of the straight lines that connect MA with point T and are tangential to the obstacle(s) at the MA's current position.

    The questions being asked are:

    - Does AlgX converge?
    - If the answer is "yes," estimate the performance of AlgX.
    - If the answer is "no," why not? Explain and give a counterexample. Using the same idea of the tangential lines connecting MA and T, try to fix the algorithm. Your procedure must operate with finite memory. Estimate its performance.
    - Develop a test for target reachability.

    Just like the Bug1 and Bug2 algorithms, the AlgX procedure also uses the notions of (a) *hit points*, H_j, and *leave points*, L_j, on the obstacle boundaries and (b) *local directions*. Given the start S and target T points, here are some necessary details:

- Point P becomes a *hit point* when MA, while moving along the ST line, encounters an obstacle at P.
- Point P can become a *leave point* if and only if (1) it is possible for MA to move from P toward T and (2) there is a straight line that is tangential to the obstacle boundary at P and passes through T. When a leave point is encountered for the first time, it is called *open*; it may be *closed* by MA later (see the procedure).
- A *local direction* is the direction of following an obstacle boundary; it can be either *left* or *right*. In AlgX the current local direction is inverted whenever MA passes through an open leave point; it does not change when passing through a closed leave point.
- A *local cycle* is formed when MA visits some points of its path more than once.

The idea behind the algorithm AlgX is as follows. MA starts by moving straight toward point T. Every time it encounters an obstacle, it inverts its local direction, the idea being that this will keep it reasonably close to the straight line (S, T). If a local cycle is formed, MA blames it on the latest turn it made at a hit point. Then MA retraces back to the latest visited leave point, closes it, and whence takes the opposite turn at the next hit point. If that turn leads again to a local cycle, then the turn that led to the current leave point is to be blamed. And so on. The procedure operates as follows:

Initialization. Set the current local direction to "right"; set $j = 0$, $L_j = S$.

Step 1. Move along a straight line from the current leave point toward point T until one of the following occurs:

    a. Target T is reached; the procedure terminates.

    b. An obstacle is encountered; go to Step 2.

Step 2. Define a hit point H_j. Turn in the current local direction and move along the obstacle boundary until one of the following occurs:

    a. Target T is reached; the procedure terminates.

    b. The current velocity vector (line tangential to the obstacle at the current MA position) passes through T, and this point has not been defined previously as a leave point; then, go to Step 3.

    c. MA comes to a previously defined leave point L_i, $i \leq j$ (i.e., a local cycle has been formed). Go to Step 4.

Step 3. Set $j = j + 1$; define the current point as a new open leave point; invert the current local direction; go to Step 1.

Step 4. Close the open leave point L_k visited immediately before L_i. Invert the local direction. Retrace the path between L_i and L_k. (During retracing, invert the local direction when passing through an open leave point, but do not close those points; ignore closed leave points.) Now MA is at the closed leave point L_k. If L_i is *open*, go to Step 1. If L_i is *closed*, execute

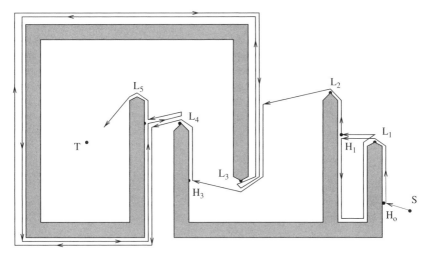

Figure 3.E.2

the second part of Step 2, "... move along ... until ...," using the current local direction.

In Figure 3.E.2, L_1 is self-closed because when it is passed for the second time, the latest visited open leave point is L_1 itself; L_4 is closed when L_3 is passed for the second time; no other leave points are closed. When retracing from L_3 to L_4, the leave point L_3 causes inversion of the local direction, but does not close any leave points.

3. Design two examples that would result in the best-case and the worst-case performance, respectively, of the Bug2 algorithm. In both examples the same three C-shaped obstacles should be used, and an M-line that connects two distinct points S and T and intersects all three obstacles. An obstacle can be mirror image reversed or rotated if desired. Obstacles can touch each other, in which case they become one obstacle; that is, a point robot will not be able to pass between them at the contact point(s). Evaluate the algorithm's performance in each case.

CHAPTER 4

Accounting for Body Dynamics: The Jogger's Problem

> Let me first explain to you how the motions of different kinds of matter depend on a property called inertia.
>
> — *Sir William Thomson (Lord Kelvin), The Tides*

4.1 PROBLEM STATEMENT

As discussed before, motion planning algorithms usually adhere to one of the two paradigms that differ primarily by their assumptions about input information: *motion planning with complete information* (*Piano Mover's problem*) and *motion planning with incomplete information* (*sensor-based motion planning, SIM paradigm*, see Chapter 1). Strategies that come out of the two paradigms can be also classified into two groups: *kinematic approaches*, which consider only kinematic and geometric issues, and *dynamic approaches*, which take into account the system dynamics. This classification is independent from the classification into the two paradigms. In Chapter 3 we studied kinematic sensor-based motion planning algorithms. In this chapter we will study dynamic sensor-based motion planning algorithms.

What is so dynamic about dynamic approaches? In strategies that we considered in Chapter 3, it was implicitly assumed that whatever direction of motion is good for the robot's next step from the standpoint of its goal, the robot will be able to accomplish it. If this is true, in the terminology of control theory such a system is called a *holonomic* system [78]. In a holonomic system the number of control variables available is no less that the problem dimensionality. The system will also work as intended in situations where the above condition is not satisfied, but for some reason the robot dynamics can be ignored. For example, a very slowly moving robot can turn on a dime and hence can execute any sharp turn if prescribed by its motion planning software.

Most of existing approaches to motion planning (including those within the Piano Mover's model) assume, first, that the system is holonomic and, second,

Sensing, Intelligence, Motion, by Vladimir J. Lumelsky
Copyright © 2006 John Wiley & Sons, Inc.

that it will behave as a holonomic system. Consequently, they deal solely with the system kinematics and ignore its dynamic properties. One reason for this state of affairs is that the methods of motion planning tend to rely on tools from geometry and topology, which are not easily connected to the tools common to control theory. Although system dynamics and sensor-based motion control are clearly tightly coupled in many, if not most, real-world systems, little attention has been paid to this connection in the literature.

The robot is a body; it has a mass and dimensions. Once it starts moving, it acquires velocity and acceleration. Its dynamics may now prevent it from making sharp, and sometimes even relatively shallow, turns prescribed by the planning algorithm. A sharp turn reasonable from the standpoint of reaching the target position may not be physically realizable because of the robot's inertia. In control theory terminology, this is a *nonholonomic* system [78]. A classical example of a nonholonomic control problem is the car parallel parking task: Because the driver does not have enough control means to execute the parking in one simple translational motion, he has to wiggle the car back and force to bring it to the desired position.

Given the insufficient information about the surroundings, which is central to the sensor-based motion planning paradigm, the lack of control means to execute any desired motion translates into a safety issue: One needs a guarantee of a *stopping path* at any time, in case a sudden obstacle makes it impossible to continue on the intended path.

Theoretically, there is a simple way out: We can make the robot stop every time it intends to turn, let it turn, and resume the motion as needed. Not many applications will like such a stop-and-go motion pattern. For a realistic control we want the robot to make turns on the move, and not stop unless "absolutely necessary," whatever this means. That is, in addition to the usual problem of "where to go" and how to guarantee the algorithm convergence in view of incomplete information, the robot's mass and velocity bring about another component of motion planning, body dynamics. Furthermore, we will see that it will be important to incorporate the constraints of robot dynamics into the very motion planning algorithm, together with the constraints dictated by collision avoidance and algorithm convergence requirements.

We call the problem thus formulated the *Jogger's Problem*, because it is not unlike the task a human jogger faces in an urban setting when going for a morning run. Taking a run involves continuous on-line control and decision-making. Many decisions will be made during the run; in fact, many decisions are made within each second of the run. The decision-making apparatus requires a smooth collaboration of a few mechanisms. First, a global planning mechanism will work on ensuring arrival at the target location in spite of all deviations and detours that the environment may require. Unless a "grand plan" is followed, arrival at the target location—what we like to call convergence—may not be guaranteed.

Second, since an instantaneous stop is impossible due to the jogger's inertia, in order to maintain a reasonable speed the jogger needs at any moment an "insurance" option of a safe *stopping path*. This mechanism will relate the jogger's mass and speed to the visible field of view. It is better to slow down at the corner—who knows what is behind the corner?

Third, when the jogger starts turning the street corner and suddenly sees a pile of sand right on the path that he contemplated (it was not there last time), some quick local planning must occur to take care of collision avoidance. The jogger's speed may temporarily decrease and the path will smoothly divert from the object. The jogger will likely want to locally optimize this path segment, in order to come back to his preplanned path quicker or along a shorter path. Other options not being feasible, the jogger may choose to "brake" to a halt and start a detour path.

As we see, the jogger's speed, mass, and quality of vision, as well as the speed of reaction to sudden changes—which represents the quality of his control system—are all tied together in a certain relationship, affecting the real-time decision-making process. The process will go on nonstop, all the time; the jogger cannot afford to take his eyes off the route for more than a fraction of a second. Sensing, local planning, global planning, and actual movement are in this process taking place simultaneously and continuously. Locally, unless the right relationship is maintained between the velocity when noticing an object, the distance to it, and the jogger's mass, collision may occur. A bigger mass may dictate better (farther) sensing to maintain the same velocity. Myopic vision may require reducing the speed.

Another interesting detail is that in the motion planning strategies considered in Chapter 3, each step of the path could be decided independently from other steps. The control scheme that takes into account robot dynamics cannot afford such luxury anymore. Often a decision will likely affect more than calculation of the current step. Consider this: Instead of turning on the dime as in our prior algorithms, the robot will be likely moving along relatively smooth curves. How do we know that the smooth path curve dictated by robot dynamics at the current step will not be in conflict with collision considerations at the next step? Perhaps at the current step we need to think of the next step as well. Or perhaps we need to think of more than one step ahead. Worse yet, what if a part of that path curve cannot be checked because it is bending around the corner of a nearby obstacle and hence is invisible?

These questions suggest that in a planning/control system with included dynamics a path step cannot be planned separately from at least a few steps that will follow it. The robot must make sure that the step it now contemplates will not result in some future steps where the collision is inevitable. How many steps look-ahead is enough? This is one thing that we need to figure out.

Below we will study the said effects, with the same objective as before—to design provably correct sensor-based motion planning algorithms. As before, the presence of uncertainty implies that no global optimality of the path is feasible. Notice, however, that given the need to plan for a few steps ahead, we can attempt local optimization. While improving the overall path, sometimes dramatically, in general a path segment that is optimal within the robot's field of vision says nothing about the path global optimality.

By the way, which optimization criterion is to be used? We will consider two criteria. The salient feature of one criterion is that, while maintaining the maximum velocity allowed by its dynamics, the robot will attempt to maximize its instantaneous turning angle toward the required direction of motion. This will allow it to finish every turning maneuver in minimum time. In a path with many turns, this should save a good deal of time. In the second strategy (which also assumes the maximum velocity compatible with collision avoidance), the robot will attempt a time-optimal arrival at its (constantly shifting) intermediate target. Intermediate targets will typically be on the boundary of the robot's field of vision.

Similar to the model used in Chapter 3, our mobile robot operates in two-dimensional physical space filled with a locally finite number of unknown stationary obstacles of arbitrary shapes. Planning is done in small steps (say, 30 or 50 times per second, which is typical for real-world robotics), resulting in continuous motion. The robot is equipped with sensors, such as vision or range finders, which allow it to detect and measure distances to surrounding objects within its sensing range. Robot vision works within some limited or unlimited *radius of vision*, which allows some steps look-ahead (say, 20 or 30 or more). Unless obstacles occlude one another, the robot will see them and use this information to plan appropriate actions. Occlusions effectively limit a robot's input information and call for more careful planning.

Control of body dynamics fits very well into the feedback nature of the SIM (Sensing–Intelligence–Motion) paradigm. To be sure, such control can in principle be incorporated in the Piano Mover's paradigm as well. One way to do this is, for example, to divide the motion planning process into two stages: First, a path is produced that satisfies the geometric constraints, and then this path is modified to fit the dynamic constraints [79], possibly in a time-optimal fashion [80–82].

Among the first attempts to explicitly incorporate body dynamics into robot motion planning were those by O'Dunlaing [83] for the one-dimensional case and by Canny et al. [84] in their *kinodynamic planning* approach for the two-dimensional case. In the latter work the proposed algorithm was shown to run in exponential time and to require space that is polynomial in the input. While the approach operates in the context of the Piano Mover's paradigm, it is somewhat akin to the approach considered in this chapter, in that the control strategy adheres to the L_∞-norm; that is, the velocity and acceleration components are assumed bounded with respect to a fixed (absolute) reference system. This allows one to decouple the equations of robot motion and treat the two-dimensional problem as two one-dimensional problems.[1]

---

[1]Though comparisons between algorithms belonging to the two paradigms are difficult, one comparison seems to apply here. Using the L_∞-norm will result, both in Ref. 84 and in the strategy described here, in a less efficient use of control resources and a "less optimal" time of path execution. Since planning with complete information is a one-time computation, this loss in efficiency is likely to be significant, due to a large deviation over the whole path of the robot's moving reference system from the absolute (world) system. In contrast, in the sensor-based approaches the decoupling of controls occurs again and again, at every step of the motion: The reference system is fixed only for the duration of one step, and so the resulting loss in efficiency should be less.

Within the Piano Mover's paradigm, a number of kinematic *holonomic* strategies make use of the *artificial potential field*. They usually require complete information and analytical presentation of obstacles; the robot's motion is affected by (a) the "repulsive forces" created by a potential field associated with obstacles and (b) "attractive forces" associated with the goal position [85]. A typical convergence issue here is how to avoid possible local minima in the potential field. Modifications that attempt to solve this problem include the use of repulsive fields with elliptic isocontours [86], introduction of special global fields [87], and generation of a numerical field [88]. The vortex field method [89] allows one to avoid undesirable attraction points, while using only local information; the repulsive actions are replaced by the velocity flows tangent to the isocontours, so that the robot is forced to move around obstacles. An approach called active reflex control [90] attempts to combine the potential field method with handling the robot dynamics; the emphasis is on local collision avoidance and on filtering out infeasible control commands generated by the motion planner.

Among the approaches that deal with incomplete information, a good number of holonomic techniques originate in maze-search strategies (see the previous chapters). When applicable, they are usually fast, can be used in real time, and guarantee convergence; obstacles in the scene may be of arbitrary shape.

There is also a group of *nonholonomic* motion planning approaches that ignore the system dynamics. These also require analytical representation of obstacles and assume complete [91–93] or partial input information [94]. The schemes are essentially open-loop, do not guarantee convergence, and attempt to solve the planning problem by taking into account the effect of nonholonomic constraints on obstacle avoidance. In Refs. 91 and 92 a two-stage scheme is considered: First, a holonomic planner generates a complete collision-free path, and then this path is modified to account for nonholonomic constraints. In Ref. 93 the problem is reduced to searching a graph representing the discretized configuration space. In Ref. 94, planning is done incrementally, with partial information: First, a desirable path is defined and then a control is found that minimizes the error in a least-squares sense.

To design a provably correct *dynamic* algorithm for sensor-based motion planning, we need a single control mechanism: Separating it into stages is likely to destroy convergence. Convergence has two faces: Globally, we have to guarantee finding a path to the target if one exists. Locally, we need an assurance of collision avoidance in view of the robot inertia. The former can be borrowed from kinematic algorithms; the latter requires an explicit consideration of dynamics.

Notice an interesting detail: In spite of sufficient knowledge about the piece of the scene that is currently within the robot's sensing range, even in this area it would make little sense at a given step to address the planning task as one with complete information, as well as to try to compute the whole subpath within the sensing range. Why not? Computing this subpath would require solving a rather difficult optimal motion control problem—a computationally expensive task. On the other hand, this would be largely computational waste, because only the first step of this subpath would be executed: As new sensing data appears at the next step, in general a path adjustment would be required. We will therefore attempt to plan only as many steps that immediately follow the current one as is needed to guarantee nonstop collision-free motion.

The general approach will be as follows: At its current position C_i, the robot will identify a visible intermediate target point, T_i, that is guaranteed to lie on a convergent path and is far enough from the robot—normally at the boundary of the sensing range. If the direction toward T_i differs from the current velocity vector V_i, moving toward T_i may require a turn, which may or may not be possible due to system dynamics.

In the first strategy that we will consider, if the angle between V_i and the direction toward T_i is larger than the maximum turn the robot can make in one step, the robot will attempt a fast smooth maneuver by turning at the maximum rate until the directions align; hence the name *Maximum Turn Strategy*. Once a step is executed, new sensing data appear, a new point T_{i+1} is sought, and the process repeats. That is, the actual path and the path that contains points T_i will likely be different paths: With the new sensory data at the next step, the robot may or may not be passing through point T_i.

In the second strategy, at each step, a *canonical solution* is found which, if no obstacles are present, would bring the robot from its current position C_i to the intermediate target T_i with zero velocity and in minimum time. Hence the name *Minimum Time Strategy*. (The minimum time refers of course to the current local piece of scene.) If the canonical path crosses an obstacle and is thus not feasible, a *near-canonical solution* path is found which is collision-free and satisfies the control constraints. We will see that in this latter case only a small number of options needs be considered, at least one of which is guaranteed to be collision-free.

The fact that no information is available beyond the sensing range dictates caution. To guarantee safety, the whole stopping path must not only lie inside the sensing range, it must also lie in its visible part. No parts of the stopping path can be occluded by obstacles. Moreover, since the intermediate target T_i is chosen as the farthest point based on the information currently available, the robot needs a guarantee of stopping at T_i, even if it does not intend to do so. Otherwise, what if an obstacle lurks right beyond the vision range? That is, each step is to be planned as the first step of a trajectory which, given the current position, velocity, and control constraints, would bring the robot to a halt at T_i. Within one step, the time to acquire sensory data and to calculate necessary controls must fit into the step cycle.

4.2 MAXIMUM TURN STRATEGY

4.2.1 The Model

The robot is a *point mass*, of mass m. It operates in the plane; the scene may include a locally finite number of static obstacles. Each obstacle is bounded by a simple closed curve of arbitrary shape and of finite length, such that a straight line will cross it in only a finite number of points. Obstacles do not touch each other; if they do, they are considered one obstacle. The total number of obstacles in the scene need not be finite.

The robot's sensors provide it with information about its surroundings within the *sensing range (radius of vision)*, a disc of radius r_v centered at its current location C_i. The sensor can assess the distance to the nearest obstacle in any direction within the sensing range. The robot input information at moment t_i includes its current velocity vector $\mathbf{V}_i$, coordinates of point C_i and of the target point T, and possibly few other points of interest that will be discussed later.

The task is to move, collision-free, from point S (start) to point T (target) (see Figure 4.1). The robot's control means include two components (p, q) of the acceleration vector $\mathbf{u} = \frac{f}{m} = (p, q)$, where m is the robot mass and f is the force applied. Though the units of (p, q) are those of acceleration, by normalizing to $m = 1$ we can refer to p and q as control forces, each within its fixed range $|p| \leq p_{max}$, $|q| \leq q_{max}$. Force p controls forward (or backward when braking) motion; its positive direction coincides with the velocity vector $\mathbf{V}$. Force q is perpendicular to p, forming a right pair of vectors, and is equivalent to the steering control (rotation of vector $\mathbf{V}$) (Figure 4.2). Constraints on p and q imply a constraint on the path curvature. The point mass assumption implies that the robot's rotation with respect to its "center of mass" has no effect on the system dynamics. There are no external forces acting on the robot except p and q. There is no friction; for example, values $p = q = 0$ and $\mathbf{V} \neq 0$ will result in a straight-line constant velocity motion.[2]

Robot motion is controlled in steps i, $i = 0, 1, 2, \dots$. Each step takes time $\delta t = t_{i+1} - t_i = const$. The step's length depends on the robot's velocity within

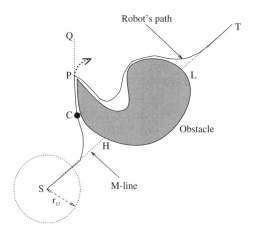

Figure 4.1 An example of a conflict between the performance of a kinematic algorithm (e.g., VisBug-21, the solid line path) and the effects of dynamics (the dotted piece of trajectory at P).

---

[2]If needed, other external forces and constraints can be handled within this model, using for example the technique described in Ref. 95.

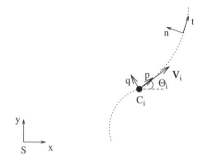

Figure 4.2   The path coordinate frame $(\mathbf{t}, \mathbf{n})$ is used in the analysis of dynamic effects of robot motion. The world frame $(\mathbf{x}, \mathbf{y})$, with its origin at the start point S, is used in the obstacle detection and path planning analysis.

the step. Steps i and $i + 1$ start at times t_i and t_{i+1}, respectively; $C_0 = S$. While moving toward location C_{i+1}, the robot computes necessary controls for step $i + 1$ using the current sensory data, and it executes them at C_{i+1}. The finite time necessary within one step for acquiring sensory data, calculating the controls, and executing the step must fit into the step cycle (more details on this can be found in Ref. 96). We define two coordinate systems (follow Figure 4.2):

- The *world coordinate frame*, $(\mathbf{x}, \mathbf{y})$, fixed at point S.
- The *path coordinate frame*, $(\mathbf{t}, \mathbf{n})$, which describes the motion of point mass at any moment $\tau \in [t_i, t_{i+1})$ within step i. The frame's origin is attached to the robot; axis $\mathbf{t}$ is aligned with the current velocity vector $\mathbf{V}$; axis $\mathbf{n}$ is normal to $\mathbf{t}$; that is, when $\mathbf{V} = 0$, the frame is undefined. One may note that together with axis $\mathbf{b} = \mathbf{t} \times \mathbf{n}$, the triple $(\mathbf{t}, \mathbf{n}, \mathbf{b})$ forms the known *Frenet trihedron*, with the plane of $\mathbf{t}$ and $\mathbf{n}$ being the *osculating plane* [97].

4.2.2   Sketching the Approach

Some terms and definitions here are the same as in Chapter 3; material in Section 3.1 can be used for more rigorous definitions. Define *M-line (Main line)* as the straight-line segment (S, T) (Figure 4.1). The M-line is the robot's desired path. When, while moving along the M-line, the robot senses an obstacle crossing the M-line, the crossing point on the obstacle boundary is called a *hit point*, H. The corresponding M-line point "on the other side" of the obstacle is a *leave point*, L.

The planning procedure is to be executed at each step of the robot's path. Any provable maze-searching algorithm can be used for the kinematic part of the algorithm that we are about to build, as long as it allows distant sensing. For specificity only, we use here the *VisBug* algorithm (see Section 3.6; either VisBug-21 or VisBug-22 will do). VisBug algorithms alternate between these two operations (see Figure 4.1):

1. Walk from point S toward point T along the M-line until, at some point C, you detect an obstacle crossing the M-line, say at point H.

2. Using sensing data, define the farthest visible *intermediate target* T_i on the obstacle boundary and on a convergent path; make a step toward T_i; iterate Step 2 until you detect the M-line; go to Step 1.

To this process we add a control procedure for handling dynamics. It is clear already that from time to time dynamics will prevent the robot from carefully following an obstacle boundary. For example, in Figure 4.1, while trying to pass the obstacle from the left, under a VisBug procedure the robot would make a sharp turn at point P. Such motion is not possible in a system with dynamics.

At times the current intermediate target T_i may go out of the robot's sight, because of the robot inertia or because of occluding obstacles. In such cases the robot will be designating *temporary intermediate targets* and use them until it can spot the point T_i again. The final algorithm will also include mechanisms for checking the target reachability and for local path optimization.

Safety Considerations.   Dynamics affects safety. Given the uncertainty beyond the distance r_v from the robot (or even closer to it in case of occluding obstacles), a guaranteed *stopping path* is the only way to ensure collision-free motion. Unless this last resort path is available, new obstacles may appear in the sensing range at the next step, and collision may be imminent. A stopping path implies a safe direction of motion and a safe velocity value. We choose the stopping path as a straight-line segment along the step's velocity vector. A candidate for the next step is "approved" by the algorithm only if its execution would guarantee a stopping path. In this sense our planning procedure is based on a one-step analysis.[3]

As one will see, the procedure for a detour around a suddenly appearing obstacle operates in a similar fashion. We emphasize that the stopping path does not mean stopping. While moving along, at every step the robot just makes sure that *if a stop is suddenly necessary, there is always a guarantee for it.*

Allowing for a straight-line stopping path with the stop at the sensing range boundary implies the following relationship between the velocity $\mathbf{V}$, mass m, and controls $\mathbf{u} = (p, q)$:

$$V \leq \sqrt{2pd} \tag{4.1}$$

where d is the distance from the current position C to the stop point. So, for example, an increase in the radius of vision r_v would allow the robot to raise the maximum velocity, by the virtue of providing more information farther along the path. Some control ramifications of this relationship will be analyzed in Section 4.2.3.

Convergence.   Because of the effect of dynamics, the convergence mechanism borrowed from a kinematic algorithm—here VisBug—needs some modification.

---

[3]A deeper multistep analysis can in principle produce locally shorter paths. It would not add to safety, however, and is not likely to justify the steep rise in computational expenses.

VisBug assumes that the intermediate target point is either on the obstacle boundary or on the M-line and is visible. However, the robot's inertia may cause it to
move so that the intermediate target T_i will become invisible, either because it
goes outside the sensing range r_v (as after point P, Figure 4.1) or due to occluding obstacles (as in Figure 4.6). The danger of this is that the robot may lose
from its sight point T_i—and the path convergence with it. One possible solution
is to keep the velocity low enough to avoid such overshoots—a high price in
efficiency to pay. The solution we choose is to keep the velocity high and, if the
intermediate target T_i does go out of sight, modify the motion locally until T_i is
found again (Section 4.2.6).

4.2.3 Velocity Constraints. Minimum Time Braking

By substituting p_{max} for p and r_v for d into (4.1), one obtains the maximum velocity, V_{max}. Since the maximum distance for which sensing information is available is r_v, the sensing range boundary, an emergency stop should
be planned for that distance. We will show that moving with the maximum
speed—certainly a desired feature—actually guarantees a minimum-time arrival
at the sensing range boundary. The suggested braking procedure, developed fully
in Section 4.2.4, makes use of an optimization scheme that is sketched briefly in
Section 4.2.4.

Velocity Constraints. It is easy to see (follow an example in Figure 4.3) that
in order to guarantee a safe stopping path, under discrete control the maximum
velocity must be less than V_{max}. This velocity, called *permitted maximum velocity*,
$V_{p\,max}$, can be found from the following condition: If $V = V_{p\,max}$ at point C_2 (and
thus also at C_1), we can guarantee the stop at the sensing range boundary (point
B_1, Figure 4.3). Recall that velocity V is generated at C_1 by the control force p.
Let $|C_1 C_2| = \delta x$; then

$$\delta x = V_{p\,max} \cdot \delta t$$

$$V_{B_1} = V_{p\,max} - p_{max} t$$

Since we require $V_{B_1} = 0$, then $t = V_{p\,max}/p_{max}$. For the segment $|C_2 B_1| = r_v - \delta x$, we have

$$r_v - \delta x = V_{p\,max} \cdot t - \frac{p_{max} t^2}{2}$$

From these equations, the expression for the maximum permitted velocity $V_{p\,max}$
can be obtained:

$$V_{p\,max} = \sqrt{p_{max}^2 \delta t^2 + 2 p_{max} r_v} - p_{max} \delta t$$

As expected, $V_{p\,max} < V_{max}$ and converges to V_{max} with $\delta t \to 0$.

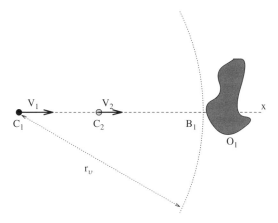

Figure 4.3   With the sensing radius equal to r_v, obstacle O_1 is not visible from point C_1. Because of the discrete control, velocity V_1 commanded at C_1 will be constant during the step interval (C_1, C_2). Then, if $V_1 = V_{max}$ at C_1, then also $V_2 = V_{max}$, and the robot will not be able to stop at B_1, causing collision with obstacle O_1. The permitted velocity thus must be $V_{p\,max} < V_{max}$.

4.2.4   Optimal Straight-Line Motion

We now sketch the optimization scheme that will later be used in the development of the braking procedure. Consider a dynamic system, a moving body whose behavior is described by a second-order differential equation $\ddot{x} = p(t)$, where $\|p(t)\| \le p_{max}$ and $p(t)$ is a scalar control function. Assume that the system moves along a straight line. By introducing state variables x and V, the system equations can be rewritten as $\dot{x} = V$ and $\dot{V} = p(t)$. It is convenient to analyze the system behavior in the *phase space* (V, x).

The goal of control is to move the system from its initial position $(x(t_0), V(t_0))$ to its final position $(x(t_f), V(t_f))$. For convenience, choose $x(t_f) = 0$. We are interested in an optimal control strategy that would execute the motion in minimum time t_f, arriving at $x(t_f)$ with zero velocity, $V(t_f) = 0$. This optimal solution can be obtained in closed form; it depends upon the above/below relation of the initial position with respect to two parabolas that define the switch curves in the phase plane (V, x):

$$x = -\frac{V^2}{2p_{max}}, \qquad V \ge 0 \tag{4.2}$$

$$x = \frac{V^2}{2p_{max}}, \qquad V \le 0 \tag{4.3}$$

This simple result in optimal control (see, e.g., Ref. 98) is summarized in the control law that follows, and it is used in the next section in the development of the braking procedure for emergency stopping. The procedure will guarantee robot safety while allowing it to cruise with the maximum velocity (see Figure 4.4):

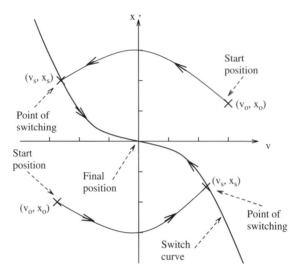

Figure 4.4   Depending on whether the initial position (V_0, x_0) in the phase space (V, x) is above or below the switch curves, there are two cases to consider. The optimal solution corresponds to moving first from point (V_0, x_0) to the switching point (V_s, x_s) and then along the switch line to the origin.

Control Law: *If in the phase space the initial position (V_0, x_0) is above the switch curve (4.2), first move along the parabola defined by control $\hat{p} = -p_{max}$ toward curve (4.2), and then with control $\hat{p} = p_{max}$ move along the curve to the origin. If point (V_0, x_0) is below the switch curve, move first with control $\hat{p} = p_{max}$ toward the switch curve (4.3), and then move with control $\hat{p} = -p_{max}$ along the curve to the origin.*

The Braking Procedure. We now turn to the calculation of time it will take the robot to stop when it moves along the stopping path. It follows from the argument above that if at the moment when the robot decides to stop its velocity is $V = V_{p\,max}$, then it will need to apply maximum braking all the way until the stop. This will define uniquely the time to stop. But, if $V < V_{p\,max}$, then there are a variety of braking strategies and hence of different times to stop.

Consider again the example in Figure 4.3; assume that at point C_2, $V_2 < V_{p\,max}$. What is the *optimal braking strategy*, the one that guarantees safety while bringing the robot in minimum time to a stop at the sensing range boundary? While this strategy is not necessarily the one we would want to implement, it is of interest because it defines the limit velocity profile the robot can maintain for safe braking. The answer is given by the solution of an optimization problem for a single-degree-of-freedom system. It follows from the Control Law above that the optimal solution corresponds to at most two curves, I and II, in the phase space (V, x) (Figure 4.5a) and to at most one control switch, from $\hat{p} = p_{max}$ on line I to $\hat{p} = -p_{max}$ on line II, given by (4.2) and (4.3). For example, if braking

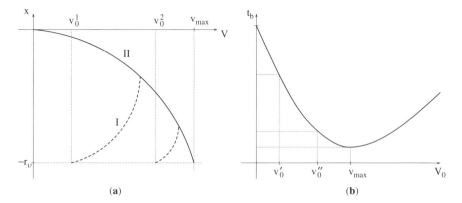

Figure 4.5 (a) Optimal braking strategy requires at most one switch of control. (b) The corresponding time–velocity relation.

starts with the initial values $x = -r_v$ and $0 \le V_0 < V_{max}$, the system will first move, with control $\hat{p} = p_{max}$, along parabola I to parabola II (Figure 4.5a),

$$x(V) = \frac{V^2 - V_0^2}{2p_{max}} - r$$

and then, with control $\hat{p} = -p_{max}$, toward the origin, along parabola II,

$$x(V) = \frac{V^2}{2p_{max}}$$

The optimal time t_b of braking is a function of the initial velocity V_0, radius of vision r_v, and the control limit p_{max},

$$t_b(V_0) = \frac{\sqrt{2V_0^2 + 4p_{max}r_v} - V_0}{p_{max}} \tag{4.4}$$

Function $t_b(V_0)$ has a minimum at $V_0 = V_{max} = \sqrt{2p_{max}r_v}$, which is exactly the upper bound on the velocity given by (4.1); it is decreasing on the interval $V_0 \in [0, \ V_{max}]$ and increasing when $V_0 > V_{max}$ (see Figure 4.5b). For the interval $V_0 \in [0, \ V_{max}]$, which is of interest to us, the above analysis leads to a somewhat counterintuitive conclusion:

Proposition 1. *For the initial velocity V_0 in the range $V_0 \in [0, \ V_{max}]$, the time necessary for stopping at the boundary of the sensing range is a monotonically decreasing function of V_0, with its minimum at $V_0 = V_{max}$.*

Notice that this result (see also Figure 4.5) leaves a comfortable margin of safety: Even if at the moment when the robot sees an obstacle on its way it moves with the maximum velocity, it can still stop safely before it reaches the obstacle. If the robot's velocity is below the maximum, it has more control options for braking, including even one of speeding up before actual braking. Assume, for example, that we want the robot to stop in minimum time at the sensing range boundary (the origin in Figure 4.5a); consider two initial positions: (i) $x = -r_v$, $V = V_0^1$ and (ii) $x = -r_v$, $V = V_0^2$; $V_0^2 > V_0^1$. Then, according to Proposition 1, in case (i) this time is bigger than in case (ii). Note that because of the discrete control it is the permitted maximum velocity, $V_{p\max}$, that is to be substituted into (4.4) to obtain the minimum time. (More details on the braking procedure can be found in Ref. 99).

4.2.5 Dynamics and Collision Avoidance

The analysis in this section consists of two parts. First we incorporate the control constraints into the model of our mobile robot and develop a transformation from the moving path coordinate frame to the world frame (see Section 4.2.1). Then the Maximum Turn Strategy is produced, an incremental decision-making mechanism that determines forces p and q at each step.

Transformation from Path Frame to World Frame. The remainder of this section refers to the time interval $[t_i, t_{i+1})$, and so index i can be dropped. Let $(x, y) \in \mathcal{R}^2$ be the robot's position in the world frame, and let θ be the (slope) angle between the velocity vector $\mathbf{V} = (V_x, V_y) = (\dot{x}, \dot{y})$ and x axis of the world frame (see Figure 4.2). The planning process involves computation of controls $\mathbf{u} = (p, q)$, which for every step defines the velocity vector and eventually the path, $\mathbf{x} = (x, y)$, as a function of time. The normalized equations of motion are

$$\ddot{x} = p \cos \theta - q \sin \theta$$
$$\ddot{y} = p \sin \theta + q \cos \theta$$

The angle θ between vector $\mathbf{V}$ and x axis of the world frame is found as

$$\theta = \begin{cases} \arctan\left(\frac{V_y}{V_x}\right), & V_x \geq 0 \\ \arctan\left(\frac{V_y}{V_x}\right) + \pi, & V_x < 0 \end{cases}$$

To find the transformation from path frame to the world frame $(\mathbf{x}, \mathbf{y})$, present the velocity in the path frame as $\mathbf{V} = V\mathbf{t}$. Angle θ is defined as the angle between $\mathbf{t}$ and the positive direction of x axis. Given that control forces p and q act along the $\mathbf{t}$ and $\mathbf{n}$ directions, respectively, the equations of motion with respect to the path frame are

$$\dot{V} = p$$
$$\dot{\theta} = q/V$$

Since the control forces are constant over time interval $[t_i, t_{i+1})$, within this interval the solution for $V(t)$ and $\theta(t)$ becomes

$$V(t) = V_0 + pt$$

$$\theta(t) = \theta_0 + \frac{q \log(1 + \frac{tp}{V_i})}{p} \tag{4.5}$$

where θ_0 and V_0 are constants of integration and are equal to the values of $\theta(t_i)$ and $V(t_i)$, respectively. By parameterizing the path by the value and direction of the velocity vector, the path can be mapped onto the world frame using the vector integral equation

$$\mathbf{r}(t) = \int_{t_i}^{t_{i+1}} \mathbf{V} \cdot dt \tag{4.6}$$

Here $\mathbf{r}(t) = (x(t), y(t))$ and $\mathbf{V} = (V \cdot \cos(\theta), V \cdot \sin(\theta))$ are the projections of vector $\mathbf{V}$ onto the world frame $(\mathbf{x}, \mathbf{y})$. After integrating Eq. (4.6), we obtain a set of solutions in the form

$$x(t) = \frac{2p \cos \theta(t) + q \sin \theta(t)}{4p^2 + q^2} V^2(t) + A$$

$$y(t) = -\frac{q \cos \theta(t) - 2p \sin \theta(t)}{4p^2 + q^2} V^2(t) + B \tag{4.7}$$

where terms A and B are

$$A = x_0 - \frac{V_0^2(2p \cos(\theta_0) + q \sin(\theta_0))}{4p^2 + q^2}$$

$$B = y_0 + \frac{V_0^2(q \cos(\theta_0) - 2p \sin(\theta_0))}{4p^2 + q^2}$$

Equations (4.7) are directly defined by the control variables p and q; $V(t)$ and $\theta(t)$ therein are given by Eq. (4.5).

In general, Eqs. (4.7) describe a spiral curve. Note two special cases: When $p \neq 0, q = 0$, Eqs. (4.7) describe a straight-line motion along the vector of velocity; when $p = 0$ and $q \neq 0$, Eqs. (4.7) produce a circle of radius $V_0^2/|q|$ centered at the point (A, B).

Selection of Control Forces. We now turn to the control law that guides the selection of forces p and q at each step i, for the time interval $[t_i, t_{i+1})$. To ensure a reasonably fast convergence to the intermediate target T_i, those forces are chosen such as to align, as fast as possible, the direction of the robot's motion with the direction toward T_i. First, find a solution among the controls (p, q) such that

$$(p, q) \in \{(p, q) : p \in [-p_{max}, +p_{max}], \quad q = \pm q_{max}\} \tag{4.8}$$

where $q = +q_{max}$ if the intermediate target T_i lies in the left semiplane, and $q = -q_{max}$ if T_i lies in the right semiplane with respect to the vector of velocity. That is, force p is chosen so as to keep the maximum velocity allowed by the surrounding obstacles. To this end, a discrete set of values p is tried until a step that guarantees a collision-free stopping path is found. At a minimum, the set should include values $-p_{max}$, 0, and $+p_{max}$. The greater the number of values that are tried, the closer the resulting velocity is to the maximum sought. Force q is chosen on the boundary, to produce a maximum turn in the appropriate direction. On the other hand, if because of obstacles no adequate controls in the range (4.8) can be chosen, this means that maximum braking should be applied. Then the controls are chosen from the set

$$(p, q) \in \{(p, q) : p = -p_{max}, \quad q \in (\pm q_{max}, \ 0]\} \tag{4.9}$$

where q is found from a discrete set similar to p in (4.8). Note that sets (4.8) and (4.9) always include at least one safe solution: By the algorithm design, the straight-line motion with maximum braking, $(p, q) = (-p_{max}, 0)$, is always collision-free (for more detail, see Ref. 96).

4.2.6 The Algorithm

The resulting algorithm consists of three procedures:

- *Main Body.* This defines the motion within the time interval $[t_i, t_{i+1})$ toward the intermediate target T_i.
- *Define Next Step.* This chooses the forces p and q.
- *Find Lost Target.* This handles the case when the intermediate target goes out of the robot's sight.

Also used in the algorithm is a procedure called *Compute T_i*, from the VisBug algorithms (Section 3.6), which computes the next intermediate target T_{i+1} and includes a test for target reachability. Vector $\mathbf{V}_i$ is the current vector of velocity, and T is the robot's target location. The term "safe motion" below refers to the mechanism for determining the next robot's position C_{i+1} such as to guarantee a stopping path (Section 4.2.2).

Main Body. The procedure is executed at each step's time interval $[t_i, t_{i+1})$ and makes use of two procedures, *Define Next Step* and *Find Lost Target* (see further below):

- M1: Move in the direction specified by *Define Next Step*, while executing *Compute T_i*. If T_i is visible, do the following: If $C_i = T$, the procedure stops; else if T is unreachable, the procedure stops; else if $C_i = T_i$, go to M2. Otherwise, use *Find Lost Target* to make T_i visible. Iterate M1.
- M2: Make a step along vector $\mathbf{V}_i$ while executing *Compute T_i*: If $C_i = T$, the procedure stops; else if the target is unreachable, the procedure stops; else if $C_i \neq T_i$, go to M1.

Define Next Step. This procedure covers all cases of generation of a single motion step. Its part D1 corresponds to motion along the M-line; D2 corresponds to a simple turn when the directions of vectors $\mathbf{V}_i$ and (C_i, T_i) can be aligned in one step; D3 is invoked when the turn requires multiple steps and can be done with the maximum speed; D4 is invoked when turning must be accompanied by braking:

- D1: If vector $\mathbf{V}_i$ coincides with the direction toward T_i, do the following: If $T_i = T$, make a step toward T; else make a step toward T_i.
- D2: If vector $\mathbf{V}_i$ does not coincide with the direction toward T_i, do the following: If the directions of $\mathbf{V}_{i+1}$ and (C_i, T_i) can be aligned within one step, choose this step. Else go to D3.
- D3: If a step with the maximum turn toward T_i and with maximum velocity is safe, choose it. Else go to D4.
- D4: If a step with the maximum turn toward T_i and some braking is possible, choose it. Else, choose a step along $\mathbf{V}_i$, with maximum braking, $p = -p_{\max}, q = 0$.

Find Lost Target. This procedure is executed when the intermediate target T_i becomes invisible. The last position C_i where T_i was visible is kept in the memory until T_i becomes visible again. A very simple fix would be this: Once T_i becomes occluded by an obstacle, in order to immediately initiate a stopping path, move back to C_i, and then move directly toward T_i. This would be quite inefficient, however. Instead, the procedure operates as follows: Once the robot loses T_i, it keeps moving ahead while defining temporary intermediate targets on the visible part of the line segment (C_i, T_i), and continuing looking for T_i. If it finds T_i, the procedure terminates, the control returns to the *Main Body*, and the robot moves directly toward T_i (see Figure 4.6a). Otherwise, if the whole segment (C_i, T_i) becomes invisible, the robot brakes to a stop, returns to C_i, the procedure terminates, and so on (see Figure 4.6b). Together these two pieces ensure that the intermediate target T_i will not be lost. The procedure is as follows:

- F1: If segment (C_i, T_i) is visible, define on it a temporary intermediate target T_i^t and move toward it while looking for T_i. If the current position is at T, exit; else if C_i lies in the segment (C_i, T_i), exit. Else go to F2.
- F2: If segment (C_i, T_i) is invisible, initiate a stopping path and move back to C_i; exit.

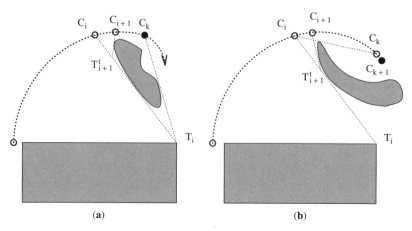

Figure 4.6   Because of its inertia, immediately after its position C_i the robot temporarily "loses" the intermediate target T_i. (**a**) The robot keeps moving around the obstacle until it spots T_i, and then it continues toward T_i. (**b**) When because of an obstacle the whole segment (C_i, T_i) becomes invisible at point C_{k+1}, the robot stops, returns back to C_i, and then moves toward T_i along the line (C_i, T_i).

Convergence.   To prove convergence of the described procedure, we need to show the following:

(i) At every step of the path the algorithm guarantees collision-free motion.
(ii) The set of intermediate targets T_i is guaranteed to lie on the convergent path.
(iii) The planning strategy guarantees that the current intermediate target will not be lost.

Together, (ii) and (iii) assure that a path to the target position T will be found if one exists. Condition (i) can be shown by induction; condition (ii) is provided by the VisBug procedure (see Section 3.6), which also includes the test for target reachability. Condition (iii) is satisfied by the procedure *Find Lost Target* of the Maximum Turn Strategy. The following two propositions hold:

Proposition 2.   *Under the Maximum Turn Strategy algorithm, assuming zero velocity,* $\mathbf{V}_S = 0$, *at the start position S, at each step of the path there exists at least one stopping path.*

By design, the stopping path is a straight-line segment. Choosing the next step so as to guarantee existence of a stopping path implies two requirements: There should be at least one safe direction of motion and the value of velocity that would allow stopping within the visible area. The latter is ensured by the choice of system parameters [see Eq. (4.1) and the safety conditions, Section 4.2.2]. As to the existence of safe directions, proceed by induction: We need to show that if a safe direction exists at the start point and at an arbitrary step i, then there is a safe direction at the step $(i + 1)$.

Since at the start point S the velocity is zero, $\mathbf{V}_S = 0$, then any direction of motion at S will be a safe direction; this gives the basis of induction. The induction proceeds as follows. Under the algorithm, a candidate step is accepted for execution if only its direction guarantees a safe stop for the robot if needed. Namely, at point C_i, step i is executed only if the resulting vector $\mathbf{V}_{i+1}$ at C_{i+1} will point in a safe direction. Therefore, at step $(i + 1)$, at the least this very direction presents a safe stopping path.

Remark: Proposition 2 will hold for $\mathbf{V}_S \neq 0$ as well if the start point S is known to possess at least one stopping path originating in it.

Proposition 3. *The Maximum Turn Strategy is convergent.*

To see this, note that by design of the VisBug algorithm (see Section 3.6.3), each intermediate target T_i lies on a convergent path and is visible at the moment when it is generated.

That is, the only way the robot can get lost is if at the following step(s) point T_i becomes invisible due to the robot's inertia or an obstacle occlusion: This would make it impossible to generate the next intermediate target, T_{i+1}, as required by VisBug. However, if point T_i does become invisible, the procedure *Find Lost Target* is invoked, a set of temporary intermediate targets T_{i+1}^t are defined, each with a guaranteed stopping path, and more steps are executed until point T_i becomes visible again (see Figure 4.6). The set T_{i+1}^t is finite because of finite distances between every pair of points in it and because the set must lie within the sensing range of radius r_v. Therefore, the robot always moves toward a point which lies on a path that is convergent to the target T.

4.2.7 Examples

Examples shown in Figures 4.7a to 4.7d demonstrate performance of the Maximum Turn Strategy in a computer simulated environment. Generated paths are shown by thicker lines. For comparison, also shown by thin lines are paths produced under the same conditions by the VisBug algorithm. Polygonal shapes are chosen for obstacles in the examples only for the convenience of generating the scene; the algorithms are oblivious to the obstacle shapes.

To understand the examples, consider a simplified version of the relationship that appears in Section 4.2.3, $V_{max} = \sqrt{2 r_v p_{max}} = \sqrt{2 r_v \cdot f_{max}/m}$. In the simulations, the robot's mass m and control force f_{max} are kept constant; for example, an increase in sensing radius r_v would "raise" the velocity V_{max}. Radius r_v is the same in Figures 4.7a and 4.7b. In the more complex scene (b), because of three additional obstacles (three small squares) the robot's path cannot curve as freely as in scene (a). Consequently, the robot moves more "cautiously," that is, slower; the path becomes tighter and closer to the obstacles, allowing the robot

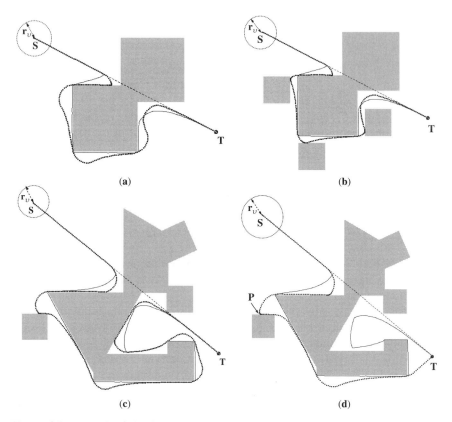

(a)

(b)

(c)

(d)

Figure 4.7 In each of the four examples shown, one path (indicated by the thin line) is produced by the VisBug algorithm, and the other path (a thicker line) is produced by the Maximum Turn Strategy, which takes into account the robot dynamics. The circle at point S indicates the radius of vision r_v.

to squeeze between obstacles. Accordingly, the time to complete the path is 221 units (steps) in (a) and 232 units in (b), whereas the path in (a) is longer than that in (b).

Figures 4.7c and 4.7d refer to a more complex environment. The difference between these two situations is that in (d) the radius of vision r_v is 1.5 times larger than that in (c). Note that in (d) the path produced by the Maximum Turn Strategy is noticeably shorter than the path generated by the VisBug algorithm. This has happened by sheer chance: Unable to make a sharp turn (because of its inertia) at the last piece of its path, the robot "jumped out" around the corner and hence happened to be close enough to T to see it, and this eliminated a need for more obstacle following.

Note the stops along the path generated by the Maximum Turn Strategy; they are indicated by sharp turns. These might have been caused by various reasons: For example, in Figure 4.7a the robot stopped because its small sensing radius

r_v was insufficient to see the obstacle far enough to initiate a smooth turn. In Figure 4.7d, the stop at point P was probably caused by the robot's temporarily losing its current intermediate target.

4.3   MINIMUM TIME STRATEGY

We will now consider the second strategy for solving the *Jogger's Problem*. The same model of the robot, its environment, and its control means will be used as in the Maximum Turn Strategy (see Section 4.2.1).

The general strategy will be as follows: At the given step i, the kinematic motion planning procedure chosen—we will use again VisBug algorithms—identifies an intermediate target point, T_i, which is the farthest visible point on a convergent path. Normally, though not always, T_i is defined at the boundary of the sensing range r_v. Then a single step that lies on a time-optimal trajectory to T_i is calculated and executed; the robot moves from its current position C_i to the next position C_{i+1}, and the process repeats.

Similar to the Maximum Turn Strategy, the fact that no information is available beyond the robot's sensing range dictates a number of requirements. There must be an emergency *stopping path*, and it must lie inside the current sensing area. Since parts of the sensing range may be occupied or occluded by obstacles, the stopping path must lie in its visible part. Next, the robot needs a guarantee of stopping at the intermediate target T_i, even if it does not intend to do so. That is, each step is to be planned as the first step of a trajectory which, given the robot's current position, velocity, and control constraints, would bring it to a halt at T_i (though, again, this will be happening only rarely).

The step-planning task is formulated as an optimization problem. It is the optimization criterion and procedure that will make this algorithm quite different from the Maximum Turn Strategy. At each step, a *canonical solution* is found which, if no obstacles are present, would bring the robot from its current position C_i to its current intermediate target T_i with zero velocity and in minimum time. If the canonical path happens to be infeasible because it crosses an obstacle, a collision-free *near-canonical solution* path is found. We will show that in this case only a small number of path options need be considered, at least one of which is guaranteed to be collision-free.

By making use of the L_∞-norm within the duration of a single step, we decouple the two-dimensional problem into two one-dimensional control problems and reduce the task to the bang-bang control strategy. This results in an extremely fast procedure for finding the time-optimal subpath within the sensing range. The procedure is easily implementable in real time. Since only the first step of this subpath is actually executed—the following step will be calculated when new sensor information appears after this (first) step is executed—this decreases the error due to the control decoupling. Then the process repeats. One special case will have to be analyzed and incorporated into the procedure—the case when the intermediate target goes out of the robot's sight either because of the robot inertia or because of occluding obstacles.

4.3.1   The Model

To a large extent the model that we will use in this section is similar to the model used by the Maximum Turn Strategy above. There are also some differences. For convenience we hence give a complete model description here.

As before, the scene is two-dimensional physical space $\mathcal{W} \equiv (x, y) \subset \mathfrak{R}^2$; it may include a finite set of locally finite static obstacles $\mathcal{O} \in \mathcal{W}$. Each obstacle $\mathcal{O}_k \in \mathcal{O}$ is a simple closed curve of arbitrary shape and of finite length, such that a straight line will cross it in only a finite number of points. Obstacles do not touch each other; if they do, they are considered one obstacle.

The robot is a *point mass*, of mass m. Its vision sensor allows it to detect any obstacles and the distance to them within its *sensing range* (radius of vision)—a disk $D(C_i, r_v)$ of radius r_v centered at its current location C_i. At moment t_i, the robot's input information includes its current velocity vector $\mathbf{V}_i$ and coordinates of C_i and of target location T.

The robot's means to control its motion are two components of the acceleration vector $\mathbf{u} = f/m = (p, q)$, where m is the robot mass and f the force applied. Controls $\mathbf{u}$ come from a set $\mathbf{u}(\cdot) \in U$ of measurable, piecewise continuous bounded functions in $\mathfrak{R}^2$, $U = \{\mathbf{u}(\cdot) = (p(\cdot), q(\cdot))/p \in [-p_{max}, p_{max}], q \in [-q_{max}, q_{max}]\}$. By taking mass $m = 1$, we can refer to components p and q as control forces, each within a fixed range $|p| \leq p_{max}$, $|q| \leq q_{max}$; $p_{max}, q_{max} > 0$. Force p controls the forward (or backward when braking) motion; its positive direction coincides with the robot's velocity vector $\mathbf{V}$. Force q, the steering control, is perpendicular to p, forming a right pair of vectors (Figure 4.8). There is no friction: For example, given velocity $\mathbf{V}$, the control values $p = q = 0$ will result in a constant-velocity straight-line motion along the vector $\mathbf{V}$.

Without loss of generality, assume that no external forces except p and q act on the system. Note that with this assumption our model and approach can still handle other external forces and constraints using, for example, the technique suggested in Ref. 95, whereby various dynamic constraints such as curvature, engine force, sliding, and velocity appear in the inequality describing the limitations on the components of acceleration. The set of such inequalities defines a convex region of the $\ddot{x}\ddot{y}$ space. In our case the control forces act within the intersection of the box $[-p_{max}, p_{max}] \times [-q_{max}, q_{max}]$, with the half-planes defined by those inequalities.

The task is to move in $\mathcal{W}$ from point S (start) to point T (target) (see Figure 4.1). The control of robot motion is done in steps i, $i = 0, 1, 2, \ldots$. Each step i takes time $\delta t = t_{i+1} - t_i = \text{const}$; the path length within time interval δt depends on the robot velocity $\mathbf{V}_i$. Steps i and $i + 1$ start at times t_i and t_{i+1}, respectively; $C_0 = S$. Control forces $\mathbf{u}(\cdot) = (p, q) \in U$ are constant within the step.

We define three coordinate systems (follow Figure 4.8):

- The *world frame*, $(\mathbf{x}, \mathbf{y})$, is fixed at point S.
- The *primary path frame*, $(\mathbf{t}, \mathbf{n})$, is a moving (inertial) coordinate frame. Its origin is attached to the robot; axis $\mathbf{t}$ is aligned with the current velocity

vector **V**, axis **n** is normal to **t**. Together with axis **b**, which is a cross product **b** = **t** × **n**, the triple (**t**, **n**, **b**) forms the *Frenet trihedron*, with the plane of **t** and **n** forming the *osculating plane* [97].

- The *secondary path frame*, (ξ_i, η_i), is a coordinate frame that is fixed during the time interval of step i. The frame's origin is at the intermediate target T_i; axis ξ_i is aligned with the velocity vector $\mathbf{V}_i$ at time t_i, and axis η_i is normal to ξ_i.

For convenience we combine the requirements and constraints that affect the control strategy into a set, called Ω. A solution (a path, a step, or a set of control values) is said to be Ω-*acceptable* if, given the current position C_i and velocity $\mathbf{V}_i$,

(i) it satisfies the constraints $|p| \leq p_{\max}$, $|q| \leq q_{\max}$ on the control forces,
(ii) it guarantees a stopping path,
(iii) it results in a collision-free motion.

4.3.2 Sketching the Approach

The algorithm that we will now present is executed at each step of the robot path. The procedure combines the convergence mechanism of a kinematic sensor-based motion planning algorithm with a control mechanism for handling dynamics, resulting in a single operation. As in the previous section, during the step time interval i the robot will maintain within its sensing range an *intermediate target* point T_i, usually on an obstacle boundary or on the desired path. At its current position C_i the robot will plan and execute its next step toward T_i. Then at C_{i+1} it will analyze new sensory data and define a new intermediate target T_{i+1}, and so on. At times the current T_i may go out of the robot's sight because of its inertia or due to occluding obstacles. In such cases the robot will rely on *temporary intermediate targets* until it can locate point T_i again.

The Kinematic Part. In principle, any maze-searching procedure can be utilized here, so long as it allows an extension to distant sensing. For the sake of specificity, we use here a VisBug algorithm (see Section 3.6; either VisBug-21 or VisBug-22 will do). Below, *M-line* (*Main line*) is the straight-line connecting points S and T; it is the robot's desired path. When, while moving along the M-line, the robot encounters an obstacle, the M-line, the intersection point between M-line and the obstacle boundary is called a *hit point*, denoted as H. The corresponding complementary intersection point between the M-line and the obstacle "on the other side" of the obstacle is a *leave point*, denoted L. Roughly, the algorithm revolves around two steps (see Figure 4.1):

1. Walk from S toward T along the M-line until detect an obstacle crossing the M-line, say at point H. Go to Step 2.

2. Define a farthest visible intermediate target T_i on the obstacle boundary in the direction of motion; make a step toward T_i. Iterate Step 2 until detect M-line. Go to Step 1.

The actual algorithm will include additional mechanisms, such as a finite-time target reachability test and local path optimization. In the example shown in Figure 4.1, note that if the robot walked under a kinematic algorithm, at point P it would make a sharp turn (recall that the algorithm assumes holonomic motion). In our case, however, such motion is not possible because of the robot inertia, and so the actual motion beyond point P would be something closer to the dotted path.

The Effect of Dynamics. Dynamics affects three algorithmic issues: safety considerations, step planning, and convergence. Consider those separately.

Safety Considerations. Safety considerations refer to collision-free motion. The robot is not supposed to hit obstacles. Safety considerations appear in a number of ways. Since at the robot's current position no information about the scene is available beyond the distance r_v from it, guaranteeing collision-free motion means guaranteeing at any moment at least one "last resort" stopping path. Otherwise in the following steps new obstacles may appear in the sensing range, and collision will be imminent no matter what control is used. This dictates a certain relationship between the velocity $\mathbf{V}$, mass m, radius r_v, and controls $\mathbf{u} = (p, q)$. Under a straight-line motion, the range of safe velocities must satisfy

$$V \leq \sqrt{2pd} \tag{4.10}$$

where d is the distance from the robot to the stop point. That is, if the robot moves with the maximum velocity, the stop point of the *stopping path* must be no further than r_v from the current position C. In practice, Eq. (4.10) can be interpreted in a number of ways. Note that the maximum velocity is proportional to the acceleration due to control, which is in turn directly proportional to the force applied and inversely proportional to the robot mass m. For example, if mass m is made larger and other parameters stay the same, the maximum velocity will decrease. Conversely, if the limits on (p, q) increase (say, due to more powerful motors), the maximum velocity will increase as well. Or, an increase in the radius r_v (say, due to better sensors) will allow the robot to increase its maximum velocity, by the virtue of utilizing more information about the environment.

Consider the example in Figure 4.1. When approaching point P along its path, the robot will see it at distance r_v and will designate it as its next intermediate target T_i. Along this path segment, point T_i happens to stay at P because no further point on the obstacle boundary will be visible until the robot arrives at P. Though there may be an obstacle right around the corner P, the robot needs not to slow down since at any point of this segment there is a possibility of a stopping path ending somewhere around point Q. That is, in order to proceed with maximum velocity, the availability of a stopping path has to be ascertained at every step i. Our stopping path will be a straight-line path along the corresponding vector V_i. If a candidate step cannot guarantee a stopping path, it is discarded.[4]

Step Planning. Normally the stopping path is not used; it is only an "insurance" option. The actual step is based on the *canonical solution*, a path which, if fully executed, would bring the robot from C_i to T_i with zero velocity and in minimum time, assuming no obstacles. The optimization problem is set up based on Pontryagin's optimality principle. We assume that within a step time interval $[t_i, t_{i+1})$ the system's controls (p, q) are bounded in the L_∞-norm, and apply it with respect to the secondary coordinate frame (ξ_i, η_i). The result is a fast computational scheme easily implementable in real time. Of course only the very first step of the canonical path is explicitly calculated and used in the actual motion. At the next step, a new solution is calculated based on the new sensory information that arrived during the previous step, and so on. With such a step-by-step execution of the optimization scheme, a good approximation of the globally time-optimal path from C_i to T_i is achieved. On the other hand, little computation is wasted on the part of the path solution that will not be utilized.

If the step suggested by the canonical solution is not feasible due to obstacles, a close approximation, called the *near-canonical solution*, is found that is both feasible and Ω-acceptable. For this case we show, first, that only a finite number of path options need be considered and, second, that there exists at least one path solution that is Ω-acceptable. A special case here is when the intermediate target goes out of the robot's sight either because of the robot's inertia or because of occluding obstacles.

Convergence. Once a step is physically executed, new sensing information appears and the process repeats. If an obstacle suddenly appears on the robot's intended path, a detour is arranged, which may or may not require the robot to stop. The detour procedure is tied to the issue of convergence, and it is built similar to the case of normal motion. Because of the effect of dynamics, the convergence mechanism borrowed from a kinematic algorithm—here VisBug—will need some modification. The intermediate target points T_i produced by VisBug lie either on the boundaries of obstacles or on the M-line, and they are visible from the corresponding robot's positions.

However, the robot's inertia may cause it to move so that T_i will become invisible, either because it goes outside of the sensing range r_v (as after point P, Figure 4.1) or due to occluding obstacles (as in Figure 4.11). This may endanger path convergence. A safe but inefficient solution would be to slow down or to keep the speed small at all times to avoid such overshoots. The solution chosen (Section 4.3.6) is to keep the velocity high and, if the intermediate target T_i goes out of sight, modify the motion locally until T_i is found again.

---

[4]A deeper, multistep analysis would be hardly justifiable here because of high computational costs, though occasionally it could produce locally shorter paths.

4.3.3 Dynamics and Collision Avoidance

Consider a time sequence $\sigma_t = \{t_0, t_1, t_2, \ldots, \}$ of the starting moments of steps. Step i takes place within the interval $[t_i, t_{i+1})$, $(t_{i+1} - t_i) = \delta t$. At moment t_i the robot is at the position C_i, with the velocity vector $\mathbf{V}_i$. Within this interval, based on the sensing data, intermediate target T_i (supplied by the kinematic planning algorithm), and vector $\mathbf{V}_i$, the control system will calculate the values of control forces p and q. The forces are then applied to the robot, and the robot executes step i, finishing it at point C_{i+1} at moment t_{i+1}, with the velocity vector $\mathbf{V}_{i+1}$. Then the process repeats.

Analysis that leads to the procedure for handling dynamics consists of three parts. First, in the remainder of this section we incorporate the control constraints into the robot's model and develop transformations between the primary path frame and world frame and between the secondary path frame and world frame. Then in Section 4.3.4 we develop the *canonical solution*. Finally, in Section 4.3.5 we develop the *near-canonical solution*, for the case when the canonical solution would result in a collision. The resulting algorithm operates incrementally; forces p and q are computed at each step. The remainder of this section refers to the time interval $[t_i, t_{i+1})$ and its intermediate target T_i, and so index i is often dropped.

Denote $(x, y) \in \Re^2$ the robot's position in the world frame, and denote θ the (slope) angle between the velocity vector $\mathbf{V} = (V_x, V_y) = (\dot{x}, \dot{y})$ and $\mathbf{x}$ axis of the world frame (Figure 4.8). The planning process involves computation of the controls $\mathbf{u} = (p, q)$, which for every step define the velocity vector and eventually

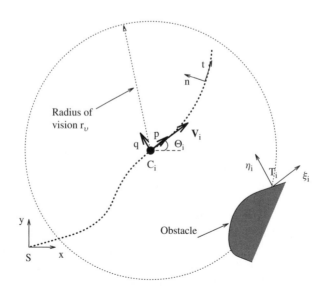

Figure 4.8 The coordinate frame $(\mathbf{x}, \mathbf{y})$ is the *world frame*, with its origin at S; $(\mathbf{t}, \mathbf{n})$ is the *primary path frame*, and (ξ_i, η_i) is the *secondary path frame* for the current robot position C_i.

the path, $(x(t), y(t))$, as a function of time. Taking mass $m = 1$, the equations of motion become

$$\ddot{x} = p \cos\theta - q \sin\theta$$
$$\ddot{y} = p \sin\theta + q \cos\theta$$

The angle θ between vector $\mathbf{V} = (V_x, V_y)$ and x axis of the world frame is found as

$$\theta = \begin{cases} \arctan\left(\frac{V_y}{V_x}\right), & V_x \geq 0 \\ \arctan\left(\frac{V_y}{V_x}\right) + \pi, & V_x < 0 \end{cases}$$

The transformations between the world frame and secondary path frame, from (x, y) to (ξ, η) and from (ξ, η) to (x, y), are given by

$$\begin{pmatrix} \xi \\ \eta \end{pmatrix} = R \begin{pmatrix} x - x_T \\ y - y_T \end{pmatrix} \tag{4.11}$$

and

$$\begin{pmatrix} x \\ y \end{pmatrix} = R' \begin{pmatrix} \xi \\ \eta \end{pmatrix} + \begin{pmatrix} x_T \\ y_T \end{pmatrix} \tag{4.12}$$

where

$$R = \begin{pmatrix} \cos\theta & \sin\theta \\ -\sin\theta & \cos\theta \end{pmatrix}$$

R' is the transpose matrix of the rotation matrix between the frames (ξ, η) and $(\mathbf{x}, \mathbf{y})$, and (x_T, y_T) are the coordinates of the (intermediate) target in the world frame $(\mathbf{x}, \mathbf{y})$.

To define the transformations between the world frame (x, y) and the primary path frame $(\mathbf{t}, \mathbf{n})$, write the velocity in the primary path frame as $\mathbf{V} = V\mathbf{t}$. To find the time derivative of vector $\mathbf{V}$ with respect to the world frame $(\mathbf{x}, \mathbf{y})$, note that the time derivative of vector $\mathbf{t}$ in the primary path frame (see Section 4.3.1) is not equal to zero. It can be defined as the cross product of angular velocity $\omega = \dot{\theta}\mathbf{b}$ of the primary path frame and vector $\mathbf{t}$ itself: $\dot{\mathbf{t}} = \omega \times \mathbf{t}$, where angle θ is between the unit vector $\mathbf{t}$ and the positive direction of $\mathbf{x}$ axis. Given that the control forces p and q act along the $\mathbf{t}$ and $\mathbf{n}$ directions, respectively, the equations of motion with respect to the primary path frame are

$$\dot{V} = p$$
$$\dot{\theta} = q/V$$

Since p and q are constant over the time interval $t \in [t_i, t_{i+1})$, the solution for $V(t)$ and $\theta(t)$ within the interval becomes

$$V(t) = pt + V_0$$

$$\theta(t) = \theta_0 + \frac{q \log(1 + tp/V_i)}{p} \qquad (4.13)$$

where θ_0 and V_0 are constants of integration and are equal to the values of $\theta(t_i)$ and $V(t_i)$, respectively. By parameterizing the path by the value and direction of the velocity vector, the path can be mapped into the world frame $(\mathbf{x}, \mathbf{y})$ using the vector integral equation

$$\mathbf{r}(t) = \int_{t_i}^{t_{i+1}} V \cdot \mathbf{t} \cdot dt \qquad (4.14)$$

Here $\mathbf{r}(t) = (x(t), y(t))$, and $\mathbf{t}$ is a unit vector of direction $\mathbf{V}$, with the projections $\mathbf{t} = (\cos(\theta), \sin(\theta))$ onto the world frame $(\mathbf{x}, \mathbf{y})$. After integrating Eq. (4.14), obtain the set of solutions

$$x(t) = \frac{2p \cos \theta(t) + q \sin \theta(t)}{4p^2 + q^2} V^2(t) + A$$

$$y(t) = -\frac{q \cos \theta(t) - 2p \sin \theta(t)}{4p^2 + q^2} V^2(t) + B \qquad (4.15)$$

where terms A and B are

$$A = x_0 - \frac{V_0{}^2 \ (2p \ \cos(\theta_0) + q \ \sin(\theta_0))}{4p^2 + q^2}$$

$$B = y_0 + \frac{V_0{}^2 \ (q \ \cos(\theta_0) - 2p \ \sin(\theta_0))}{4p^2 + q^2}$$

and $V(t)$ and $\theta(t)$ are given by (4.13).

Equations (4.15) describe a spiral curve. Note two special cases: When $p \neq 0$ and $q = 0$, Eqs. (4.15) describe a straight-line motion under the force along the vector of velocity; when $p = 0$ and $q \neq 0$, the force acts perpendicular to the vector of velocity, and Eqs. (4.15) produce a circle of radius $V_0^2/|q|$ centered at the point (A, B).

4.3.4 Canonical Solution

Given the current position $C_i = (x_i, y_i)$, the intermediate target T_i, and the velocity vector $\mathbf{V}_i = (\dot{x}_i, \dot{y}_i)$, the canonical solution presents a path that, assuming no obstacles, would bring the robot from C_i to T_i with zero velocity and in minimum time. The L_∞-norm assumption allows us to decouple the bounds on accelerations in ξ and η directions, and thus treat the two-dimensional problem as a set of two one-dimensional problems, one for control p and the other for control q. For details on obtaining such a solution and the proof of its sufficiency, refer to Ref. 99.

The optimization problem is formulated based on the Pontryagin's optimality principle [100], with respect to the secondary frame (ξ, η). We seek to optimize a criterion F, which signifies time. Assume that the trajectory being sought starts at time $t = 0$ and ends at time $t = t_f$ (for "final"). Then, the problem at hand is

$$F(\xi(\cdot), \eta(\cdot), t_f) = t_f \rightarrow \inf$$
$$\ddot{\xi} = p, \quad \|p\| \le p_{\max}$$
$$\ddot{\eta} = q, \quad \|q\| \le q_{\max}$$
$$\xi(0) = \xi_0, \quad \eta(0) = \eta_0, \quad \dot{\xi}(0) = \dot{\xi}_0, \quad \dot{\eta}(0) = \dot{\eta}_0$$
$$\eta(t_f) = \eta(t_f) = \dot{\xi}(t_f) = \dot{\eta}(t_f) = 0$$

Analysis shows (see details in the Appendix in Ref. 99) that the optimal solution of each one-dimensional problem corresponds to the "bang-bang" control, with at most one switching along each of the directions ξ and η, at times $t_{s,\xi}$ and $t_{s,\eta}$ ("s" stands for "switch"), respectively.

The switch curves for control switchings are two connected parabolas in the phase space $(\xi, \dot{\xi})$,

$$\xi = -\frac{\dot{\xi}^2}{2p_{\max}}, \qquad \dot{\xi} > 0$$

$$\xi = \frac{\dot{\xi}^2}{2p_{\max}}, \qquad \dot{\xi} < 0 \qquad (4.16)$$

and in the phase space $(\eta, \dot{\eta})$, respectively (see Figure 4.9),

$$\eta = -\frac{\dot{\eta}^2}{2q_{\max}}, \qquad \dot{\eta} > 0$$

$$\eta = \frac{\dot{\eta}^2}{2q_{\max}}, \qquad \dot{\eta} < 0 \qquad (4.17)$$

The time-optimal solution is then obtained using the bang-bang strategy for ξ and η, depending on whether the starting points, $(\xi, \dot{\xi})$ and $(\eta, \dot{\eta})$, are above or below their corresponding switch curves, as follows:

$$\hat{p}(t) = \begin{cases} \alpha_1 \cdot p_{\max}, & 0 \le t \le t_{s,\xi} \\ \alpha_2 \cdot p_{\max}, & t_{s,\xi} < t \le t_f \end{cases}$$

$$\hat{q}(t) = \begin{cases} \alpha_1 \cdot q_{\max}, & 0 \le t \le t_{s,\eta} \\ \alpha_2 \cdot q_{\max}, & t_{s,\eta} < t \le t_f \end{cases} \qquad (4.18)$$

where $\alpha_1 = -1, \alpha_2 = 1$ if the starting point, $(\xi, \dot{\xi})_s$ or $(\eta, \dot{\eta})_s$, respectively, is above its switch curves, and $\alpha_1 = 1, \alpha_2 = -1$ if the starting point is below its

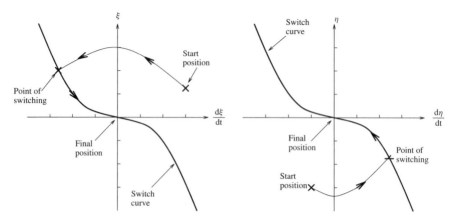

Figure 4.9 **(a)** The start position in the phase space $(\xi, \dot{\xi})$ is above the switch curves. **(b)** The start position in the phase space $(\eta, \dot{\eta})$ is under the switch curves.

switch curves. For example, if the initial conditions for ξ and η are as shown in Figure 4.9, then

$$
\hat{p}(t) = \begin{cases} -p_{max}, & 0 \le t \le t_{s,\xi} \\ +p_{max}, & t_{s,\xi} < t \le t_f \end{cases}
$$

$$
\hat{q}(t) = \begin{cases} +q_{max}, & 0 \le t \le t_{s,\eta} \\ -q_{max}, & t_{s,\eta} < t \le t_f \end{cases} \tag{4.19}
$$

where the caret sign (^) refers to the parameters under optimal control. The time, position, and velocity of the control switching for the ξ components are described by

$$
t_{s,\xi} = \frac{\sqrt{\frac{(\dot{\xi}_0)^2}{2} + \xi_0 p_{max}} + \dot{\xi}_0}{p_{max}}
$$

$$
\xi_s = \frac{\dot{\xi}_0^2}{4 p_{max}} + \frac{\xi_0}{2}
$$

$$
\dot{\xi}_s = -\sqrt{\frac{\dot{\xi}_0^2}{2 p_{max}} + \xi_0 p_{max}}
$$

and those for the η components are described by

$$
t_{s,\eta} = \frac{\sqrt{\frac{(\dot{\eta}_0)^2}{2} - \eta_0 q_{max}} - \dot{\eta}_0}{q_{max}}
$$

$$\eta_s = -\frac{\dot{\eta}_0^2}{4q_{max}} + \frac{\eta_0}{2}$$

$$\dot{\eta}_s = \sqrt{\frac{\dot{\eta}_0^2}{2q_{max}} - \eta_0 q_{max}}$$

The number, time, and locations of switchings can be uniquely defined from the initial and final conditions. It can be shown (see Appendix in Ref. 99) that for every position of the robot in the $\Re^4$ ($\xi, \eta, \dot{\xi}, \dot{\eta}$) the control law obtained guarantees time-optimal motion in both ξ and η directions, as long as the time interval considered is sufficiently small. Substituting this control law in the equations of motion (4.15) produces the canonical solution.

To summarize, the procedure for obtaining the first step of the canonical solution is as follows:

1. Substitute the current position/velocity ($\xi, \eta, \dot{\xi}, \dot{\eta}$) into the equations (4.16) and (4.17) and see if the starting point is above or below the switch curves.
2. Depending on the above/below information, take one of the four possible bang-bang control pairs p, q from (4.18).
3. With this pair (p, q), find from (4.15) the position C_{i+1} and from (4.13) the velocity $\mathbf{V}_{i+1}$ and angle θ_{i+1} at the end of the step. If this step to C_{i+1} crosses no obstacles and if there exists a stopping path in the direction $\mathbf{V}_{i+1}$, the step is accepted; otherwise, a near-canonical solution is sought (Section 4.3.5).

Note that though the canonical solution defines a fairly complex multistep path from C_i to T_i, only one—the very first—step of that path is calculated explicitly. The switch curves (4.16) and (4.17), as well as the position and velocity equations (4.15) and (4.13), are quite simple. The whole computation is therefore very fast.

4.3.5 Near-Canonical Solution

As discussed above, unless a step that is being considered for the next moment guarantees a stopping path along its velocity vector, it will be rejected. This step will be always the very first step of the canonical solution. If the stopping path of the candidate step happens to cross an obstacle within the distance found from (4.10), the controls are modified into a *near-canonical solution* that is both Ω-acceptable and reasonably close to the canonical solution. The near-canonical solution is one of the nine possible combinations of the bang-bang control pairs ($k_1 \cdot p_{max}, k_2 \cdot q_{max}$), where k_1 and k_2 are chosen from the set $\{-1, 0, 1\}$ (see Figure 4.10).

Since the canonical solution takes one of those nine control pairs, the near-canonical solution is to be chosen from the remaining eight pairs. This set is guaranteed to contain an Ω-acceptable solution: Since the current position has been chosen so as to guarantee a stopping path, this means that if everything

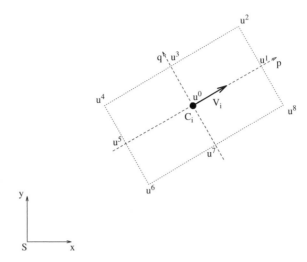

Figure 4.10 Near-canonical solution. Controls (p, q) are assumed to be L_∞-norm bounded on the small interval of time. The choice of (p, q) is among the eight "bang-bang" solutions shown.

else fails, there is always the last resort path back to the current position—for example, under control $(-p_{max}, 0)$.

Furthermore, the position of the intermediate target T_i relative to vector $\mathbf{V}_i$—in its left or right semiplane—suggests an ordered and thus shorter search among the control pairs. For step i, denote the nine control pairs $\mathbf{u}_i^j$, $j = 0, 1, 2, \ldots, 8$, as shown in Figure 4.10. If, for example, the canonical solution is $\mathbf{u}_i^2$, then the near-canonical solution will be the first Ω-acceptable control pair $\mathbf{u}^j = (p, q)$ from the sequence $(\mathbf{u}^3, \mathbf{u}^1, \mathbf{u}^4, \mathbf{u}^0, \mathbf{u}^8, \mathbf{u}^5, \mathbf{u}^7, \mathbf{u}^6)$. Note that $\mathbf{u}^5$ is always Ω-acceptable.

4.3.6 The Algorithm

The complete motion planning algorithm is executed at every step of the path, and it generates motion by computing canonical or near-canonical solutions at each step. It includes four procedures:

(i) The *Main Body* procedure monitors the general control of motion toward the intermediate target T_i. In turn, *Main Body* makes use of three procedures:

(ii) Procedure *Define Next Step* chooses the controls (p, q) for the next step.

(iii) Procedure *Find Lost Target* deals with the special case when the intermediate target T_i goes out of the robot's sight.

(iv) *Main Body* also uses the procedure called *Compute T_i*, taken directly from the kinematic algorithm (for example, VisBug-21 or VisBug-22,

Section 3.6), which computes the next intermediate target T_{i+1} so as to guarantee the global convergence, and also performs the test for target reachability.

As before, S is the starting point, T—the robot target position; at step i, C_i is the robot's current position, vector V_i—the current velocity vector. Initially, $i = 0, C_i = T_i = S$.

Procedure *Main Body*. At each step i:

If $C_i = T$, stop.
Find T_i from *Compute T_i*.
If T is found unreachable, stop.
If T_i is visible, find C_{i+1} from *Define Next Step*; make a step toward C_{i+1}; iterate; else,
Use *Find Lost Target* to produce T_i visible; iterate.

Procedure *Define Next Step*. This procedure consists of two steps:

S1: Find the canonical solution (the switch curves and controls (p, q)) using Eqs. (4.16), (4.17), and (4.18). If it is Ω-acceptable, exit; else go to S2.
S2: Find the near-canonical solution as in Section 4.3.5; exit.

Procedure *Find Lost Target*. This procedure is executed when T_i becomes invisible. The last position C_i where T_i was visible is then stored until T_i becomes visible again. Various heuristics can be used here as long as convergence is preserved. One simple strategy would be to come to a halt using a stopping path, then come back to C_i with zero velocity, and then move directly toward T_i. This may add stops that could be avoided. The procedure chosen below is somewhat smarter in that the robot does not stop unnecessarily: If the robot loses T_i, it keeps moving ahead while defining temporary intermediate targets T_i^t on the visible part of the line segment (C_i, T_i) and continues looking for T_i. If it locates T_i, it turns directly toward it without stopping (Figure 4.11a). Otherwise, if the whole segment (C_i, T_i) becomes invisible, the robot brakes to a stop, returns to C_i with zero velocity, and then moves directly toward T_i (Figure 4.11b). *Find Lost Target* operates as follows:

S1: While at $C_k, k > i$, find on the segment (C_i, T_i) the visible point closest to T_i; denote it T_k^t. If there is no such point [i.e., the whole segment (C_i, T_i) is not visible], go to S2. Else, using *Define Next Step*, compute and execute the next step using T_k^t as the temporary intermediate target; iterate.
S2: Initiate a stopping path, then go back to C_i; exit.

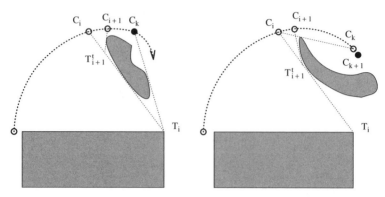

Figure 4.11 In these examples, because of the system inertia the robot temporarily "loses" the intermediate target point T_i. **(a)** The robot keeps moving forward until at C_k it sees T_i. **(b)** At C_k the robot initiates a stopping path, stops at C_{k+1}, returns back to C_i, and moves toward T_i along the line (C_i, T_i).

4.3.7 Convergence. Computational Complexity

Convergence. The collision-free motion along the path is guaranteed by the design of the canonical and near-canonical solutions. To prove convergence, we need to show that the algorithm will find a path to the target position T if one exists, or will infer in finite time that T is not reachable if true. This is guaranteed by the convergence properties of the kinematic algorithm (Section 3.6). The following two statements, Claim 1 and Claim 2, hold:

Claim 1. *Under the Minimum Time Strategy, assuming zero velocity* $\mathbf{V}_S = 0$ *at the starting position S, at each step of the path there exists at least one stopping path.*

To see this, recall that according to our model the stopping path lies along a straight line. Guaranteeing a stopping path implies two requirements: a safe direction and the velocity value that will allow a stop within the visible area. Because the latter is ensured by the choice of the system parameters in (4.10), we focus now on the existence of safe directions. Proceed by induction: We have to show that if a safe direction exists at the start point and on an arbitrary step i, then there is a safe direction on the step $(i + 1)$.

Since at the start point S the velocity is zero, any direction of motion at S will be a safe direction. This gives the basis of induction. The induction step is as follows. Under the algorithm, a candidate step is accepted for execution only if its direction guarantees a safe stop for the robot if needed. Namely, at point C_i, step i is executed only if the resulting vector $\mathbf{V}_{i+1}$ at C_{i+1} will point in a safe direction. Therefore, at step $(i + 1)$, at the least this very direction can be used for a stopping path.

Remark: Claim 1 will hold for $\mathbf{V}_S \neq 0$ as well if there exists at least one stopping path originating at the start point S.

In practical terms, this is a reasonable condition. If for some reason the robot did not start at S and was passing through it on the fly—which is already strange enough—it is hard to imagine that point S happened to be so bad that it could not even provide a stopping path.

Claim 2. *The Minimum Time Strategy guarantees convergence.*

To see this, note that at each step i at its current position C_i, the robot uses its sensing to generate the next intermediate target point T_i. That point T_i is known to lie on a convergent path of the kinematic algorithm (Section 3.6.3). At the moment when T_i is generated, it is visible. Hence, the only way that the robot can get lost is if at the next step C_{i+1} point T_i becomes invisible due to the robot inertia or obstacle occlusion: This would make it impossible to generate the intermediate target T_{i+1} as required by the kinematic algorithm. But, if indeed point T_i becomes invisible, the *Find Lost Target* procedure is invoked and a set of temporary intermediate targets T_{i+1}^t and associated steps are executed until point T_i becomes visible again (see Figure 4.11a). Thus the robot always moves toward a point that lies on a convergent path and itself converges to the target T.

Computational Complexity. As with other on-line sensor-based algorithms, it would not be very informative to try to assess the algorithm complexity the way it is usually done for algorithms with complete information, as a function of the number of vertices of approximated obstacles (see Chapter 1). As one reason, algorithms with complete information deal with one-time computation, whereas in sensor-based algorithms the important complexity measure is the amount of computations at each step; the total computation time is simply a linear function of the path length.

As shown in Section 4.3.4, though the canonical solution found by the algorithm at each path step is the solution of a fairly complex time-optimal problem, its computational cost is remarkably low, thanks to the (optimal) bang-bang control. This computation includes (a) substituting the initial conditions $(\xi, \eta, \ddot{\xi}, \ddot{\eta})$ into the equations for parabolas [Eqs. (4.16) and (4.17)] to see if the start point is above or below the corresponding parabola and (b) simply taking the corresponding control pair $(\hat{p}, \hat{q})$ from the four choices in (4.18). The parabola equations themselves are found beforehand, only once. The near-canonical solution, when needed, is similar and as fast. Note that a single-step computation is of constant time: Though the canonical solution represents the whole multistep trajectory within the sensing range of radius r_v, the computation time is independent of the value r_v and of the length of path within the sensing range.

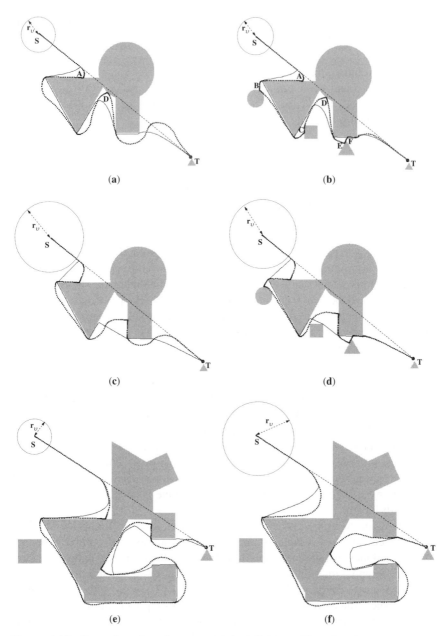

Figure 4.12 Examples of performance of the Minimum Time Strategy. Parts **(a)** and **(b)** differ in that more obstacles are added in **(b)**. Parts **(c)** and **(d)** relate to the same scenes, respectively, and have a larger radius of vision r_v. The radius of vision in **(f)** is significantly larger than that in **(e)**. Note the stopping points along the paths.

4.3.8 Examples

The performance of Minimum Time Strategy algorithm is illustrated in Figure 4.12. The examples shown are computer simulations. The robot mass m and constraints on control parameters p and q are the same for all examples: $p_{max} = q_{max}$, $p_{max}/m = 1$. The generated paths are shown in thicker lines. For the purpose of comparison, also shown (in a thin line) are paths produced under the same conditions by a kinematic algorithm VisBug.

The radius of vision r_v is the same in both Figures 4.12a and 4.12b. The difference is in the environment: In Figure 4.12b there are additional obstacles; that is, the robot suddenly uncovers them at a close distance when turning around a corner. Note that in Figure 4.12b the path becomes tighter and shorter, though it takes longer: Measured in the number of steps, the path in Figure 4.12a takes 242 steps, and the path in Figure 4.12b takes 278 steps. One can say the robot becomes more cautious in Figure 4.12b.

A similar pair of examples shown in Figures 4.12c and 4.12d illustrates the effect of the radius of vision r_v: Here it is twice as big as the radius of vision in Figures 4.12a and 4.12b. The times to execute the path here are 214 and 244 steps, respectively, shorter than in the corresponding examples in Figures 4.12a and 4.12b. The examples thus demonstrate that better sensing (larger r_v) results in shorter time to complete the task: More crowded space results in longer time, though possibly in shorter paths.

Note that in some points the algorithm found it necessary to make use of the stopping path; those points are usually easy to recognize from the sharp turns in the path. For example, in Figure 4.12a the robot came to a halt at points A and D, and in Figure 4.12b it stopped at points A–F. The algorithm's performance in a more complicated environment is shown in Figures 4.12e and 4.12f. In Figure 4.12f the radius of vision r_v is significantly larger than that in Figure 4.12e. Note again that richer input information provided by a larger sensing range is likely to translate into shorter paths.

CHAPTER 5

Motion Planning for Two-Dimensional Arm Manipulators

If we imagine constructions to be made with rigid rods ... we should find that
different laws hold for these from those resulting on the basis of Euclidean plane
geometry. The surface is not a Euclidean continuum with respect to the rods,
and we cannot define Cartesian co-ordinates in the surface.

— *Albert Einstein,*
Relativity: The Special and General Theory

5.1 INTRODUCTION

In Chapter 3 we have developed the foundations of the SIM (Sensing–Intelligence–Motion) paradigm (called also sensor-based robot motion planning). Basic
algorithms were developed for the simplest case of a point robot that possesses
tactile sensing and operates in a two-dimensional scene populated with obstacles
of arbitrary shapes. The algorithms were then extended to richer sensing such as
vision, as well as to algorithm versions that take into account robot dynamics.
When the robot starts on its journey, it knows nothing about the shapes, locations,
and number of individual obstacles in the scene. It acquires information about
its surroundings from its sensors—much the way we humans see and listen and
smell when moving in the physical world. The robot's only goal is to arrive at
its target location. This means that when it arrives there, it may still know very
little about the scene. In other words, knowing more about the scene is not the
objective. In a way, the less the robot knows about the scene at the end—that is,
the less it wonders around trying to locate its target—the better its performance.

This chapter is devoted to developing sensor-based motion planning for *robot
arm manipulators*. This work will be heavily based on the developments in
Chapter 3. While the basic strategies developed there can be used for real-world mobile robots, their one very important use is to serve as a foundation
for motion planning strategies for robot arm manipulators. The importance of
this area for practice is underscored by today's situation in the field of robotics:
In terms of actual utilization, overall investment, and engineering complexity

robot arm manipulators are way more important than mobile robots. According to the UNECE (United Nations Economic Commission for Europe) report "World Robotics 2003" [101], by 2003 about 1,000,000 industrial robots had been used worldwide. By far most of these robots are arm manipulators.

And yet, while at least some commercial mobile robots come today with rudimental means for sensor-based motion planning, by and large no such means exist for arm manipulators. Exceptions do exist, but as a rule they relate to motion planning of the robot end effector, the tool, rather than the whole robot. This is certainly not because of lack of interest. If available, such systems would find an immediate and wide use—even in the same industries where arm manipulators are used today—by helping decrease the cost of systems. In a typical industrial system, the cost of the robot is a fraction—perhaps 20% or so—of the total cost of the work cell. Much of the rest are means to compensate for the robot's inability to avoid collisions with its surroundings on its own.

Motion planning systems would also allow robot manufacturers to move their products into new domains—agriculture (to pick op fruits and berries and other crops), nursing homes (to help move and feed patients), homes of the elderly (to help them handle various home chores), outer space (to assemble large structures, such as telescopes and space stations)—in short, to a whole slew of applications that are good candidates for automation but could not be automated so far because of the high level of uncertainly characteristic of such tasks.

There are two major reasons as to why commercial robots intended for a high level of uncertainty are not here yet. First, appropriate theory and algorithms are just beginning to appear. Second, the sensing technology that is required for such algorithms to operate is also at the development stage. (The issues of sensing hardware is addressed in Chapter 8.)

Whatever research has been done on motion planning for arm manipulators, its lion share relates to motion planning for arm hands and grippers. Collision avoidance for the rest of the robot body has been largely left out. Again, this is not because of the lack of need. A quick glance at a layout of a typical robot cell with a robot arm manipulator will show how crowded those cells are. The problem of handling potential collisions for the whole body of an arm manipulator is acute.

Works that focus on robot hands' collision avoidance do of course advance the general progress in robotics. One can imagine applications where the designer makes sure that potential collisions can occur only near the arm hand. On a robot welding line in an automotive plant, the operation can be designed and scheduled so that no objects would endanger (or would be endangered by) the robot body. Because the robot tool must be close to the parts to be welded, this cannot be avoided, and so the robot hand will be the only part of the robot body that can come close to other objects. This way, unpredictable events can happen only at the tool: Parts to be welded may be positioned slightly off, their dimensions may deviate slightly, one part may be slightly bent, and so on.

Hence the designer of such a system will seek collision avoidance procedures that will take care of the arm's hand only. In our example, such algorithms would lead the gun of a welding robot arm clear of the parts being welded. Providing the assurance that only the arm hand can be in danger of collisions is expensive and can be justified only in a well-controlled environment, of which an automotive factory floor is a good example. In general the practical use of such algorithms is limited. They would not be of much use in tasks with a reasonable level of uncertainty—as for example, outdoors.

As in the case of mobile robots, both exact (provable) and heuristic motion planning algorithms have been explored for arm manipulators. It is important to note that while good human intuition can sometimes justify the use of heuristic motion-planning procedures for mobile robots, no such intuition exists for arm manipulators. As we will see in Chapter 7, more often than not human intuition fails in motion planning tasks for arm manipulators. One can read these results as a promise that a heuristic automatic procedure will likely produce unpleasant surprises. Theoretical assurances of algorithms' convergence becomes a sheer necessity.

Similar to the situation with mobile robots (see Chapter 3), historically motion planning for arm manipulators has received most attention in the context of the paradigm with complete information (Piano Mover's model). Both exact and heuristic approaches have been explored [15, 16, 18, 20–22, 24, 25, 102]. Little work has been done on motion planning with uncertainty [54].

In this and the next chapters, sensor-based motion planning will be applied to the whole robot body. No point of the robot body should be in danger of a collision. But bodies of robot arm manipulators are very complex. Parts move relative to each other, and shapes are elaborate; it would not be feasible in practice to supply a collision avoidance algorithm with an exact description of the robot body. Our objective will be to make the algorithms immune to specifics of arm geometry.

Similar to how we approached the problem in Chapter 3, we will first consider simple systems, namely, planar arm manipulators. These results may already have some limited use in applications; for example, in terms of programming and motion planning, a class of arm manipulators called SCARA (which stands for Selective Compliant Articulated Robot Arm) consists of essentially plane-oriented devices; they are used widely in tasks where the "main action" takes place in a plane (such as assembly on a conveyer belt), and the third dimension plays a secondary role. However, the main motivation behind the simpler cases considered in this chapter is to develop a theoretical framework that will be used in the next chapter to develop motion planning strategies for three-dimensional (3D) robot arms of various kinematics.

The same as with mobile robots, the uncertainty of the robot surroundings precludes a sensor-based algorithm from promising an optimal path for an arm manipulator. Instead, the objective is to generate a "reasonable" path for the arm (if one exists), or to conclude that the target position cannot be reached if that happen to be so. We will discover that for the arm manipulators considered here a purely local sensory feedback is sufficient to guarantee reaching a global objective—that is, to guarantee algorithm convergence.

We will do the necessary analysis using the simplest tactile sensing and simplified shapes for the robot. Since such simplifications often cause confusion as to algorithms' applicability, it is worthwhile to repeat these points:

Types of Sensing and Robot Geometry Versus Algorithms. Here and elsewhere in this text, when we develop motion planning algorithms based on tactile sensing and on a simplified shapes of robots (say, a point mobile robot or a stick line arm manipulator links), this does not imply that tactile sensors or simplified shapes are the only, or the recommended, modalities for an algorithm at hand: (a) Any type of sensing (tactile, proximal, vision, etc.) can be used with such algorithms, either directly or with small easily realizable modifications, provided that sensing covers every point of the robot body. (b) The algorithm will work with the robots or arm manipulator links of any shapes. See Section 1.2 and later in Section 5.1.1.

Major and Minor Linkage. Following Ref. 103, we use the notion of a *separable* arm, which is an arm manipulator that can be naturally divided into (a) the *major linkage* responsible for the arm's *position planning* (or *gross motion*) and (b) the *minor linkage* responsible for the *orientation* of the arm's end effector (its hand) in the arm workspace. As a rule, existing arm manipulators are separable, and so are the limbs of humans and animals (although theoretically this does not have to be so).

The notions of major and minor linkages are tied to the notion of a *minimal configuration.* For the major linkage a minimal configuration is the minimum number of links (and joints) that the arm needs to be able to reach any point in its workspace. For the two-dimensional (2D) case the minimal configuration includes two links (and two joints). Looking ahead, in the three-dimensional (3D) case the minimal configuration for the major linkage includes three links (and three joints). In the algorithmic approach considered here, motion planning is limited to the gross motion—that is, to the major linkage. The implicit assumption is that the motion of the end effector (i.e., the minor linkage) can be planned separately, after the arm's major linkage arrives in the vicinity of the target position. For all but very unusual applications, this is a plausible assumption. Although theoretically this does not have to be so, providing orientation for the minor linkage is usually significantly simpler than for the major linkage, simply because the hand is small.

Types of Two-Link Arm Manipulators. These kinematic pairs are called two-dimensional (2D) arms, to reflect the fact that the end effector of any such arm moves on a two-dimensional surface—or, in topological terms, in a surface *homeomorphic to a plane*. With this understanding, we can call these arms *planar arms*, as they are often called, although the said surface may or may not be a plane. Only revolute and sliding joints will be considered, the two types that are primary joints used in practical manipulators. Other types of joints appear very rarely and are procedurally reducible to these two [103].

A *revolute* joint between two links is similar to the human elbow: One link rotates about the other, and the angle between the links describes the joint value at any given moment (see Figure 5.1a). In a *sliding* joint the link slides relative to the other link; the linear displacement of sliding is the corresponding joint value (see Figure 5.1b). Sliding joints are also known in the literature on kinematics as *prismatic joints*.

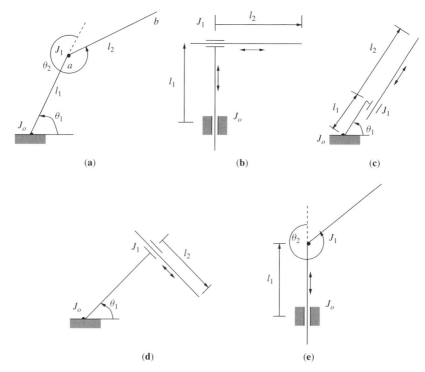

Figure 5.1 Five kinematically distinct two-link planar robot arms manipulators: **(a)** RR arm (revolute–revolute, or articulated); **(b)** PP arm (prismatic–prismatic, or Cartesian); **(c)** RP parallel arm (revolute–prismatic parallel); **(d)** RP perpendicular arm (revolute–prismatic perpendicular); **(e)** PR arm (prismatic–revolute).

In this chapter, no constraints will be imposed on the shapes of arm links (see below). This is very helpful: It means, for example, that the minor linkage can be "frozen" during the gross motion, thus effectively making the minor linkage a part of the outer link of the major linkage. It also means that the robot does not have to worry about its own shape when planning collision avoidance motion.

We obtain the list of distinct planar arm configurations with revolute and prismatic joints from the classification of kinematic configurations of three-dimensional major linkages with revolute and prismatic joints. Out of 36 possible configurations described in Ref. 103, 12 are listed as kinematically useful and distinct. (We will return to those when considering three-dimensional arms in Chapter 6.) The rest are not admissible, either because they degenerate into one- or two-dimensional cases or because they are equivalent to some other configurations. Of these 12 types, only five are meaningful for the case of a planar major linkage. These are as follows:

- Arm manipulator with two revolute joints; it is often called an *articulated arm* or *revolute–revolute arm*, or *RR arm* for short (Figure 5.1a).

- Arm with two prismatic joints, typically referred to as a *Cartesian arm* or a *PP arm* (Figure 5.1b).
- *RP (revolute–prismatic) arm*, which has a revolute joint followed by a parallel prismatic link (Figure 5.1c).
- Another *RP arm*—a revolute link followed by a perpendicular prismatic link (Figure 5.1d).
- *PR (prismatic–revolute) arm*—prismatic joint followed by a revolute joint (Figure 5.1e).

In the next section, a general model of the arm manipulator and of the environment in which it operates will be outlined, along with necessary definitions. Any modifications that the model may require for a specific arm configuration will appear in the corresponding sections. Next we will consider in detail the first of the five two-link arms, the RR (revolute–revolute) arm (Figure 5.1a). We will study interactions between the arm and the obstacles in the arm's workspace, eventually deriving a path planning algorithm with guaranteed convergence.

In the sections that follow the RR arm study, we will study in a similar fashion, except in a more brisk pace, each of the remaining four arms depicted in Figure 5.1. When developing the corresponding motion planning procedures, we will observe that the algorithmic issues for these arms turn out to be simpler compared to the RR arm.

For a reader familiar with the Piano Mover's techniques, it will come perhaps as a surprise that in principle each of the two-link arms shown in Figure 5.1 will require its own version of the sensor-based motion planning algorithm. While a detailed reasoning as to why that is so will come later, here the reader is invited to accept this fact as a law of nature. Indeed, nature operates in a similar fashion: Constraints imposed on animals' motion by the kinematics of their limbs and bodies make each species move differently from others. Each species' "algorithm" for obstacle avoidance differs from other species' algorithms. Humans are no exception: A person who lost his leg in an accident will have to re-learn the use of legs. He will learn to walk around or step over objects in ways dramatically different from how he handled this task before the accident.

On the other hand, there will be much in common between motion planning algorithms for different arm manipulators. Based on this observation, in Section 5.8.4 we will attempt to build a unified theory of planar manipulators. This will allow us to derive planar arm algorithms as a special case of one general strategy. We will see, in particular, that the arm configurations b, c, d, and e in Figure 5.1 are in some sense special cases of the RR (revolute–revolute) arm.

As a concluding remark in this introductory section, one should keep in mind that the planar two-link arms shown in Figure 5.1 can be arranged in more kinematic arrangements, and hence more geometries, beyond those shown in the figure. Note, for example, that in the RR arm planar arm shown in Figure 5.1a the axes of both joints are parallel. A number of other configurations of RR arms can be obtained by manipulating the mutual arrangement of those axes. We will consider this variety and its effect on motion planning later, in Section 5.3. It suffices to say here that the workspace of what we call "planar arms" is not necessarily planar, but it remains two-dimensional.

5.1.1 Model and Definitions

The Robot Arm. The arm body consists of two links, l_1 and l_2, and two joints, J_0 and J_1. Joint J_0 is fixed and is the origin of the reference system. See different arm configurations in Figures 5.1a to 5.1e. Solely for presentation purposes we represent the links as straight-line segments, of lengths l_1 and l_2, respectively.

The arm's *configuration* is defined by its *kinematics*—that is, by the way its joints connect the links together. In the arms in Figure 5.1 a link may be rotating about its joint, in which case the joint is a *revolute joint*, or it may be sliding in it, in which case the joint is a *prismatic joint*.

Accordingly, in the case of a revolute joint the corresponding link is of a constant length, and in case of a prismatic joint the corresponding link is of a variable length. An *arm solution* for a given point P in the arm workspace (W-space)—equivalent terms that we may use are *arm position, arm coordinates*, and *link positions*—is defined by a pair of variables called *joint values*. Joint values are either angles (as in Figure 5.1a) or linear translations (as in Figure 5.1b), or a combination of both (as in Figures 5.1c, 5.1d, 5.1e). An equivalent presentation for the same solution P is given, for example, by Cartesian coordinates (x, y) of the link endpoints, a_p and b_p; b_p also designates the arm endpoint position at point P. Positive directions and zero positions for joint values of the five arms are shown in Figures 5.1a to 5.1e.

As said above, for the considered class of path planning algorithms, the shape of arm links and of obstacles—for example, the fact of their convexity or concavity—is of no importance. Without loss of generality, and solely for better visualization and material presentation, line segment links and circular obstacles are used in most figures of this section.

The arm is capable of performing the following actions:

1. Moving the arm endpoint through a prescribed simple curve (called *main line* or *M-line*) that connects the arm's start (S) and target (T) positions.
2. When the arm body hits an obstacle, identifying the point(s) of the arm body that is in contact with the obstacle.
3. Following the obstacle boundary.

The first of these operations implies that the arm is capable of computing coordinates of consecutive points along the M-line and, if necessary, transforming them into the corresponding joint values (using, for example, appropriate procedures of inverse kinematics [8]).

The sole purpose of the second operation is to provide the local information needed to pass around the obstacle. At any moment, when at least one point of the arm body is in contact with an obstacle, the arm identifies coordinates of the points of contact on the arm body relative to the arm's internal reference system. Note that such identification is a local operation that does not require global information about the environment. (As an example of this ability, recall that a blindfolded person can easily indicate a point of his body that touches an object). We do not discuss here specific ways in which this capability can be realized. For our purposes, assume that the arm body is covered with sensors such that when one sensor contacts an obstacle, the point of contact on the arm body is known.

For the third operation, following the obstacle boundary, imagine that while being in contact with an obstacle, the arm follows the obstacle boundary while a weak constant force pushes it against the obstacle. (This situation is similar to a blindfolded person walking around a building while keeping his finger on the wall.) At any given moment during the motion, there is a variable point of contact between the obstacle boundary and the arm body. In the algorithm, the arm will plan its next step along the obstacle boundary in such a way that, after the step has been made, the arm is still in contact with the obstacle. Again, we will not discuss here how this important capability can be realized in the physical system. Note that if the arm endpoint follows an obstacle boundary up to the W-space boundary—for example, the arm is fully outstretched—it is not clear whether on the boundary the arm is still in contact with the obstacle. To avoid this limit case, we assume that no point of the W-space boundary may be a point of contact between an obstacle and the arm endpoint.

Passing around an obstacle is a continuous motion of the arm during which the arm is in constant contact with some obstacle. Because of the arm and obstacle interaction, some areas of the W-space not occupied by the actual obstacle may be inaccessible by the arm endpoint. Such areas create a *shadow* of the obstacle; for the arm endpoint, a shadow presents as real an obstacle as points of the actual obstacle. The actual obstacle and its shadow(s) constitute the *virtual obstacle* of a given obstacle. When the arm is passing around an actual obstacle, the arm endpoint follows the *virtual line*, which is the boundary of the virtual obstacle. Below we will study this phenomenon in more detail.

Input Information. Sensing.   At the start, the only information available to the arm are coordinates of its current position S and its target position T.

When moving, the arm obtains its information about the surrounding world from its *sensors*. Sensors can be of any type—tactile, proximal, vision, range sensing, and so on—as long as they provide sufficient input information. As explained in Section 1.2.3, richer sensing will often result in better efficiency (for example, shorter paths), but will not guarantee algorithm convergence or produce better performance bounds. A property of robot sensing that is absolutely necessary for the planning algorithms to operate successfully is that sensing should encompass the whole robot body; that is, it should allow the robot arm to detect a potential collision at any point of its body. No blind spots are allowed. To develop motion planning algorithms, we will first assume whole-body tactile sensing.

W-Space. The arm operates in its *workspace (W-space)*, which is defined in the plane. The *W*-space *boundaries* depend on the arm configuration and dimensions of its links. The arm's *position* in *W*-space is defined by positions of its joints and links.

    W-space may contain obstacles. Each obstacle is a simple closed curve of finite size. There may be only a finite number of obstacles present in *W*-space. Formally, this means that the boundary of any obstacle is homeomorphic to a circle and that any circle of a limited radius or a straight line passing through *W*-space will intersect with a finite set of obstacles. Being rigid bodies, obstacles cannot intersect one another. Two or more obstacles may touch each other, in which case for the arm they effectively present one obstacle. At any position of the arm with respect to a set of obstacles, at least some arm motion is assumed to be feasible.

The Task. Our objective for the arm is to move it from the starting position S to the target position T, or to (correctly) conclude in finite time that no path from S to T exists. Only continuous motion of the links is allowed. Both positions S and T lie within the *W*-space. A position of the arm is said to be feasible if, when the arm's endpoint is in this position, the arm's links and joints intersect no obstacles. Position S is known to be feasible. Because of obstacles in the scene, position T may or may not be feasible. Even if position T is feasible, it may or may not be reachable from position S.

C-Space. The *configuration space (C-space)* is a representation space in which the arm is shrunk to a point. In our case, *C*-space is the space of arm joint variables, which happen to be the arm's independent control variables. Every path and every virtual obstacle has its corresponding image in *C*-space. A combination of the virtual line with the corresponding arm solutions defines the *virtual boundary* of the obstacle. The virtual boundary is a curve that forms the boundary of the obstacle image in *C*-space. The transformation from *C*-space to *W*-space is unique. As we will discuss later, depending on the arm configuration, the transformation from *W*-space to *C*-space may or may not be unique. We will soon see that for all the arms shown in Figure 5.1 the corresponding *C*-space presents a two-dimensional manifold.

    One should not confuse the dimensionality of the manifold in question with the dimensionality of space in which the manifold appears. For example, later in this text we will deal with the surfaces of a common torus or a sphere. While each of these is a two-dimensional manifold, they appear in the three-dimensional space. In general, *C*-space is a k-dimensional manifold in a Euclidean space whose dimension is higher than k. Accordingly, the metric in a manifold in question may or may not be Euclidean. We will see later, for instance, that unlike what occurs in a Euclidean space, up to four distinct shortest routes between two points may appear on the surface of the torus.

Sketching the Approach. To develop a path planning procedure, the problem of motion planning for a planar arm will be first reduced to that of moving a point (robot) around simple closed curves. The key property will then be deduced: For a two-link arm, no matter how complex the arm motion around an actual physical obstacle in W-space, the corresponding virtual boundary in C-space presents a simple curve—that is, a curve with no self-intersections and double points. This will be shown to be true for each of the arms in Figure 5.1.

With this property in hand, by transforming the motion planning problem from W-space to C-space, we will effectively make our problem similar to the one that was tackled in Chapter 3 for mobile robots. In fact, on a certain level of generalization, both problems look identical. The actual algorithms will differ due to a number of new issues that need to be worked out. Still, understanding the Bug family algorithms from Chapter 3 will help one grasp the algorithms for robot arms that we are about to develop.

We can now sketch the idea behind a motion planning algorithm for a planar robot arm manipulator. It is easier to describe the operation in C-space; the actual operation in W-space proceeds accordingly. As one will notice, the sketch sounds much like the algorithm Bug2; deviations and complexities will be added later.

At the beginning, the C-space arm image point moves along a simple M-line, which is a desired path from point S to point T, an equivalent of the straight-line M-line for the mobile robot (Section 3.3). During this motion, when (in W-space) some point of the arm body meets an obstacle, in C-space this corresponds to the image of M-line intersecting the obstacle's *virtual boundary*. The point of intersection is said to define a *hit point*, H_j, where j is the running index enumerating such points.

We will show below that the virtual boundary is a simple curve, a curve with no self-intersections or double points. This being so, at the hit point the arm has a simple choice: to walk along the virtual boundary in one or the opposite direction along the curve. Since no information is available beforehand as to which of the two directions is better, one direction, called the *local direction*, will be chosen once and for all.

While following the obstacle virtual boundary, the arm may meet the M-line again. If it does, and if this occurs at a distance (measured appropriately along the M-line) from point T shorter than the distance from the latest hit point H_j to T, the arm will define a *leave point*, L_j. Hit and leave points will play an important role in the path planning procedure. We will see below that these points come in pairs, (H_j, L_j), $j = 1, 2, \ldots$. For convenience, denote $L_o = S$, Start, with no corresponding H_o. The motion planning algorithm proper, the proof of its convergence, and the test for target reachability will emerge from our analysis of the described scheme and of C-space properties.

Similar to the mobile robot case (Chapter 3), under our scheme the arm will need no beforehand information about the obstacles in order to move properly. The C-space presentation is used primarily for the analysis, the algorithm development, and the proof of convergence. No explicit mapping of any kind from W-space to C-space and no explicit calculation of C-space will ever take place before or during the actual arm motion.

5.2 PLANAR REVOLUTE–REVOLUTE (RR) ARM

Let us reiterate, with a bit more specifics of the RR-arm, the arm's model given in Section 5.1.1. The arm consists of two links, l_1 and l_2, and two revolute joints, J_0 and J_1 (Figure 5.2). Joint J_0 is fixed. Strictly for better visualization, links will be drawn as line segments. (As mentioned above, the shape of the arm links, or the fact of their being smooth or convex or concave, will be of no importance to the planning algorithm.) Link $l_i, i = 1, 2$, is hence a straight-line segment of length l_i. It can rotate indefinitely about the corresponding joint producing an angle θ_i, called the *joint value*. If W-space (workspace) is free of obstacles, the arm endpoint b can reach any point within the W-space boundaries.

The arm's W-space is formed by a circle of radius $(l_1 + l_2)$ (the outer circle, Figure 5.2) and by a circular "dead zone" (the inner circle, Figure 5.2) of radius $|l_1 - l_2|$. The middle circle in Figure 5.2 is a locus of points reachable by joint J_1. For a given position P of the arm endpoint in W-space, the corresponding pair of values (θ_1^p, θ_2^p), or the set of Cartesian coordinates of the link endpoints a_p and b_p, represent an *arm solution (arm position)* for P. It is easy to see that, in general, any position of the arm endpoint in W-space, except for points along the W-space boundaries, corresponds to two arm solutions.

An obstacle in W-space is a closed curve of finite length homeomorphic to a circle; that is, it cannot have self-intersections or double points. This also means

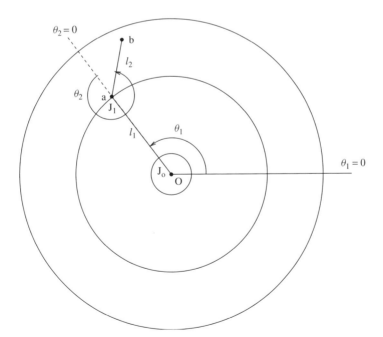

Figure 5.2   Revolute–revolute (RR) arm. l_0 and l_1 are joints; θ_1 and θ_2 are joint values; b is the arm endpoint.

that for all practical purposes an obstacle can be of arbitrary shape. There can be only a finite number of obstacles present in W-space. Any disk or a straight line passing through W-space intersects a finite set of obstacles. Being rigid bodies, obstacles cannot intersect. Two or more obstacles may touch each other, in which case the arm will treat them as one obstacle. Only such configurations of sets of obstacles are considered for which, at any position of the arm, at least some arm motion is possible. Only continuous motion of robot links is allowed.

At any given moment, the arm knows its current coordinates θ_1 and θ_2, as well as coordinates of the target position T. The starting position S is known to be reachable; that is, when the arm is in the position S, no arm links intersect any obstacles. It is not known whether position T is reachable and, if so, whether T can be reached from S. The arm is said to be moving in *free space* when it has no contacts with obstacles. Repeating the description given in the section above, the arm is assumed to be capable of the following actions:

1. Moving the arm endpoint through a prescribed simple curve (called *main line* or *M-line*) that connects points S and T.
2. Identifying the point(s) of contact on the arm body when the arm hits an obstacle.
3. Following the obstacle boundary.

The first operation implies that the arm is capable of computing coordinates of consecutive points along the M-line and transforming them into the corresponding pairs (θ_1, θ_2).

The purpose of the second operation is to provide information needed to pass around an obstacle. This is done with the help of the arm's tactile sensing. When at least one point of an arm link is in contact with an obstacle, relative coordinates of the point(s) of contact can be identified in the link reference system. Note that the identification is a local operation that does not require any additional information about the environment. Assume, for example, that the arm is covered with a "skin" with densely spaced tactile sensors, so that when a sensor contacts an obstacle, the point of contact on the arm body is known.

For the third operation, imagine that, while in contact with an obstacle, the arm follows the obstacle boundary as if some weak force pushes it against the obstacle. Therefore, at any moment during such motion, there is a variable point or points of contact between the obstacle boundary and the arm body.[1]

If the arm endpoint follows the obstacle up to the W-space boundary—for example, points on the outer circle in Figure 5.2—it is not clear whether the arm is still in contact with the obstacle on the boundary. To avoid an ambiguity, assume that no point of the W-space boundary can be a point of contact between an obstacle and the arm.

---

[1]A similar ability is considered in works on *compliance control* of robot wrists (see, e.g., Ref. 104).

5.2.1 Analysis

Here we will expand to our RR arm manipulator the theory developed in Section 3.3 for mobile robots. One important part of that theory is making use of distinctive topology of obstacles—namely, the fact that any obstacle is a simple closed curve. Exploiting this fact resulted in elegant motion planning algorithms with guaranteed convergence. We now intend to establish a similar characteristic of obstacles faced by our RR arm—namely, that the arm's complete passing around an obstacle presents some sort of simple closed curves.

As we will soon observe, this is not so in the arm workspace. Simple examples will show that paths produced by the arm endpoint when moving around even simple obstacles are complex and self-intersecting. We will also see, however, that the said property holds for all virtual obstacles in C-space. It will further be shown that the number of such closed curves per obstacle is limited—a fact that is important for the algorithm completeness. These facts will become the basis of the algorithm design. We will then study the nonuniqueness of choices for the M-line caused by peculiarities of the arm kinematics, and establish a criterion for choosing appropriate M-lines. Finally, we will address one side effect of the developed motion planning procedure, which can sometimes cause the arm to repeat parts of its path.

Obstacles in W-Space. Consider an example of the arm interaction with obstacles in the arm workspace (W-space). The formal underpinnings of our observations will become clearer in the subsequent analysis of C-space.

We begin with a simple circular obstacle A in the arm's workspace (Figure 5.3). Starting at position S, the arm moves its endpoint along the M-line (S, T) toward the target position T. In this example the M-line happens to be a straight line. Denote by (a_i, b_i) the ends of link l_2, where point b_i is the arm endpoint. After traveling for a while in free space, at some moment the arm will contact obstacle A, at which time the link l_2 position is (a_2, b_2), and the point of contact on A is b'. Now the arm will attempt to pass around the obstacle in order to continue its motion along the M-line.

Observe that here the arm has two options for maneuvering around the obstacle while maintaining a contact with it. With option 1, starting at the link l_2 position (a_2, b_2), the arm endpoint moves along the curve $b_2, b_3, \ldots, b_6, b_7, b_8$. Soon thereafter (between points b_8 and b_9), the arm endpoint encounters the M-line (S, T) and can continue moving along it toward T. When at T, the position of link l_2 is (a'_T, T).

With option 2, starting again at point b_2, the arm endpoint passes through the curve $b_2, b_{14}, b_{13}, b_{12}$. At point b_{12} the arm endpoint will encounter the M-line and then continue along it toward T. When at T, the position of link l_2 will be (a''_T, T).

In other words, depending on the option taken, the arm endpoint may encounter the M-line at different points, and the arm may consequently arrive at point T with different positions of its links. Notice that we can accommodate this discrepancy: For example, when moving under option 1, after passing point b_8 and reaching

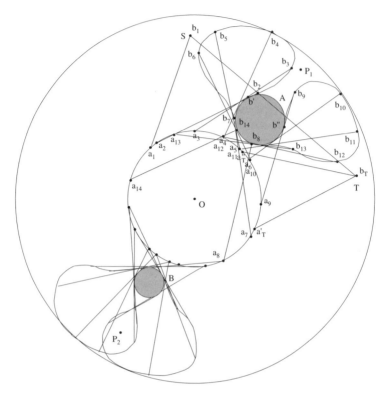

Figure 5.3 Obstacles A and B form "shadows"; the arm endpoint cannot reach points inside a shadow. For example, point P_1 is in the shadow of the circular obstacle A and thus cannot be reached. The shadow of the circular obstacle B forms two disconnected "subshadows."

the M-line, the arm endpoint can continue through points b_9, b_{10}, b_{11}, meeting the M-line again at point b_{12} as under option 2.

In fact, when starting at point b_2 under any of the two options, if one continues "rotating" the arm around obstacle A while keeping in contact with it, the arm endpoint will make a complete closed curve, passing through the points $b_2, b_3, b_4 \ldots, b_8, b_9 \ldots, b_{13}, b_{14}$ and eventually arriving at the same point b_2. This indicates that the paths produced under both options are complementary to each other: When added together, they form a closed curve.

Regarding this curve, consider the area whose curvilinear boundary passes through points b', b_2, b_3, b_4, then the segment b_4, b_{10} of the workspace boundary, then points b_{10}, b_9, b'' of our curve, and finally the smaller part of the obstacle A boundary between points b'' and b'. This area is called the *shadow* of obstacle A: Though this is a part of free space, no point (such as P_1) inside this area can be reached by the arm endpoint.

This suggests that an obstacle shadow will be perceived by the arm as an obstacle, as real as an actual physical obstacle. The arm cannot penetrate either

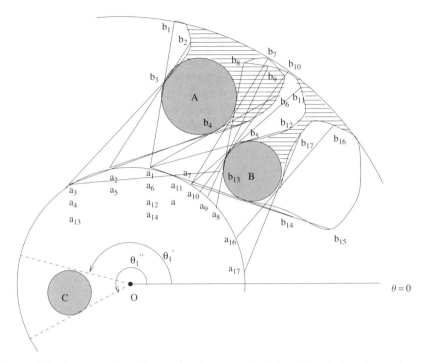

Figure 5.4 An example of interaction between obstacles. The shadow (shaded area) behind obstacle A is the result of interaction between obstacles A and B. If the arm moves through the positions (a_1, b_1), (a_2, b_2), ..., (a_{17}, b_{17}), at any moment it is in contact with either obstacle A or B. This means that the arm will perceive these two obstacles as one obstacle. Because of obstacle C, link l_1 cannot realize any angle values θ_1 in the range $\theta_1' < \theta_1 < \theta_1''$.

of them. The shape of a shadow depends on the shape, size, and position in W-space of the corresponding actual obstacle that creates the shadow, as well as on the arm links' shapes and dimensions. An obstacle can form disconnected shadows, as in the case of obstacle B (Figure 5.3). Or, obstacles can interact in forming shadows; this happens, for example, when two or more points of the arm body touch two or more actual obstacles simultaneously, as at position (a_8, b_8) in Figure 5.4.

Definition 5.2.1. *A virtual obstacle X is an area (or areas) in W-space, no points of which can be reached by the arm endpoint because of the arm's possible interference with the actual obstacle X.*

Thus a virtual obstacle consists of the corresponding actual obstacles and their shadows. In W-space a virtual obstacle forms one or more compact areas (see Figure 5.4). Whereas topologically this combination presents little of interest in W-space, we will see below that it possesses interesting properties in the arm's C-space.

Definition 5.2.2. *Passing around an obstacle presents a continuous motion of the arm, during which the arm is constantly in contact with the corresponding physical obstacle(s).*

It is clear from Figure 5.4 that two or more actual obstacles may be interpreted by the arm as a single virtual obstacle. In Figure 5.4, at any position from the set (a_1, b_1), (a_2, b_2), ..., (a_{17}, b_{17}) the arm is in contact with at least one of the actual obstacles A and B. Hence the two obstacles will be interpreted as one.

Definition 5.2.3. *A virtual line is a curve in W-space that the arm endpoint follows when passing around an obstacle. The virtual line forms the boundary of a virtual obstacle in W-space.*

A virtual line is not necessarily a smooth curve. For example, if the arm endpoint follows a sharp corner on an obstacle, or if the arm contacts some obstacle while passing around another obstacle [as in the link position (a_8, b_8), Figure 5.4], the virtual line may form sharp turns. Nor is a virtual line necessarily a non-self-intersecting curve (see virtual boundary of obstacle B, Figure 5.3), differing in this respect from the boundaries of physical two-dimensional objects. We will discuss this issue later, when analyzing the arm C-space properties.

Points of contact on the arm may undergo a discontinuous jump when passing around obstacles. This can happen because of the shapes of obstacles and arm links involved, or because of the arm–obstacle interaction. In Figure 5.4, for example, during link l_2 motion through positions (a_1, b_1), (a_2, b_2), and so on, an instant before position (a_8, b_8) link l_2 is in contact with obstacle A; an instant after position (a_8, b_8) the link is in contact with obstacle B. Accordingly, in this short period the contact point on the arm jumps from a point of contact on one side of link l_2 to a completely different point on the link's other side.

Note, however, that even in such cases there will be no discontinuity in the virtual curve.[2] For example, in the area of point b_8, which corresponds to the jump of the contact point mentioned above (Figure 5.4), the virtual line remains continuous. There will be more on the virtual line continuity in our analysis of the arm C-space.

Observe also that some distinct pieces of the virtual line may be associated with the same physical curve. Such is, for example, a part of the virtual line (b_{14}, b_8) (Figure 5.3), which is a part of obstacle A boundary. When trying to do a complete "rotation" by the arm around A, the arm endpoint will follow the curve segment (b_{14}, b_8) twice, once in each of the two directions.

The requirement of continuous contact while passing around the obstacle is equivalent to adding a constraint on the arm motion. In general, the arm's position relative to obstacles is described by one of these three situations:

1. No contact with obstacles takes place; the motion is unconstrained, and all points in the vicinity of the arm endpoint are available for its next position.

---

[2]Given the physics of the underlying phenomenon, this is not surprising: Physical motion is continuous, so the arm endpoint must be moving through a continuous curve.

2. One degree of freedom of the system (not necessarily one arm link) is constrained by an obstacle boundary; then only points along the virtual line—that is, a one-dimensional curve—are available for the next positions of the arm endpoint.

3. Two degrees of freedom of the system are constrained: No motion is possible.

Because of our model's assumption that some motion is always possible, case 3 is impossible. Case 2 thus includes all cases of interaction between the arm and obstacles.

Obstacles in C-Space.   *Configuration space (C-space)* of our RR arm manipulator is presented as the surface of a common two-dimensional torus defined by two independent angular variables, θ_1 and θ_2 [57]. Values of these variables are the arm joint values, respectively. An arm position P with coordinates (joint values) θ_1^P and θ_2^P in W-space corresponds to a point P with the same coordinates on the surface of the C-space torus. Continuity is preserved in this mapping: A small change in the position of arm links in W-space translates into a small displacement of the corresponding image point in C-space. A closed curve in W-space has its closed curve counterpart in C-space [105]. For an M-line in W-space, there is an M-line image in C-space (Figure 5.5).

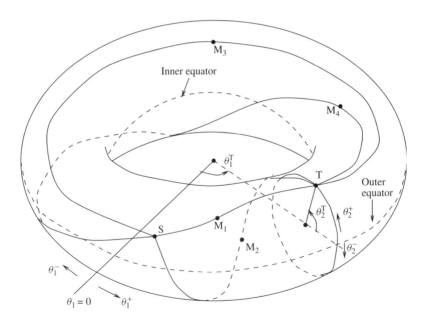

Figure 5.5 C-space torus. Zeroes and positive and negative directions for both angles θ_1 and θ_2 are shown. For a given θ_1, the point $\theta_2 = 0$ lies at the corresponding point of the torus's outer equator. For example, coordinates of point T are (θ_1^T, θ_2^T). Points M_1, M_2, M_3, and M_4 are the middle points of four M-lines, the four "straight line" routes between points S and T.

A geodesic curve connecting points S and T on the surface of the C-space torus corresponds to a straight line in the plane of variables (θ_1, θ_2). This geodesic curve can therefore be used as the "shortest" M-line between positions S and T. Because of the torus topology, in general, four such "shortest" M-lines can appear. Shown in Figure 5.5 are these four M-lines, their middle points, and positive and negative directions and zero points for both variables (θ_1, θ_2). With appropriate positioning of points S and T on the torus, all four M-lines can be made indeed equal. Otherwise, each M-line presents the "shortest" curve for a given set of directions of change of variables (θ_1, θ_2).

Since in general every position of the arm endpoint corresponds to two positions of the arm, defining uniquely the image of a virtual line in C-space will require some additional information about the corresponding arm positions.

Definition 5.2.4. *A virtual boundary is a curve in C-space that represents the image of the corresponding virtual line.*

Clearly, the virtual boundary corresponds to one out of two sets of arm positions tied to the virtual line. Where is the other set? The other set is physically unrealizable: In each such position the arm links would cross through the corresponding obstacle.

The virtual boundary separates an area of C-space occupied by the virtual obstacle from the rest of C-space. A finite number of actual obstacles in W-space produce a finite number of virtual obstacles in C-space. Each intersection of the M-line with the virtual line in W-space has its counterpart intersection of the M-line image with the virtual boundary. Unlike virtual lines, virtual boundaries cannot form self-intersections or double points. This means that at any point during the motion along the virtual boundary in C-space, there is one and only one possible direction for continuing the motion. Therefore, the following statement holds.

Lemma 5.2.1. *A virtual boundary can consist of only simple curves.*

To define the virtual boundary corresponding to the virtual line of obstacle A in Figure 5.3, points a_1 to a_{14} have to be added, coordinates of the endpoint of link l_1; the respective positions (a_i, b_i) of link l_2 are shown in the figure. Note that the coinciding points on the virtual line correspond to different positions of link l_2. That is, in C-space all points of the virtual boundary are distinct. The same is true for obstacle B.

Theorem 5.2.1. *A virtual boundary can consist of only simple closed curves. (See the proof in the Appendix to this chapter.)*

This statement will be pivotal in the design of the motion planning algorithm for an RR arm. Formally, the statement means that no matter what direction is chosen for following the virtual boundary, eventually the whole curve will be explored, and the arm will return to the position where it started. The theorem does not tell us how many simple closed curves may constitute a given virtual boundary. Can a virtual boundary consist of two, three, or an infinite number of simple closed curves? We will need to address this question, because it is very important from the algorithmic standpoint.

Following a simple virtual boundary is as advantageous as it was for the Bug family algorithms in Section 3.3 to follow the simple closed curves of obstacles, and for the same reason: When the arm meets an obstacle, one of only two possible directions for passing around it will have to be chosen. Since no information about the obstacles is available, neither of the two directions is preferable to another.

Definition 5.2.5. *A local direction is a predefined direction for passing around a virtual obstacle in C-space; it can be either right or left.*

Here "right" and "left" are defined in the same natural way as we did with the Bug algorithms. Looking at the scene from above, going left means going along the curve clockwise, that is having the obstacle to one's right, and going right means going along the curve counterclockwise—that is, having the obstacle to one's left.

The motion planning algorithm (which is still to be formulated) will proceed as follows. The arm's endpoint starts moving along the M-line from its starting position S toward the target position T. (In C-space the arm and both positions S and T are points.) When during this motion the arm encounters a virtual boundary—which means the arm contacts an obstacle—it defines on it a *hit point* H. The arm then starts passing around the obstacle using the chosen local direction. Since in doing so it follows a simple closed curve, it will eventually either reach point T, or return to the hit point H, or meet the M-line again. In the latter case, if the distance, as measured along the M-line, between the point where the arm meets the M-line and T is shorter than that between the hit point H and T, the arm defines this point as a *leave point L*. In Figure 5.3 the arm position (a_2, b_2) is the hit point H_1, and position (a_{12}, b_{12}) is the leave point L_1. (As mentioned above, depending on the chosen local direction and the way of passing around the obstacle, some other point might be defined as the leave point; this option will be discussed further later.)

We now turn to the question of the maximum number of simple closed curves that may form a virtual obstacle. Unlike some special two-dimensional *nonorientable* surfaces, such as the *Moebius strip* and *Kline bottle*, the surface of the common torus is topologically an *orientable* surface [57]. By continuously moving on one side of an orientable surface, a point robot will never find itself on the other side of the surface (which can happen on nonorientable surfaces). This fact follows from the Jordan Curve Theorem [57, 105], according to which any closed curve homeomorphic to a circle drawn around and in the vicinity of a given point on an orientable surface divides the surface into two separate domains, for which the curve is their common boundary [57].

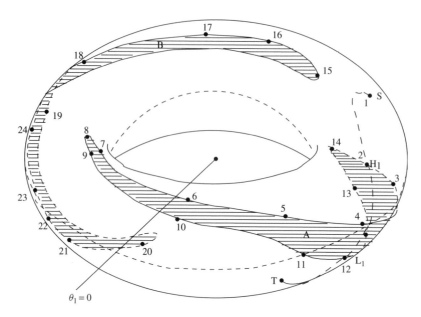

Figure 5.6   The C-space representation of obstacles A and B of Figure 5.3. Unlike in W-space, the boundary of each obstacle is a single closed curve—even for obstacle B, which forms two disconnected "subshadows" in Figure 5.3. The curve (S, T) is the image of the straight line (S, T) of Figure 5.3. Line (S, T) intersects the virtual boundary A in two points, the hit point H_1 (it coincides with point 2) and the leave point L_1 (it coincides with point 12).

This suggests that at least some virtual obstacles can be formed by a single closed curve. Obstacles A and B in Figure 5.3 are examples of obstacles (in W-space) whose images (virtual boundaries) in C-space are single closed curves. Those images are shown in Figure 5.6. Note also that although the virtual obstacle B includes two separate subshadows in W-space, in C-space B becomes one area separated from the rest of the torus by a single closed curve. For our algorithm we need to know if these examples exhaust all possible cases, and if not, what other options are there.

A very different example, of an obstacle virtual boundary formed by two closed curves, is shown in Figures 5.7. To understand the example, the reader may find it helpful to try to follow the arm motion as it passes around the obstacle A. In W-space (Figure 5.7a), the arm starts at the position (a_1, b_1), and the arm endpoint goes through a closed curve defined by points $b_1, b_2, \ldots, b_{10}, b_1$. The part b_4 to b_8 of the curve is an arc of a circle of radius l_2 centered at a point defined by indices a_4 to a_8. Starting from position (a_{11}, b_{11}) would result in a symmetric but different closed curve (in order not to complicate the picture, it is not shown in Figure 5.7a). The image of the corresponding virtual obstacle is shown in Figure 5.7b. As one can see, the virtual boundary forms an annulus, a band-like formation, on the torus.

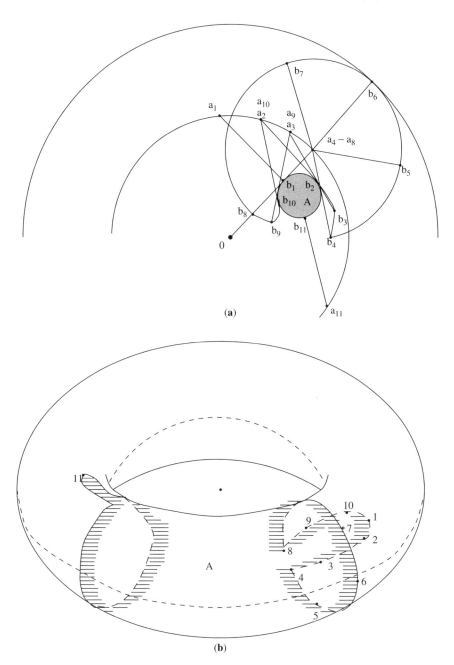

(a)

(b)

Figure 5.7   (a) As the arm passes around the obstacle A, starting with link l_2 position (a_1, b_1), the arm endpoint goes through a closed curve indicated in W-space by points b_1 to b_{10}. With the starting position (a_{11}, b_{11}), a similar but distinct closed curve would be formed (not shown here). **(b)** In C-space these two closed curves form the boundary of the band-like virtual obstacle (partially shaded). No point inside this area can be reached by the arm endpoint.

Recall that an ability to explore the whole obstacle boundary is an important function exploited in the Bug family algorithms (Section 3.3). The robot may rarely use it, but it should be there: Bug algorithms need it for assuring convergence and for the target reachability test. We intend to bring this same mechanism into the process of motion planning for arm manipulators. The example in Figure 5.7, where two simple closed curves form the virtual obstacle, raises a question: How many more simple closed curves can a virtual obstacle have? Unless the robot knows this, it will not know whether it explored the whole obstacle or there is still something unexplored. And, if the robot does know that number, how would it know if it has explored the whole obstacle if that were its goal? The maximum number of simple closed curves in a virtual boundary is given by the following lemma.

Lemma 5.2.2. *For the RR arm, a virtual boundary of an obstacle can be formed by no more than two closed curves. (See the proof in the Appendix to this chapter.)*

This is a good news.[3] One conclusion from Lemma 5.2.2 is that if the arm endpoint completes a full circle on its way around an obstacle, this does not necessarily mean that the whole virtual boundary has been traversed. There may be another, yet unobserved, closed curve which limits the virtual obstacle "from the other side" of the torus. On the other hand, if the robot explored both closed curves of a virtual boundary, this definitely means the robot has explored the whole obstacle. We classify obstacles into two types according to topology of their virtual boundaries.

Definition 5.2.6. *The virtual boundary of an obstacle of Type I is formed by a single closed curve. The virtual boundary of an obstacle of Type II is formed by two closed curves. No obstacle can be of both types. Type I and Type II are complementary and together cover all possible virtual obstacles.*

For the path planning algorithm, it would be important to know whether a closed curve traversed by the arm thus far belongs to a Type I or a Type II obstacle. If such inference is possible, it would allow us to produce a test that the algorithm can use to plan further robot motion. Namely, if the curve traversed thus far belongs to an obstacle of Type I, the robot would know that it has explored that obstacle completely. And, if the curve traversed thus far belongs to an obstacle of Type II, the robot would know that somewhere out there is still another unexplored closed curve of the same virtual boundary. The following discussion helps produce such a test.

A C-space image of an obstacle is an area on the surface of the C-space torus separated from the rest of the torus by the obstacle virtual boundary. Taking into account Lemma 5.2.2 and allowing for any continuous deformations of obstacle

---

[3]In principle, there are more complex arms with rather unusual kinematics that have more than two simple closed curves per virtual boundary. They are not used in practice and are not discussed in this text.

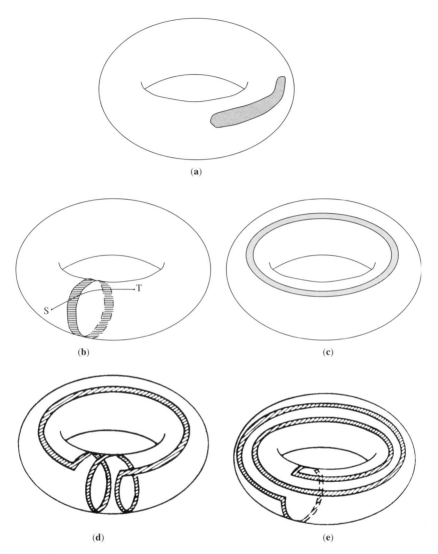

Figure 5.8 These five cases exhaust all possible ways to separate an area on the torus from the rest of its surface. Let C_i be the integral of the angle θ_i, $i = 1, 2$, taken along an obstacle boundary closed curve. Then: **(a)** $C_i = 0$, $C_{3-i} = 0$; **(b)** and **(c)** $C_i = 0$, $C_{3-i} = 2\pi$; **(d)** and **(e)** $C_i = 2\pi$, $C_{3-i} = n \cdot 2\pi$, $n = 1, 2, \ldots$; $i = 1, 2$.

boundaries, all possible ways to separate an area on the torus from the rest of its surface can be reduced to five cases shown in Figure 5.8. The case in Figure 5.8a corresponds to a Type I obstacle; the four remaining cases correspond to Type II obstacles. The cases in Figure 5.8b and 5.8c are topologically equivalent; the cases in Figure 5.8d and 5.8e are equivalent as well. From the path planning standpoint, all five cases are distinct and are treated in the algorithm separately.

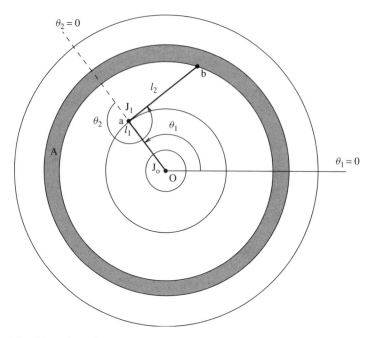

Figure 5.9   Obstacle A forms in C-space a Type II obstacle type shown in Figure 5.8c.

Figures 5.3 and 5.6 provide examples of the Type I case of Figure 5.8a. An example for the Type II case in Figure 5.8b is shown in Figure 5.7. The Type II case shown in Figure 5.8c appears, for example, when an obstacle in W-space presents a ring whose center is in the system origin and whose smaller radius is larger than l_1 (see obstacle A, Figure 5.9). One might say the example is not excessively realistic. This is true, except that with a bit more work one can come up with a rather realistic example that would still demonstrate the same phenomenon. An example for the cases in Figures 5.8d and 5.8e appears in Figure 5.10.

As these examples show, all five cases of Figure 5.8 are physically realizable, and therefore they should be accounted for in the algorithm. Consider two counters, C_1 and C_2, corresponding to the angles θ_1 and θ_2 of the arm joints, respectively. When the arm travels in free space, the content of each counter is zero. Once the arm hits an obstacle, both counters are turned on. While the arm follows a closed curve of a virtual boundary, each counter integrates the corresponding angle, taking into account the sign. As the arm completes a closed curve, the contents of each counter must be $n \cdot 2\pi$, $|n| = 0, 1, 2, \ldots$.

For a closed curve of some obstacle, the resulting values of the pair (C_1, C_2) define its *arm joints range* (or, simply, *range*). An obstacle of Type I is defined by the range of its single closed curve. For a Type I obstacle, its range is hence $(0, 0)$. For a Type II obstacle, since a closed curve of a given range cannot be reduced by topological deformation to a curve of a different range, both

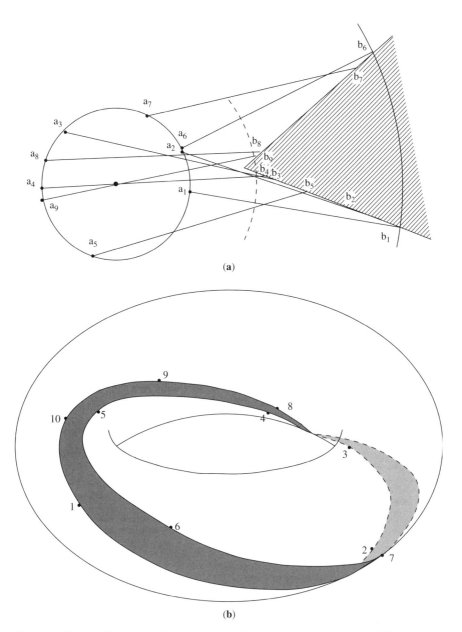

Figure 5.10 (a) W-space and (b) C-space with an obstacle whose virtual boundary in C-space is formed by two nonintersecting closed curves. The curves' ranges are $C_1 = 2\pi, C_2 = 2\pi$, making this another example of Type II cases shown in Figures 5.8d and 5.8e. Each point $P_i, i = 1, \ldots, 10$, on the obstacle virtual boundary in C-space is defined by the corresponding pair of points (a_i, b_i) in W-space.

curves of a Type II obstacle have the same range (see Figures 5.8b to 5.8e). For a given obstacle, therefore, the term "arm joints range" is equivalent to the term *obstacle range*. For a Type II obstacle, its range is $(n_1 \cdot 2\pi, n_2 \cdot 2\pi)$, with either $n_i = 0$, $|n_{3-i}| = 1$ (Figures 5.8b and 5.8c) or $|n_i| = 1$, $|n_{3-i}| = 1, 2, \ldots$ (Figures 5.8d and 5.8e); $i = 1, 2$. Numbers n_i uniquely represent the range of an obstacle.

After the arm has passed around an obstacle, the range accumulated in counters C_1, C_2 is, therefore, an indicator of the obstacle's type. If, after completing a closed curve, the contents of the counters is $(0, 0)$, then the obstacle in question is of Type I and thus the whole virtual boundary has already been traversed. If, on the other hand, the range of the closed curve is different from $(0, 0)$, then the obstacle is of Type II and hence there must be another closed curve somewhere that corresponds to the same obstacle.

This observation can be used in the reachability test that we will need in the complete motion planning algorithm. The mechanism of the test is very similar to that in the Bug2 algorithm (Section 3.3.2). Namely, in the case of a Type I obstacle, if, after having defined a leave point, the arm returns back to it without ever meeting the M-line and without having defined the next hit point, this indicates that the target position cannot be reached. This may happen, for example, if the obstacle forms a ring around the target position. A similar idea works for the Type II obstacles: If the arm traverses both closed curves of a Type II virtual obstacle without ever meeting the M-line and without defining the next hit point, the target is not reachable. These mechanism will be used in the algorithm's reachability test.

We emphasize that in no way does the above discussion imply that traversing closed curves of obstacle boundaries is a necessary part of the motion planning algorithm. More often than not, the robot will be passing only parts of obstacles that it encounters on its way to the target position. The test above is needed for algorithm completeness: If once in a while the analysis above is needed, the algorithm will provide the mechanism for it.

M-Line and Path Planning. How do we choose the M-line? If the arm encounters no obstacles on its way, the arm endpoint will proceed directly from point S to point T along the M-line. It is desirable, then, to have a simple and shorter M-line, such that one could easily determine whether a given point does or does not lie on it. In the case of planning algorithms for mobile robots (Section 3.3), we chose a straight line (S, T) for the M-line.

An M-line can be defined for the RR arm in a number of ways. One option is a straight line (S, T) in W-space (see Figures 5.3 and 5.6). Another reasonable choice for M-line would be a "uniform descent" curve in W-space, described in polar coordinates as $R = p \cdot \varphi + q$, where R is the distance from the arm endpoint to the fixed base O of the arm (Figure 5.2), and φ is the angular position of the arm endpoint relative to some zero axis passing through the base O. This is a straight line in the plane of variables (R, φ), $0 \leq \varphi \leq 2\pi$. Knowing link positions for points S and T, one can find the corresponding values R and φ (or the complement of φ to 2π) and then calculate p and q.

One disadvantage of these M-lines is that with them the motion can be non-monotonic in terms of the joint angles θ_1 and θ_2 (see Figure 5.3). Worse yet, depending on the arm positions S and T, continuous motion along such an M-line may be impossible even if the scene contains no obstacles. This happens, for example, if the M-line is a straight line passing through the arm's dead zone (Figure 5.2).

Figuring out beforehand if the nonmonotonic change in joint values or the M-line passing through the workspace dead zone is a problem in the case at hand is an extra difficulty, which can be avoided with still another choice for the M-line, the one that it preferred in this text. We choose the M-line that is a straight line in the plane of variables θ_1 and θ_2—that is, a straight line between points S and T in C-space. When no obstacles are present, this M-line results in an economical and uniform change of the arm joints from the starting to the final position. That is, motion along this M-line is monotonic in θ_1 and θ_2. On the C-space torus the model of this M-line is a geodesic curve—a tight thread connecting points S and T. In Figure 5.5 this "shortest route," M_1, corresponds to the positive change of both angles θ_1 and θ_2.

There are, however, three other ways to obtain a geodesic curve between points S and T on the torus surface. All four are obtained by using both positive and negative directions of change for each of the angles θ_1 and θ_2. These are shown in Figure 5.5, as well as on the flattened torus in Figure 5.11. We will see that from time to time the algorithm may use a combination of these routes. One of the four routes corresponds to the global minimum, and the other three—called *complementary* routes—correspond to the local minima of the distance between S and T in the plane (θ_1, θ_2), (Figure 5.5). For some special locations of points S and T, two or more of the four M-lines may become equal. For example, with points S and T located at the opposite points of the torus's outer equator, all four M-lines are of equal length.

Each M-line is characterized by its M_k-*line segment* between points S and T, $k = 1, 2, 3, 4$. (For better visibility, in Figure 5.5 M_k also indicates the middle point of the corresponding segment.) An M-line that represents a straight line in the plane θ_1, θ_2 is given by the equation

$$\theta_2 = p \cdot \theta_1 + q \tag{5.1}$$

To compute coefficients p and q for all M_k-lines—we may need this for the motion planning algorithm—"flatten" the torus by cutting it along two mutually perpendicular circles passing through point T, one of which is parallel to the large equator of the torus; this is shown in Figure 5.11. This operation produces a rectangle, with point S lying somewhere inside of it. Point T is *identified*; that is, it produces four points T_k, $k = 1, 2, 3, 4$, each sitting in the rectangle's corners and corresponding to one of the M_k-lines. Denote $\delta_i = \text{sign}(\theta_i^T - \theta_i^S)$, $i = 1, 2$,

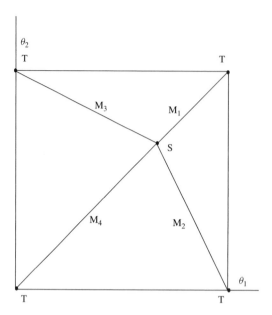

Figure 5.11 Flattened C-space torus of Figure 5.5.

where sign() takes values $+1$ or -1 depending on the sign of the argument. Then, coordinates of the endpoints S and T_k of each of the M_k-lines are as follows:

$$
\begin{aligned}
M_1 &: \ S = (\theta_1^S, \theta_2^S), \ \ T_1 = (\theta_1^T, \theta_2^T) \\
M_2 &: \ S = (\theta_1^S, \theta_2^S), \ \ T_2 = (\theta_1^T, \theta_2^T - 2\pi \cdot \delta_2) \\
M_3 &: \ S = (\theta_1^S, \theta_2^S), \ \ T_3 = (\theta_1^T - 2\pi \cdot \delta_1, \theta_2^T) \\
M_4 &: \ S = (\theta_1^S, \theta_2^S), \ \ T_4 = (\theta_1^T - 2\pi \cdot \delta_1, \theta_2^T - 2\pi \cdot \delta_2)
\end{aligned}
\tag{5.2}
$$

Substituting these into (5.1), coefficients p and q for each of the segments are found:

$$
M_1 : p = \frac{\theta_2^T - \theta_2^S}{\theta_1^T - \theta_1^S}, \qquad\qquad q = \frac{\theta_1^T \cdot \theta_2^S - \theta_2^T \cdot \theta_1^S}{\theta_1^T - \theta_1^S}
$$

$$
M_2 : p = \frac{\theta_2^T - \theta_2^S - 2\pi \cdot \delta_2}{\theta_1^T - \theta_1^S}, \ q = \frac{\theta_1^T \cdot \theta_2^S - \theta_2^T \cdot \theta_1^S + 2\pi \cdot \delta_2 \cdot \theta_1^S}{\theta_1^T - \theta_1^S}
$$

$$
M_3 : p = \frac{\theta_2^T - \theta_2^S}{\theta_1^T - \theta_1^S - 2\pi \cdot \delta_1}, \ q = \frac{\theta_1^T \cdot \theta_2^S - \theta_2^T \cdot \theta_1^S - 2\pi \cdot \delta_1 \cdot \theta_2^S}{\theta_1^T - \theta_1^S - 2\pi \cdot \delta_1}
$$

$$
M_4 : p = \frac{\theta_2^T - \theta_2^S - 2\pi \cdot \delta_2}{\theta_1^T - \theta_1^S - 2\pi \cdot \delta_1}, \ q = \frac{\theta_1^T \cdot \theta_2^S - \theta_2^T \cdot \theta_1^S - 2\pi \cdot (\delta_1 \cdot \theta_2^S - \delta_2 \cdot \theta_1^S)}{\theta_1^T - \theta_1^S - 2\pi \cdot \delta_1}
$$

$$
\tag{5.3}
$$

For algorithmic purposes, it is important to (a) detect the fact that a given point in C-space lies on the currently used M_k-line and (b) compute its relative position on the segment M_k. We will use for this the parametric description of the M-line equation [97]. For example, for M_1-line, the symmetric presentation using coordinates of points S and T is,

$$\frac{\theta_1 - \theta_1^S}{\theta_1^T - \theta_1^S} = \frac{\theta_2 - \theta_2^S}{\theta_2^T - \theta_2^S} = T \tag{5.4}$$

from which the parametric equations for θ_1 and θ_2 are found, with t as a parameter:

$$\theta_1 = \theta_1^S \cdot (1 - t) + \theta_1^T \cdot t$$
$$\theta_2 = \theta_2^S \cdot (1 - t) + \theta_2^T \cdot T \tag{5.5}$$

A pair of angles (θ_1, θ_2) is recognized as a point on the M-line if both equations in (5.5) produce the same value of t. The relative position of a point on the M-line segment is determined as follows:

$t = 0$ corresponds to point S;

$t = 1$ corresponds to point T;

$0 < t < 1$ corresponds to points inside the segment (S, T);

$t < 0$ correspond to points outside the segment and closer to S than to T;

$t > 1$ corresponds to points outside the segment and closer to T than to S.

We emphasize that in the plane (θ_1, θ_2) where the straight line $\theta_2 = f(\theta_1)$ is defined, angles θ_i and $(\theta_i - 2\pi)$ are not equivalent. In determining whether a given point lies within the segment (S, T) of a given M_k-line, care should be taken in choosing the right line from (5.2) to represent the angles in (5.5).

We order four M-line segments according to (5.2), with M_1 being the shortest segment. Here the length of a segment is the Euclidean distance between points S and T in the plane (θ_1, θ_2), with coordinates presented as in (5.2). For each angle θ_i, $i = 1, 2$, there is an interval $\delta\theta_{i,k} = |\theta_{i,k}^T - \theta_{i,k}^S|$ or $\delta\theta_{i,k} = |\theta_{i,k}^T - \theta_{i,k}^S - 2\pi|$ related to the corresponding segment M_k; $\delta\theta_{i,k} \leq 2\pi$.

Definition 5.2.7. *For two M-line segments M_k and M_m, $k, m = 1, 2, 3, 4, k \neq m$, their complementarity shows in that, for one or both angles θ_i, intervals $\delta\theta_{i,k}$ and $\delta\theta_{i,m}$ add to 2π (see Figure 5.5). These two segments are said to be complementary over the angle θ_i.*

The following discussion helps clarify how complementary M-lines are used during the path planning. Assume that M_1 has been chosen as the M-line. Imagine that the arm, while following M-line, encounters an obstacle (i.e., defines on it a hit point), tries to pass it around, returns to the hit point without ever meeting

M-line, and, after having examined the contents of the counters C_1 and C_2, concludes that it is dealing with a Type II obstacle. Now the arm needs to explore the second closed curve of the virtual boundary. This second closed curve must lie somewhere in a direction "opposite" to that in which the first closed curve had been spotted. Such "opposite" directions are provided by the complementary M_k-lines (Figure 5.5). The notion of M_k-line complementarity over a certain angle θ_i provides a way of selecting M_k-lines. No more than two complementary M-lines will have to be tried to complete the task. The outcome of processing the first closed curve of the virtual boundary is one of these three cases:

1. The accumulated range is $(0, n_2)$; $|n_2| \geq 1$–an integer. This case is shown in Figure 5.8b: The arm starts at S, moves along the line (S, T) (which happens to correspond to the change of θ_1 in the positive direction), encounters an obstacle, goes around one of its closed curves, and concludes that the further motion along M_1-line in this direction is impossible. On the other hand, using the opposite (in this example, the negative) direction of change of θ_1 guarantees reaching the second closed curve of the virtual obstacle. How would the arm choose the right M-line for this stage? According to (5.2), such direction can be realized with either an M_3-line or an M_4-line (Figure 5.5). Because no additional information is available, for example, the shortest of these will be chosen. (There is, of course, no guarantee that the chosen M-line is better than the other possible choice.)

2. The accumulated range is $(n_1, 0)$; $|n_1| \geq 1$. This case could appear, for example, in Figure 5.8c. This situation is symmetric to the previous one except that it relates to angle θ_2. The opposite direction of change for θ_2 can be realized with either M_2-line or M_4-line (Figure 5.5). The shortest of these will be chosen.

3. The accumulated range is (n_1, n_2); $|n_i| \geq 1$ (Figures 5.8d and 5.8e). Again, further motion along M_1-line is impossible. Any one of the segments M_2, M_3, or M_4 can be used as a new M-line. The shortest of these will be chosen.

Where would the arm start for the exploration of the second closed curve? Since no information about the obstacle's second closed curve has been accumulated so far, one point to start with the new M-line is S. If neither of the two M-lines brings the arm to the target—that is, if point T has not been reached after exploration of the obstacles both closed curves—the target is not reachable. Using in the worst case two M-lines instead of one represents the topological distinction of motion planning on the torus as compared to the plane. With the addition of complementary M-lines, the convergence theorems developed in Section 3.3 for the Bug family algorithms can be used to ensure that our algorithm for the revolute–revolute arm terminates.

Local Cycles. As we learned in Section 3.3, a planar sensor-based motion planning procedure for a mobile robot can create local cycles in the path. Local cycles happen in scenes with obstacles of certain shapes and with certain mutual positions of obstacles and start and target points. The procedure that we are about to formulate for the RR arm motion planning can create local cycles as well. As for mobile robots, local cycles do not affect convergence of the RR arm planning algorithm, but they do result in longer paths. For the process to work, the algorithm does not need to recognize those local cycles or undertake any specific action when they occur.

A *local cycle* is created when the arm image point on the C-space torus comes back to a previously defined hit point, and the contents of its counters C_i at this moment is different from $n_i \cdot 2\pi$, $|n_i| = 0, 1, 2, \ldots, i = 1, 2$. This may occur only if between two consecutive encounters of the same hit point the arm has defined other hit point(s). (Recall that the counters C_i are turned on when a hit point is defined.) Otherwise the contents of C_i would be exactly $n_i \cdot 2\pi$.

An example with a local cycle is shown in Figures 5.12, 5.13, and 5.14. Using this example, we will first demonstrate the process of path generation by the algorithm, indicate the place where a local cycle is created, and then present the algorithm itself. (One may want to return to this example again when reading the algorithm procedure below.) In this example the workspace includes four

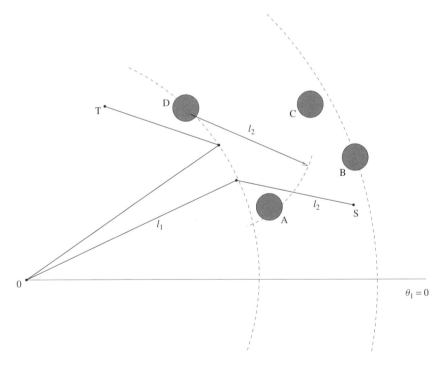

Figure 5.12 Four simple obstacles, A, B, C, and D, interfere with the arm's attempt to move from start S to target T. Note the simultaneous interaction of both arm links with obstacles: For example, when link l_1 touches obstacle D, obstacles A and C are on the way of link l_2.

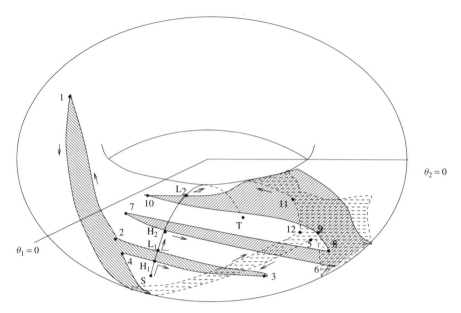

Figure 5.13   The virtual obstacle that corresponds to obstacles $A, B, C,$ and D of Figure 5.12 forms two separate closed curves in C-space. Arrows indicate the direction of motion. Note how simple disk obstacles in W-space form a rather intricate labyrinth in C-space.

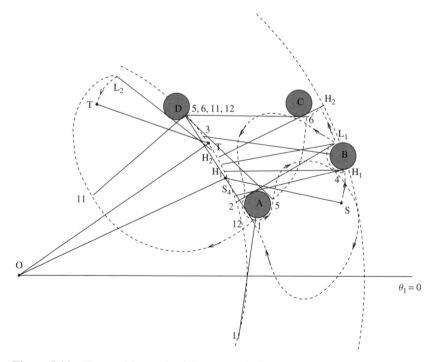

Figure 5.14   The resulting path of the arm endpoint in the problem of Figure 5.12.

simple circular obstacles, A, B, C, D (Figure 5.12). In spite of their simplicity, we will see that moving between these obstacles turns out to be tricky.

As discussed before, at times both links of the arm may interact with obstacles simultaneously, or one link may interact simultaneously with more than one obstacle. For instance, when link l_1 touches obstacle D (Figure 5.12), obstacles A and C may be on the way of link l_2. Therefore, link l_2 may be touching at least one of obstacles A or C, while at the same time link l_1 is touching obstacle D. This means that in C-space, obstacles A, C, and D form a single obstacle. Furthermore, note that if we position link l_2 between obstacles B and C and start moving link l_1 past obstacle A, at some point link l_2 will simultaneously touch obstacles B and C. In other words, in C-space all four obstacles create a complicated single obstacle, a kind of a labyrinth (Figure 5.13). As it is formed by two closed curves, this is a Type II obstacle. Also shown in Figure 5.13 is the M-line (S, T), chosen as a straight line in a "flatten" C-space.

Let us choose "right" as the local direction for the arm's passing around an obstacle. Although the sensing, the motion, and the actual algorithm procedure will be happening in W-space, to understand what happens under this procedure it is better to follow the generated path first in C-space. As we know already, the reason C-space is easier to follow is that the C-space image of the path has no self-intersections or double points, and obstacles are sets of simple closed curves.

When watching it in C-space (Figure 5.13), one will recognize in the arm's path (indicated by arrows) the "signature" of the Bug2 procedure described in Section 3.3.2. (We will see below that the resulting algorithm becomes more complex than Bug2 because it has to take into account the more complex topology of the torus compared to the plane.) Starting at S, the arm's image point moves along M-line toward T. On its way it encounters obstacle B and defines a hit point, H_1. Here the robot turns right (counterclockwise) relative to the obstacle boundary; after passing point 3, it meets M-line again and defines here the leave point L_1.

The next path segment is along M-line and is short, producing the next hit point, H_2. Turning right at H_2, the robot embarks on another path segment along the obstacle boundary, which takes it through points $H_2, 6, 5, 4$, bringing it back to point H_1. A local cycle has been created. (More detail on conditions under which local cycles are created can be found in Section 3.3.) Now the robot will pass point H_1 "on the fly," still continuing along the obstacle boundary. It is now looking for a candidate for a leave point that is closer to point T than point H_2 is to T. This path segment will take it again through points 3 and L_1 and then through points $2, 1, 12$, and 11 until it encounters M-line again and defines the leave point L_2. From then on, it directly proceeds along M-line to point T.

Note that along its way the robot will see only one of the two simple closed curves that form our virtual obstacle—and even this one only partially. The robot will never know that the other closed curve even exists. It will not know that it has dealt with a Type II obstacle. As we will see, this is not always so: Some seemingly similar cases can be more trying for the robot and may require a more elaborate exploration.

How does this all relate to the actual motion of the arm manipulator in W-space? The actual path in W-space whose reflection in C-space we just considered is shown in Figure 5.14. The path created by the arm endpoint on the way from S to T is shown in a dash line. The direction of motion is indicated by arrows; also shown are the intermediate positions of the arm that appear in Figure 5.13. It is a good exercise to try to follow the path in Figure 5.14, going through the same positions that we have followed in C-space: positions $S, H_1, 3, L_1, H_2, 6, 5, 4, H_1, 3, L_1, 2, 1, 12, 11, L_2, T$. Note that the local cycle—the path segment that includes positions $H_1, 3$, and L_1—has been passed twice.

As an exercise, try to see what kind of path would be created if the local direction happened to be "left."

To reiterate, the actual planning of the path by the motion planning algorithm is done completely in the workspace (W-space), based on the sensing data from the arm sensors. No preliminary exploration or computation of the C-space virtual obstacles takes place. If the target position cannot be reached because of the interfering obstacles, the reachability test will conclude so after some limited exploration.

5.2.2   Algorithm

With all the necessary details ready, the whole motion planning procedure for the planar arm of Figure 5.2 can now be formulated. The name for it would be the *RR-Arm Algorithm*. Looking ahead, for the reasons that will become clear later, the algorithm can also serve two-link arms with other kinematics; hence it also deserves a more general name, the *Two-Link-Arm Algorithm*. Some preliminaries:

- The hit points H_j and leave points L_j are numbered in the order of their being defined along the path; $L_o = S$.
- Any local direction, left or right, can be chosen for the arm's passing around an obstacle. For the sake of specificity, let us choose "right."
- If a new M-line is introduced, the arm starts again at point S, and the numbering starts over.
- Distance $d(P, Q)$ between the arm positions P and Q is the Euclidean distance in the plane of variables (θ_1, θ_2); the length of an M-line segment is defined similarly.
- In the case of a Type II obstacle, a flag is used to indicate that one of the two virtual boundary's closed curves has been already processed.
- The test for target reachability is explicitly built into the algorithm's Steps 5a and 5b. The test is based on the necessary and sufficient condition for target reachability described in Section 3.3.2.

The RR-Arm Algorithm procedure is as follows:

1. All four complementary M-lines, M_k, $k = 1, 2, 3, 4$, are defined according to the formulae (5.2) and (5.5) and are ordered as follows: M_1 is the shortest; M_2 complements M_1 over the angle θ_2 [as in (5.2)]; M_3 complements M_1 over θ_1; M_4 complements M_1 over both θ_1 and θ_2. Go to Step 2.
2. M_1-line is designated as M-line. Set the flag down. Set $j = 1$. Go to Step 3.
3. Counters C_1 and C_2 are set to zero. From point L_{j-1}, the arm moves along M-line until one of the following takes place:
   (a) Target T is reached. The procedure stops.
   (b) An obstacle is encountered and a hit point, H_j, is defined. Go to Step 4.
4. Counters C_1 and C_2 are turned on. The arm follows the obstacle boundary in the chosen local direction, until one of the following takes place:
   (a) Target T is reached. The procedure stops.
   (b) M-line is met at a distance d from T such that $d < d(H_j, T)$; point L_j is defined. Increment j. Go to Step 3.
   (c) The arm returns to position H_j (which completes a closed curve along the virtual boundary) without ever meeting the M-line. Go to Step 5.
5. Examine the obstacle range accumulated in counters C_1 and C_2. One of the following takes place:
   (a) The range is $(0, 0)$, which means it is a Type I obstacle. Target cannot be reached. The procedure stops.
   (b) The range is not $(0, 0)$ and the flag is up, which means it is the second closed curve of the virtual boundary of a Type II obstacle. The target cannot be reached. The procedure stops.

   The remaining three events relate to the case when the range is not $(0, 0)$ and the flag is down (which means that the current virtual boundary is the first closed curve of the virtual boundary of a Type II obstacle).
   (c) The range is $(0, n_2)$, $|n_2| \geq 1$; designate the shorter of M_3 and M_4 as the M-line. Go to Step 6.
   (d) The range is $(n_1, 0)$, $|n_1| \geq 1$; designate the shorter of M_2 and M_4 as the M-line. Go to Step 6.
   (e) The range is (n_1, n_2), $|n_1| \geq 1$, $|n_2| \geq 1$; designate the shortest of M_2, M_3, and M_4 as the M-line. Go to Step 6.
6. The arm moves back to Start S. Set the flag up. Set $j = 1$. Go to Step 3.

5.2.3 Step Planning

Physical realization of the above motion planning algorithm requires a small additional piece that addresses the following question: With the overall path being generated by the motion planning algorithm, how does one plan every step along that path? While this question is somewhat outside our main topic of sensor-based motion planning, one cannot avoid it; after all, a path is built from individual steps.

The robot control system is guided by a computer; that is, it is a discrete control system. Accordingly, motors that realize the motion of robot joints usually receive computer commands at some fixed *sampling rate*, say 20 or 50 times per second. For example, the rate of 50 commands per second translates into 20 ms per each motion cycle. Within each cycle the control computer(s) receives and processes the sensor data, calculates the next step in space based on guidance from the motion planning algorithm, translates the step into commands to individual robot joints, and sends those increments to the corresponding motors; the latter execute the step motion. Then the next 20-ms cycle starts, and so on. Special software makes sure that the robot passes each step "on the fly," without stopping, resulting in continuous motion.

Recall that the arm motion consists of either moving along an M-line or moving ("sliding") along obstacles. When moving along an M-line, the step calculation is easy: Each step is a linear function of arm coordinates on the previous step and increments $d\theta_i$ for joint values. The function's coefficients depend on the decided speed. When moving along an obstacle boundary, a step is computed along the line tangent to the corresponding C-obstacle at the point of local contact. Because this tangent has two possible directions, the direction chosen is the one that corresponds to the current local direction (see details above).

Whether at the current moment the arm is in contact with one or a few physical obstacles, in C-space this corresponds to one point of contact, and so the solution for a step is unique. We reiterate that the fact of computing each step based on the C-space representation does not mean that any beforehand C-space calculations take place. Each step is solely a function of entities in W-space, namely, the prior step θ_i coordinates and sensing data. More detail on the step calculation procedure can be found in Ref. 106.

5.2.4 Example

The workspace in this example contains four obstacles, A, B, C, D (Figure 5.15). Obstacles are convex and concave, with straight-line and curved boundaries. The dashed line (S, T) is the shortest M-line [as in (5.1) and (5.2)]. S and T are the start and target positions of the arm; also shown are some intermediate positions of link l_2 along the way: $(1, 1), (2, 2), \ldots, (12, 12)$. The accepted local direction is "right."

The arm proceeds as follows. Starting at point S, it moves toward point T along the M-line, until it senses obstacle A. The arm then starts passing around A (see, e.g., point 1) until it senses obstacle B; it proceeds along obstacle B until it meets the M-line again (point 2); it follows the M-line and very soon meets obstacle D; it moves along D until it meets obstacle C (e.g., point 3); while following C, it meets again obstacle D (point 4); it follows D (points 5, 6, 7, 8) until it meets the M-line again; and finally it proceeds along M-line to T.

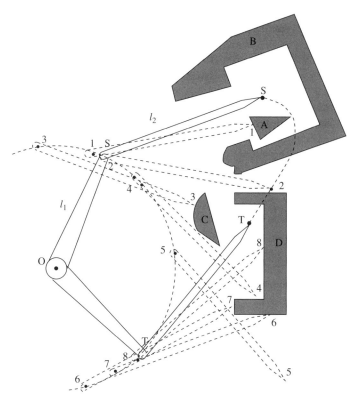

Figure 5.15   An example of the RR-Arm Algorithm performance in an emulated "industrial" work cell. The arm is required to move from the starting position (S, S) to the target position (T, T), along the M-line (S, T) (dashed line) that presents the shortest path between S and T in C-space. The arm workspace is crowded with four obstacles A, B, C, D. Shown also are some intermediate positions of link l_2, $(1, 1)$ to $(8, 8)$. No *a priori* information about the scene is given to the arm, and no preliminary calculations occur; the arm simply senses its surroundings and makes planning decisions "on the fly," in real time.

Note that although the M-line plays a crucial role in the RR-Arm Algorithm's workings, in this example the arm actually spends very little time moving along the M-line; that is, only a small part of the arm's path coincides with the M-line. Most of the time is spent passing around obstacles.

5.2.5 Motion Planning with Vision and Proximity Sensing

Similar to how in Section 3.6 VisBug algorithms were obtained from the Bug family algorithms, we would like to modify the RR-Arm Algorithm to make it compatible with proximal sensing. If that is feasible, a similar process will likely work for other arm manipulator configurations. The process is not as straight-forward as with the Bug algorithms, both on the sensing hardware level and on the algorithmic level.

Hardware. Let us consider briefly the sensing hardware requirements. (More will be said on this in Chapter 8, which is fully devoted to the issues of sensing hardware for arm manipulators.) Recall (see Section 3.6) that the primary requirement to the arm sensing is to functionally cover the whole body of the arm, so that no obstacle approaching any point of the arm body would go unde-tected. Tactile sensing, which was assumed for the arm in this section, satisfies this requirement. Less satisfactory, however, is tactile sensing's inability to han-dle system dynamics. Being made from hard unyielding materials, today's robot arm manipulators are quite heavy. Touching an object while moving creates sig-nificant instantaneous forces and accelerations, which can easily hurt the robot and the object that it touches. This is less so at the robot hand, where special measures can be taken to handle contact, but very much so at the rest of the robot body. Humans and animals are less vulnerable to such encounters because the soft muscle and fat tissues under their skin produce a softening "braking" effect. Today's robots do not have this option. The way out is proximity (distance) sensing (see Figure 5.16).

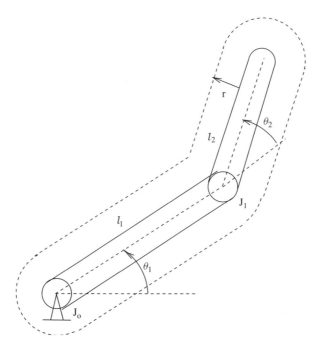

Figure 5.16 Proximity sensing of this arm manipulator forms a sensing cushion around the arm. The sensing range radius is r. An obstacle will be detected at any point of the robot body as long as the obstacle–robot distance is within r.

What kind of proximity sensing is acceptable? In Chapter 8 we will address this question in much detail. Briefly, of particular interest is vision sensing. As discussed in Section 1.2, vision is good when the scene in which the robot operates is much bigger than the robot itself. Mobile robots usually operate in this setting: An obstacle is detected way before it appears next to the robot. One way to approximate this situation formally is to say that obstacles are (almost) always detected when they are outside the robot's convex hull.

The situation changes if the size of the workspace is comparable to the robot dimensions, which is a standard case for an arm manipulator. Indeed, the arm's base is fixed, and the arm is expected to reach all areas of its work cell. Obstacles are almost always within the robot's convex hull and can appear next to any point of the robot body. In this situation the advantages of vision sensing for motion planning are diminished. Obstacles can appear from behind or from the sides, at any link of the arm. Depending on where video cameras are attached, an approaching obstacle may be occluded by cables that deliver power or materials to the arm hand, or by one of the arm's own links, or by some other piece of machinery in the work cell.

Trying to fight this problem by covering the arm (or the work cell walls) with many "eyes" will make the system awkward and hardly feasible. Another option, decreasing the number of cameras by putting them in a few strategic locations and then using auxiliary arm motion to disambiguate invisible obstacles, is even more awkward and creates other difficulties. The conclusion is that vision is useful in a limited role, such as protecting the arm's gripper when manipulating objects. To protect the whole arm body at short distances, sensing media that provides a physical coverage of the body will likely serve the task better than vision.

Various proximity sensing devices—infrared, ultrasound, capacitance, and others—fit our needs well. All of these have some limited distance of operation, and all have their own pluses and minuses. For example, ultrasound sensing has a wide sensing range that will likely reach the boundaries of the arm workspace, but its resolution is not very good. Other properties—for example, accuracy, reliability, and so on—may be important as well in the choice of sensing hardware. Without going into more specifics (see Chapter 8 for more detail), let us assume here that our revolute–revolute arm is equipped with some generic proximity sensing hardware, such that every point of the arm body can sense approaching obstacles.

Algorithmic Issues. How do we incorporate proximity sensing in the RR-Arm Algorithm or similar motion planning procedures? In the discussion on sensing versus motion planning for mobile robots (Section 3.6), we assumed that the robot has a circular sensing range, with some limited radius of sensing. For a mobile robot, this is a reasonable and natural assumption; it can be easily modified for some practical constraints, such as partial sectoral sensing. For an arm manipulator the situation is more complex. Assuming a similar limited sensing distance at each point of the robot body, in the workspace the outline of the sensing range will be a complex figure (Figure 5.16) that surrounds the arm and changes its shape as the arm links move relative to each other.

In C-space the situation is somewhat closer to the mobile robot sensing range simply because in C-space the arm becomes a point. But that's where the similarity stops. Assume that the arm's sensing allows it to sense objects in W-space within some *sensing range r* (Figure 5.16). An obstacle will be detected when it is at a distance equal to or smaller than r from the point on the arm body closest to it.

With this kind of proximity sensing, the robot's sensing range in C-space plane (θ_1, θ_2) is not a circle anymore. In fact, it is not even an entity with fixed parameters. The sensing range image will look in C-space more or less like the one in Figure 5.17 (point C is the arm's current position). As the arm links move relative to the arm base and relative to each other, the joint angles (θ_1, θ_2) change accordingly. The sensing range C-space image then moves in the plane (θ_1, θ_2),

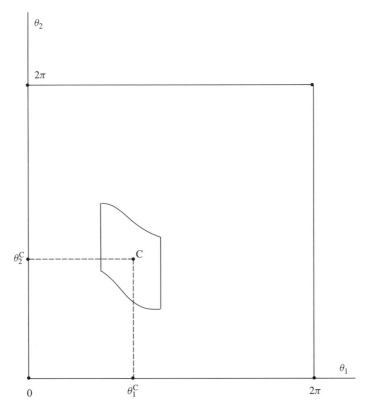

Figure 5.17 In C-space the sensing range of the revolute–revolute arm has a shape similar to the one shown here. The point in the center of the figure corresponds to the values (θ_1, θ_2) of the arm's current position. As the arm moves in its workspace, this figure moves in C-space, with its shape and dimensions changing continuously.

with its shape and dimensions shrinking and expanding from rectangle-like to ellipse-like, and with shapes in between like in Figure 5.17. Animation of this process makes for a wonderful movie: One sees a strange creature that is moving while constantly changing its shape according to some mystifying law. The extent of variability in the sensing range C-space image depends on the sensing range r and the arm's kinematics.

Calculation of the sensing range C-space image is an interesting though rather involved task; there are many details and many special cases to attend to. With good equations for the sensing range, one could improve motion planning algorithms by providing a look-ahead optimization of the arm's next few steps, or attempt algorithms that take into account the arm dynamics, similar to the work we did in Chapter 4. To my knowledge, today there are no published analyses on this topic. As a first approximation, one can start with a simplified model of the sensing range, presenting it as a circle whose radius changes as a function of the arm position (θ_1, θ_2). A conservative approximation would be to model the arm sensing by the maximum circle inscribed in the real sensing range. With this model the robot would be safe, but much sensing would be wasted: In some directions in the (θ_1, θ_2) plane the actual sensing will go much farther than the circular model will indicate.

As the arm moves, its sensing range image in C-space "breathes," shrinking and expanding as it moves in the plane (θ_1, θ_2). The extent of such changes depends on the motion. It is easy to see, for example, that if we fix angle θ_2 and let angle θ_1 change, in C-space of Figure 5.17 the sensing range figure will move horizontally, and its shape will remain the same. This is because the motion does not involve any changes in the relative position of links l_1 and l_2. Any motion involving a change in angle θ_2 will cause changes in the shape of the sensing range figure.

Except for the added calculation due to the variable sensing range in C-space, incorporating proximity sensing in the arm motion planning algorithm is similar to the analogous process for mobile robots (Section 3.6). One can combine, for example, one of the VisBug algorithms for a mobile robot (Section 3.6) with the RR-Arm Algorithm developed in this chapter. The fact that the latter is noticeably more complex than Bug algorithms calls for a careful analysis. To date, there are no published results in this area, in spite of its significant theoretical and practical potential.

How proximity sensing can affect the RR-Arm Algorithm performance can be seen in Figure 5.18. Here link l_2 happens to be attached to link l_1 not by its endpoint, as in some of our prior figures, but by some other point on the link. (This is a more realistic design; it often occurs in industrial arm manipulators.) Note how elegant and economical the arm's path becomes when the arm is provided with proximity sensing (Figure 5.18b), compared to its performance with tactile sensing (Figure 5.18a). In fact, the robot path in Figure 5.18b is almost the optimal path between the S and T locations; it could hardly be improved even by a procedure operating with complete information. This of course will

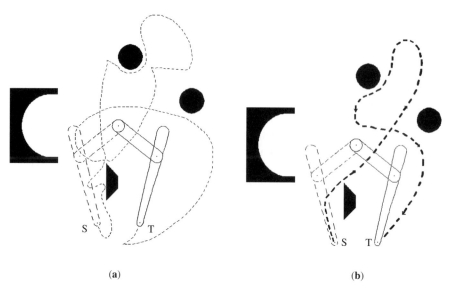

(a)   (b)

Figure 5.18   Performance of RR-Arm Algorithm in a scene with four obstacles (black objects): **(a)** with tactile sensing and **(b)** with proximity sensing.

not be always so; as we know, obtaining information about the scene "on the fly" rather than beforehand precludes one from guaranteeing the optimal solution.[4]

5.2.6   Concluding Remarks

Let us summarize here some properties of the sensor-based motion planning strategy for the revolute–revolute arm manipulator developed in this section.

- Of a pivotal importance during the development of RR-Arm Algorithm were the topological properties of the arm configuration space (C-space). These properties not only allowed us to convert the problem from moving a kinematic jointed arm to moving a point "robot" in the corresponding C-space, they also allowed us to reduce the problem from searching the whole space to searching only a tiny one-dimensional subset of space. Analysis carried out in this section sheds much light on the motion planning issues involved, and it should serve us well in studying other arm configurations in this and next chapters.

- The fact that the C-space of the RR arm is a two-dimensional manifold, namely a common torus, turned out to play an important role in the RR-Arm

---

[4]As we will see in Chapter 7, even with the benefit of seeing the whole scene and of prior training, humans are not able to compete even with the performance shown in Figure 5.18a, let alone with that in Figure 5.18b. This fact is at the heart of the argument for synergistic human–robot systems, where responsibilities between the partners are divided in accordance with their abilities.

Algorithm, not lastly because this C-space structure allows for more than one "short" route between the start and target positions, which have been used with profit by the algorithm. The analysis demonstrates that the arm kinematics can greatly influence the algorithm structure. In the following sections we will study in a similar vein the remaining four of the five configurations of planar two-link arms shown in Figure 5.1. We will conclude these studies with an attempt, in Section 5.8.4, to develop a unifying theory that will allow one to consider each of the five kinematics of Figure 5.1 as a special instant of one general case.

- Planning of arm motion with the described RR-Arm Algorithm is done completely in the workspace (W-space), based on the sensing data from the arm sensors. The above analysis and examples referring to the arm configuration space have been used solely to establish the theory and develop the algorithmic machinery.

- A similar consideration applies to the sensing used. Whereas most of our algorithm design process relied on tactile sensing, this was done only for the sake of simpler explanation. As discussed in Section 5.2.5 (and more in Chapter 8), proximity sensing, and not tactile sensing, should be used in practical arm manipulator motion planning systems.

- No preliminary exploration of obstacles and no beforehand partial or complete computation of the scene in W-space or C-space takes place or is expected by the algorithm. By the time the arm arrives at the target location, it may know very little about the space that it just traversed.

- If the desired target position is not feasible because of interfering obstacles, the reachability test built into the algorithm will make this conclusion, usually quickly enough and without exploring the whole space.

- The algorithm plans the robot arm motion better than humans do. We will discuss this interesting observation in great detail in Chapter 7. In brief, when watching the RR-Arm Algorithm in action, humans have difficulty grasping its mechanism or the rationale behind the paths it generates. A quick glance at the paths in Figures 5.14, 5.15, and 5.18 should help convince one that this is so. This is so even for simple scenes, and it is so for tactile as well as for more complex sensing. The difficulty for humans is not in that the algorithm is overly complex. With quick training, one will be able to understand and use the RR-Arm Algorithm in C-space—but not in W-space. Unfortunately, this would be a useless demonstration because C-space is not available for motion planning; remember, our primary assumption is that no information about the scene is available beforehand. On the other hand, asking human operators to use the algorithm in the workspace, the way a robot arm manipulator does it, turns out to be hopeless. And humans own strategies, whatever they are, consistently show an inferior performance compared to that of RR-Arm Algorithm (see Chapter 7).

Recall how very different our current situation is from the one we faced with mobile robot motion planning algorithms (Chapter 3). We observed there that, first, humans' own motion planning strategies work pretty well in the related tasks and, second, humans have no problem interpreting, learning, and using relevant motion planning algorithms. In comparison, motion planning for even a simple arm manipulator is a task that poses serious mental challenges to a human. This fact goes a long way in explaining difficulties that human operators exhibit when controlling real-world teleoperated robotics systems. The price for those difficulties is operators' mistakes and an overly slow operation that rules out many real-life tasks.

This suggests the need for changes in today's approaches to the design of teleoperated systems. In particular, it is highly desirable to shift the responsibility for obstacle collision avoidance from the operator's shoulders to the robot intelligence. We will return to this topic in Chapters 7 and 8.

5.3   DISTINCT KINEMATIC CONFIGURATIONS OF RR ARM

Even for the most popular revolute and sliding (prismatic) joint types, each combination of joint types of an arm manipulator can be realized in more than one kinematic configuration. The RR arm that we analyzed above (Section 5.2) is especially prolific in terms of variability of kinematic configurations. We will review here those configurations, as an example of how such variability occurs as well as to see how the theory developed above applies to them. This exercise also helps train one's spatial intuition, a useful quality in the work we do here.

Four different configurations of RR arms are shown in Figure 5.19. As we will see, different kinematic configurations result in different, sometimes quite unusual, configurations of their workspace. This fact does not change the basics of sensor-based motion planning considered above. No matter how an RR arm is configured, the following statements are true:

(a) The arm's two degrees of freedom guarantee that its endpoint moves in a two-dimensional manifold, be it a plane or some other surface.

(b) The arm's configuration space is still a common torus, and so the theory and the motion planning algorithm developed in Section 5.2 applies.

RR Arm of Figure 5.19a.   This arm is recognized easily—it is the same planar two-link RR arm manipulator that has been studied in much detail in Section 5.2 (see Figures 5.1a and 5.2). The arm lies and moves in the plane. Its two joint axes are parallel to each other and perpendicular to the arm's plane. The main difference between this arm and the remaining three arms in Figure 5.19 is that the two joint axes of those three arms are mutually perpendicular rather than parallel.[5] This changes the arm workspace rather dramatically.

---

[5]There exist arm designs where joint axes intersect at angles different from parallel or perpendicular. Some such designs have even been patented, because they provide interesting kinematic properties. No such kinematics is considered in this text.

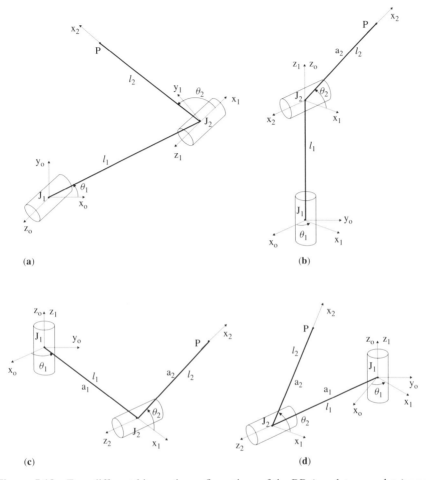

Figure 5.19 Four different kinematic configurations of the RR (revolute–revolute) arm: (a) RR arm studied in detail in Section 5.2; the arm's endpoint moves in the plane. (b) This arm's endpoint moves on the surface of a sphere. (c) This arm's endpoint moves on the surface of a torus. (d) This arm's endpoint moves on the surface of a truncated sphere.

RR Arm of Figure 5.19b. See Figure 5.20a, which shows both revolute joints of this arm, J_1 and J_2, in the same spot, with the joint axes intersecting at $90°$. Because link l_1 produces no physical displacement, we can take it as being of zero length, $l_1 = 0$. Then, only one link, l_2, is physically present in this arm; hence the arm looks like an outstretched human arm. Sometimes this mechanism is interpreted as a single *ball joint*. Since from the control standpoint this device still has two independent control variables, seeing it as two independent revolute joints, rather than one (ball) joint, is more in line with our other notation in this text.

The arm's first joint angle, θ_1, is responsible for motion in one plane; for speci-ficity, assume this is a horizontal plane. The second joint angle, θ_2, is responsible for the arm motion in the vertical plane. Together they allow the arm endpoint (end effector) to reach any point on the surface of a sphere of radius l_2. The workspace (W-space) of this arm is hence a *sphere*. Any point P on the sphere corresponds to two joint solutions, (θ_1^P, θ_2^P) and $(\pi + \theta_1^P, \pi - \theta_2^P)$.

Since the body of link l_2 moves in the three-dimensional (3D) space inside the W-space sphere, it can interact with 3D obstacles that may appear inside the workspace sphere, thus presenting a potential collision avoidance issue. The fact that one end of link l_2 is fixed (at the base J_1) and the motion of its other end is limited to the sphere surface constrains the link interaction with obstacles significantly. In terms of motion planning, this is equivalent to motion along a curve rather than around a "real" 3D object. This means that the 2D motion planning algorithms of Section 5.2 fully apply here.

Proceeding in this direction, we want to chose an M-line, the line that the arm endpoint would go through if no obstacle interfered with the arm motion. Since the C-space (configuration space) of this arm is a torus, the choice is among four M-lines (Section 5.2). These are shown in Figure 5.20. Denote the positive direction of change of angle θ_i, $i = 1, 2$, by "+" and denote the negative one by "−". Then the four M-lines are four geodesic curves, as shown in Figure 5.20b, with the corresponding joint angles changing as follows:

$$
\begin{array}{ccc}
 & \theta_1 & \theta_2 \\
M_1: & + & + \\
M_2: & - & + \\
M_3: & + & - \\
M_4: & - & -
\end{array}
$$

The choice of the M-line and the motion planning procedure will proceed accord-ing to the RR-Arm Algorithm (Section 5.2.2).

RR Arm of Figure 5.19c. A detailed picture of this arm configuration is shown in Figure 5.21a. The only difference between this configuration and the one in Figure 5.20a is that here the arm's two joints are at a distance from each other, equal to the length of the first link, l_1. Links l_1 and l_2 lie in the same plane—in general, depending on link l_1 shape, in parallel planes. The arm's endpoint moves along the surface of a torus, and so its W-space is a *torus*.[*] This torus may or may not have a hole depending on the relation between the lengths of links l_1 and l_2:

$l_2 > l_1$ produces a W-space torus with no hole;

$l_2 < l_1$ produces a W-space torus with a hole.

Projections of W-space onto the horizontal (xy) and vertical (xz or yz) planes for both cases, $l_2 > l_1$ and $l_2 < l_1$, are shown in Figures 5.21b and 5.21c, respec-tively.

---

[*]To emphasize, it is not the configuration space that forms a torus here, as we had it before, but the workspace.

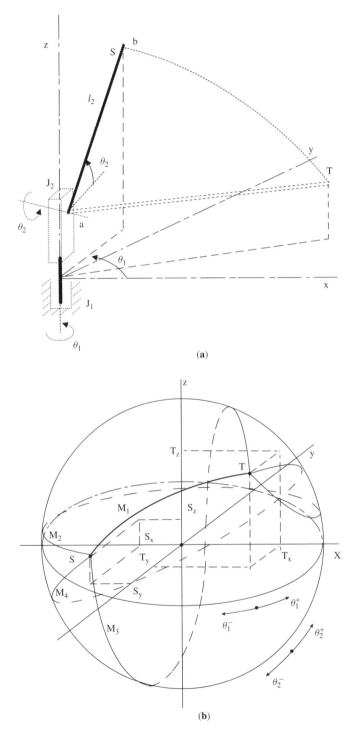

Figure 5.20   The RR arm of Figure 5.19b. (**a**) The arm design. (**b**) The arm's workspace, with four complementary M-lines shown. (The sketches are not to the same scale.)

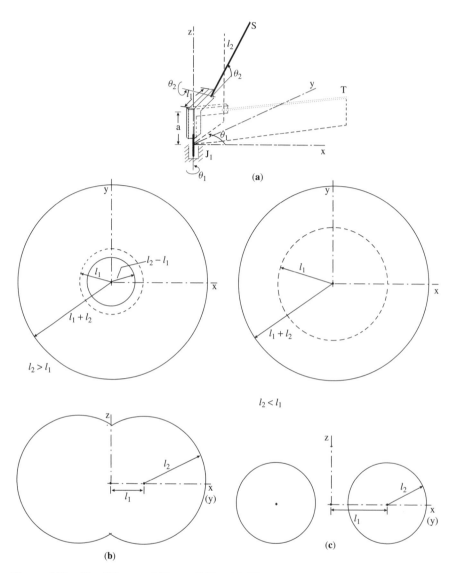

Figure 5.21   The RR arm of Figure 5.19c. (**a**) The arm design. (**b**) Two projections of the arm W-space in case $l_2 > l_1$. (**c**) Two projections of the arm W-space in case $l_2 < l_1$.

RR Arm of Figure 5.19d. Details of this arm configuration are shown in Figure 5.22a. The difference between this configuration and the one in Figure 5.21c is that here link l_2 rotates in the plane perpendicular, rather than parallel, to link l_1. In other words, the arm looks like a fan, with link l_2 being the fan propeller's only blade. This design makes for a somewhat strange workspace: The arm's endpoint moves along the surface of a truncated sphere of radius $\sqrt{l_1^2 + l_2^2}$

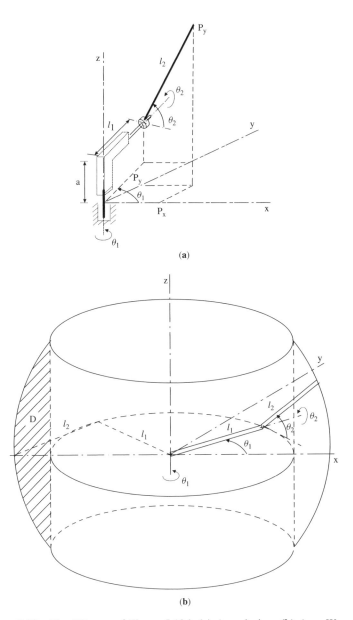

Figure 5.22   The RR arm of Figure 5.19d. **(a)** Arm design. **(b)** Arm W-space.

(Figure 5.22b). Link l_1 moves in the horizontal plane xy. Link l_2 is always inside the volume D limited by the sphere and a cylinder whose radius is l_1 and whose height is $2l_2$ (see Figure 5.22b; the cross section of the volume D is shaded). The body of link l_2 may therefore interact with 3D obstacles that happen to appear in volume D.

5.4   PRISMATIC–PRISMATIC (PP, OR CARTESIAN) ARM

This arm is the second one, arm (b), among the five arms shown in Figure 5.1. The reason it is called Cartesian is that the displacement in each of its joints translates directly in exactly the same motion of the arm endpoint in Cartesian plane (see Figure 5.23). The arm has two prismatic (sliding) joints, with joint values l_1 and l_2—hence its other name, a *PP Arm*. The boundaries of the arm's W-space are limited by the rectangle whose sides are equal to the maximum lengths of links l_1 and l_2. We assume that no obstacles outside W-space can interfere with the arm motion; hence the path planning problem is limited to the arm's W-space rectangle. The M-line is defined as a straight-line segment connecting the arm's starting and target points, S and T.

From Section 5.2.1, a *shadow* of an obstacle X is the area of workspace no point of which can be reached by the arm endpoint due to interference of X with the arm motion. Functionally, then, for the arm an obstacle shadow is the same as the physical obstacle that causes it. Observe in Figure 5.23 that for any obstacle in W-space of the PP arm, no points to the right of an obstacle can ever be reached by the arm endpoint. We will see that this property of this arm kinematics makes motion planning a rather simple task.

The boundary of an obstacle plus the boundary of the shadow that it forms produces the *virtual line* of the obstacle. The said property of the PP arm—that no points to the right of an obstacle can be reached by the arm endpoint—means

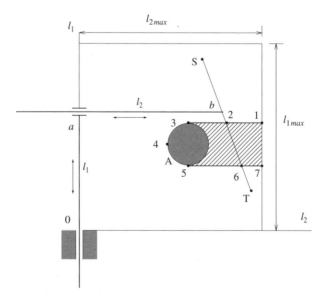

Figure 5.23   Cartesian (PP) arm. The virtual obstacle includes the actual obstacle and its shadow (shaded).

that any obstacle's virtual line is a simple curve. In Figure 5.23, the virtual line of the circular obstacle A is the line that passes through points 1, 2, 3, 4, 5, 6, 7.

Recall that in the arm configuration space (C-space), the *virtual obstacle* is the C-space image of the corresponding physical obstacle. The *virtual boundary* of the obstacle is the C-space image of the corresponding virtual line. In Figure 5.23, the virtual boundary of obstacle A is the same line (1, 2, 3, 4, 5, 6, 7). Clearly, this will be always so. Hence, any virtual boundary is a simple curve. This simple correspondence makes the PP arm's C-space practically identical to its W-space, and hence unnecessary for further analysis.

As we will see later, the above simple structure of PP arm virtual obstacles brings about one specific condition that allows a direct and simplified use for this arm of the basic planning procedure developed in Section 3.3, and ensures its convergence.

Another condition that the basic planning procedure requires—that the virtual boundary be a closed curve—does not hold for the PP arm. For example, in Figure 5.23 no point of the line segment between points 7 and 1 of the obstacle shadow can be reached by the arm endpoint, and so it is not a part of the virtual boundary. This will present no difficulty for the planning procedure, though. In general, an obstacle's virtual boundary in C-space of this arm consists of four distinct segments:

- The "left" curve corresponding to the arm endpoint's following the actual obstacle boundary; in Figure 5.23 this segment comprises points 3-4-5.
- Two mutually parallel straight line segments corresponding to those points of the arm body (other than the arm endpoint) that touch the actual obstacle; in Figure 5.23 these are lines 3-2-1 and 5-6-7.
- The straight-line segment that is a part of the W-space boundary; in Figure 5.23, this is line 7-1.

Of these, the first three segments form a simple open curve, each point of which can be reached by the arm endpoint. The fact that the fourth segment cannot be reached by the arm endpoint poses no algorithmic problems since the endpoints of the said simple open curve—in Figure 5.23, points 1 and 7—can easily be recognized from the fact that they always lie on the W-space boundary and hence correspond to the maximum value of the joint value l_2. This fact will help in showing the algorithm convergence. An important statement that helps simplify the path planning procedure of the PP arm follows directly from Figure 5.23:

Lemma 5.4.1. *For the Cartesian two-link arm, if the target point T is reachable from the starting point S, then there exists a path from S to T such that it corresponds to a monotonic change of the joint value l_1.*

This can be shown as follows. Depending on whether the difference $(l_1^S - l_1^T)$ is positive, zero, or negative, establish the *direction of change* of link l_1 motion along the M-line from S to T—positive, zero, or negative, respectively. Note that this information is known before the motion takes place, from coordinates of points S and T.

If during its motion along the M-line from S to T the arm encounters an obstacle, such a local direction is chosen for passing around the obstacle for which the corresponding change in joint value l_1 coincides with the established M-line direction of change of l_1. In the special case of l_1 being constant in the vicinity of the hit point, the local direction should be chosen as corresponding to decreasing values of l_2; otherwise the arm will not be able to pass around the obstacle.

If, while passing around an obstacle, the current value l_1 moves outside of the interval (l_1^S, l_1^T), then, clearly, point T lies in the shadow of the obstacle and cannot be reached. If the M-line direction of change of l_1 is "0"—which will be so if the M-line happen to be parallel to link l_2—and an obstacle is met along the way, then, again, point T cannot be reached because it is in the shadow of the obstacle. In other words, Lemma 5.4.1 holds, and this helps simplify the planning procedure.

While the PP arm planning procedure will work for the arm links of any shapes, it can be further simplified and made more efficient if link l_2 is assumed to present an elongated rectangle whose sides are parallel to the joint axes l_1 and l_2, respectively. If link l_2 happens to be of a more complex shape, the algorithm can replace it with a minimum rectangle that contains the link. Hence link l_2 in the example of Figure 5.23 would be treated as a rectangle of width zero.

Now the whole path planning procedure for the *PP-Arm Algorithm* can be formulated; $L_o = S$.

1. Establish the M-line direction of change of link l_1 (see above). Set $j = 1$. Go to Step 2.
2. From point L_{j-1}, the arm moves along M-line until one of the following occurs:
   (a) Target T is reached. The procedure stops.
   (b) An obstacle is encountered and a hit point, H_j, is defined. Choose the local direction using the M-line direction of change of l_1. Go to Step 3.
3. The arm follows the obstacle virtual boundary until one of the following occurs:
   (a) The target T is reached. The procedure stops.
   (b) The M-line is met at a distance d from T such that $d < d(H_j, T)$; point L_j is defined. Increment j. Go to Step 2.
   (c) The current value l_1 is outside of the interval (l_1^S, l_1^T). Target T cannot be reached. The procedure stops.

It can be shown that the algorithm will work correctly if the arm links and obstacles in the scene are of arbitrary chapes.

5.5 REVOLUTE–PRISMATIC (RP) ARM WITH PARALLEL LINKS

The revolute–prismatic parallel kinematic configuration (Figure 5.1c) is a common major linkage in commercial robot arm manipulators. The so-called Stanford manipulator robot (see, e.g., Ref. 7) is a typical example of this type. Joint values of this arm are the angle θ_1 of the first (revolute) joint and the variable length l_2 of the second (prismatic) joint; $0 \leq l_2 \leq l_{2\,\text{max}}$ (see Figure 5.24a). The outer boundary of the arm's workspace (W-space) is a circle of radius $(l_1 + l_{2\,\text{max}})$. Its inner boundary is a circle of radius l_1, which defines the *dead zone* inaccessible to the arm endpoint. Unlike in Figure 5.1c, in order to not overcrowd the picture, no dead zone appears in the arm in Figure 5.24a; that is, here $l_1 = 0$. With the arm endpoint b in some position of W-space—say, S—the position of the link's rear end, a_S, can be found by passing a line segment of length $l_{2\,\text{max}}$ from b_S through the origin O.

For specificity, let us define the M-line as a straight-line segment connecting points S and T; denote it M_1-line. In the example in Figure 5.24a, one path from S to T would be as shown, the curve $(S, 1, 2, 3, 4, 5, 6, 7, 8, 9, 10, T)$. Observe that if due to obstacles the arm cannot reach point T by following the M_1-line, it might be able to reach T "from the other side," by changing angle θ_1 in the direction opposite to that of the M_1-line. Hence, similar to how we did it with the RR arm (Section 5.2), a *complementary* M-line, the M_2-line, is introduced, defined as consisting of three parts: two straight-line segments, (S, S') and (T, T'), continuing the M_1-line segment outwards until its intersection with the W-space outer boundary, and a segment of the W-space outer boundary that corresponds to the interval of θ_1 complementing that of the M_1-line to 2π. This choice for the complementary M-line is largely arbitrary; any M_2-line will do, as long as it is uniquely defined, is computationally simple, and complements the θ_1 interval of M_1-line to 2π.

As the arm moves along the M-line, obstacles will interfere with its motion in two different ways. In our example the arm endpoint will be forced to leave M-line two times: The first time is when obstacle A interferes with the rear part of link l_2, creating the curve $(1, 2, 3, 4, 5)$, (Figure 5.24a), and the second time is when obstacle B interferes with the front part of link l_2, creating the curve $(6, 7, 8, 9, 10)$ (Figure 5.24a). Note two shadows formed during this process (shaded in Figure 5.24a): Under no circumstances the arm endpoint will be able to reach any point within the figures with boundaries $(0, 1, 2, 3, 4, 5, 0)$ and $(6, 7, 8, 9, 10, 11, 12)$.

Define a *front contact* of link l_2 as a situation where a part of the link between its front end (which is the part containing the arm endpoint) and the origin is in contact with the obstacle. A *rear contact* of the link refers to a situation where a part of the link between its rear end and the origin is in contact with the obstacle. Correspondingly, a front contact forms the *front shadow* of the obstacle, and a rear contact forms the *rear shadow* . In Figure 5.24a, the rear shadow of obstacle A is limited by the line $(1, 0, 5, 4, 3, 2, 1)$, and the front shadow of obstacle B is limited by the line $(6, 7, 13, 9, 10, 11, 12, 6)$.

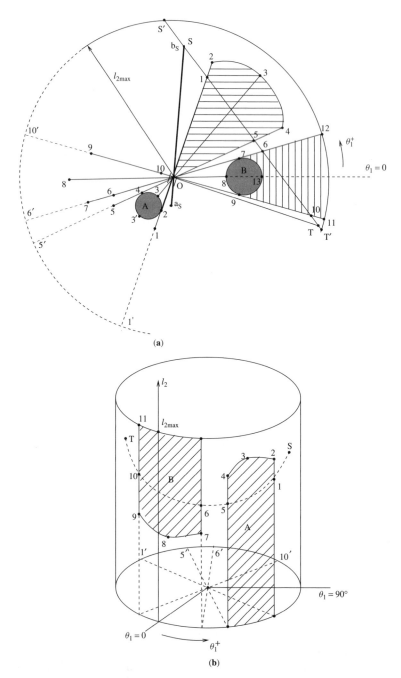

Figure 5.24   Revolute–prismatic (RP) parallel two-link arm. **(a)** W-space, with obstacles A and B. Shadows (shaded) result from interaction between the rear end of link l_2 with obstacle A and the front end of l_2 with obstacle B. **(b)** C-space images of the corresponding virtual obstacles. Line (S, T) is the image of the straight line (S, T) in W-space.

Note also that these same obstacles A and B create another two shadows, which are not of importance in our current task but might show up with some other M-line: the arm endpoint will not be able to reach any point within the figures described by points in the left part of Figure 5.24a, $(4, 5, 5', 1', 1, 2, 3', 4)$ and $(0, 10', 6', 0)$. An obstacle that extends into the arm's dead zone forms only one shadow, which extends from the obstacle all the way to the W-space outer periphery. An obstacle X that is in the shadow of another obstacle Y will never interact with the arm. If obstacle X partially intersects the shadow of obstacle Y, it will be treated by the arm as a part of Y.

Therefore, with the exception of obstacles that extend into the arm's dead zone, any virtual obstacle in the workspace of this RP arm includes the actual obstacle itself plus two shadow components: one front shadow and one rear shadow. The front shadow extends from the obstacle to the outer boundary of W-space [see the curve $(6, 7, 8, 9, 10, 11, 12, 6)$ in Figure 5.24a]; the rear shadow extends from the arm origin O into the periphery, but in general not to the boundary, of W-space [see the curve $(1, 0, 5, 4, 3, 2, 1)$ in Figure 5.24a].

Unlike the more complicated situation with the RR arm (Section 5.2), in the case of our RP arm the virtual lines of obstacles are always simple curves. In a special case when the M-line crosses the dead zone, the latter can be treated simply as an obstacle interfering with the rear end of link l_2.

Two independent variables, one angular and the other linear displacement, can be represented by the surface of a cylinder. The C-space of our arm is therefore a cylinder whose flat sides, called *base circles*, correspond to the first joint value, θ_1, and whose height corresponds to the second joint value, l_2. The C-space of the example in Figure 5.24a is shown in Figure 5.24b. Shown in the figure are the images of an M-line and of the shadow components of virtual obstacles A and B that interfere with the arm motion. In order to not overcrowd the picture, for the complementary two shadow components only their projections on the lower base circle, $(1', 5')$ and $(6', 10')$, are shown.

Unlike the RR arm, each point in W-space of our RP arm has only one arm solution. That is, there is one-to-one mapping between W-space and the corresponding C-space, as compared to the one-to-two mapping in the case of the RR arm. Because of this, and also because virtual lines in W-space are always simple curves, the virtual boundaries in C-space are also simple curves. Recall that this constitutes a necessary condition for the basic path planning procedures (Section 3.3). In general, each virtual boundary in C-space of the RP arm consists of a combination of three distinct segments:

1. A curve formed when the front or the rear end of link l_2 follows the boundary of the actual obstacle; for obstacle B in Figure 5.24b this segment passes through points 7-8-9.
2. A vertical straight-line segment formed when points of the arm body other than the arm endpoint touch the obstacle while passing around it; in Figure 5.24b this consists of lines 7-6-12 and 9-10-11).

3. A segment that is a part of one of the base circles (e.g., line 11-12, Figure 5.24b); the inside points of this segment cannot be reached by the arm.

In order to apply the basic path planning procedure to this arm, the algorithm has to reflect specifics of moving along the C-space cylinder. Similar to the RR arm, one concern in our case is whether obstacle boundaries may be formed by more than one simple curve. Recall that if a virtual boundary is formed by one simple (closed) curve, it is called a Type I obstacle, and if the virtual boundary has more than one simple (closed) curve, it is a Type II obstacle (see Section 5.2.2). Starting with one specific case, observe that if a ring-like actual obstacle appears in W-space, positioned so that it separates the arm from the W-space outer boundary, the result will be a band-like virtual obstacle in C-space—formally, a Type II obstacle. One simple closed curve of the band can be reached by the arm, whereas the other, formed by one of the base circles, is inaccessible to the arm. Because of this, and in spite of the fact that the virtual boundary has two closed curves, from the standpoint of path planning we will treat it as a Type I obstacle.

As another case, observe the arm shown in Figure 5.1c, where $l_1 \neq 0$. If an obstacle extends from W-space into the dead zone, it is easy to see that in C-space a swath-like virtual obstacle will appear, whose virtual boundary in C-space includes two separate "simple curves," plus two vertical lines each connecting the opposite base circles of the C-space cylinder. This is a real Type II obstacle. Similar to the RR arm, if during the arm motion one such curve of a Type II obstacle has been completely explored by the arm without ever meeting the M-line, it is clear that the second curve has to be explored as well. To do that, the *complementary M_2-line* will be used.

As with the Cartesian arm studied above, the choice of the local direction for following the virtual obstacle by our RP arm happens to be unique. Once the arm encounters an obstacle, one of two possible cases arises. If the contact is a front contact—that it, it corresponds to the front part of the arm contacting the obstacle—then only such a local direction is meaningful that corresponds to decreasing values of l_2. As one can see in Figure 5.24a, the opposite local direction would never bring the arm any closer to the target. If, on the other hand, the contact is a rear contact, then only such local direction should be chosen that corresponds to increasing values of l_2. The reachability test is built in a manner similar to this test for the RR arm (Section 5.2.2), taking into account the simpler structure of the RP arm's C-space (see the algorithm below).

How will the arm tell a front contact from a rear contact? By our model, the arm's sensing lets it know which point of its body contacts the obstacle. The arm also knows at all times which point of link l_2 is at the joint point of the link. This information allows the arm to always distinguish a front contact from a rear contact.

Because of the unique choice of the local direction, there is no need to investigate the whole curve of the virtual boundary. If, while passing around the obstacle in the chosen local direction, the arm reaches one of the limits of l_2, it can safely conclude that it is dealing with a Type II obstacle, so the arm should start looking for the second curve of the virtual boundary using the complementary M-line. The procedure is further simplified through the use of the following statement similar to the one in Section 5.2.2:

Lemma 5.5.1. *For the two-link revolute–prismatic (RP) arm, if position T is reachable from the starting position S, then there exists a path from S to T such that it corresponds to a monotonic change of the joint value θ_1.*

In the motion planning procedure, a flag is used to indicate processing of each of the two curves of a Type II virtual boundary. When the complementary M-line is introduced, the numbering of hit and leave points starts over; $L_o = S$. The distance used is a Euclidean distance in W-space. Assume the M_1-line is the shorter of the two complementary M-lines. The procedure *RP-Arm Algorithm* includes the following steps.

1. Establish an M_1-line as the M-line. Set the flag down. Set $j = 1$. Go to Step 2.
2. From point L_{j-1}, the arm moves along the M-line until one of the following occurs:
   (a) Target T is reached. The procedure stops.
   (b) An obstacle is encountered and a hit point, H_j, is defined. In case of a front contact, choose the local direction such that it corresponds to decreasing values of l_2. In the case of a rear contact, choose the local direction such that it corresponds to increasing values of l_2. Go to Step 3.
3. The arm follows the virtual boundary until one of the following occurs:
   (a) The target is reached. The procedure stops.
   (b) Current joint value θ_1 is outside the interval (θ_1^S, θ_1^T). The target cannot be reached. The procedure stops.
   (c) The M-line is met at a distance d from T such that $d < d(H_j, T)$. Point L_j is defined. Increment j. Go to Step 2.
   (d) The value l_2 approaches one of its limits, and the flag is down (i.e., the first curve of the virtual boundary of a Type II obstacle has been processed). Set the flag up. Set $j = 1$. Establish an M_2-line as the M-line. Move the arm back to S. Go to Step 2.
   (e) The value l_2 approaches one of its limits, and the flag is up (i.e., the second curve of the virtual boundary of a Type II obstacle has been processed). The target cannot be reached. The procedure stops.

5.6   REVOLUTE–PRISMATIC (RP) ARM WITH PERPENDICULAR LINKS

The arm is shown in Figure 5.1d. An example of the arm's interaction with two circular obstacles is shown in Figure 5.25a. C-space images of the corresponding virtual obstacles and of the M-line are shown in Figure 5.25b. The path generated under the motion planning algorithm passes through points $S, 1, 2, \ldots, 9, T$. Although the arm's configuration and its interaction with obstacles appear to be quite different from those of the RP arm with parallel links (Figure 5.1c), both arm manipulators turn out to be very similar from the standpoint of path planning. Visually this is evident when comparing Figures 5.24 and 5.25: Whereas the W-space looks quite different for both arms, their C-spaces look very similar. The reader is therefore referred to the previous section for analysis and motion planning algorithm.

5.7   PRISMATIC–REVOLUTE (PR) ARM

The PR arm is shown in Figure 5.1e. A more detailed sketch appears in Figure 5.26. The arm's first joint value is the variable length of its first link, $0 \leq l_1 \leq l_{1\,max}$. The second joint value is angle θ_2. The length of second link, l_2, is constant. The arm's W-space boundary is a combination of a rectangle with sides of length $l_{1\,max}$ and $2l_2$, respectively, and two semicircles of radius l_2 attached to the rectangle, as shown in Figure 5.26. As before, assume that no obstacles may interfere with the part of link l_1 as it moves outside of W-space.[6]

The two circles of radius l_2 centered at the limit positions O and O_1 of link l_1 are called the *limit areas* of arm's W-space. Note that any point belonging to a limit area has only one corresponding arm solution, whereas any point of W-space that is outside the limit areas has two possible arm solutions. For example, in Figure 5.26 a single arm solution exists for point P, and two arm solutions exist for point P_1. As we will see, for the path planning purposes this peculiarity makes the arm distinct from those considered so far and leads to a combination of features of the algorithms for those arms.

An obstacle is considered to be *inside the limit area* if only one arm solution exists for *all points* of the obstacle virtual line. An obstacle is *outside the limit area* if two arm solutions exist for *all points* of the obstacle virtual line. An intermediate situation, referred to as *partially inside the limit area*, is when some points of the obstacle virtual line have one solution and some other points have two solutions.

Consider three circular obstacles A, B, and C (Figure 5.27a), positioned outside, partially inside, and fully inside the limit areas of the arm W-space, respectively. Positions of the link endpoints at some points of the corresponding virtual

---

[6]Incorporating that part of the arm body into the algorithm is very easy because link l_1 moves always along the same line.

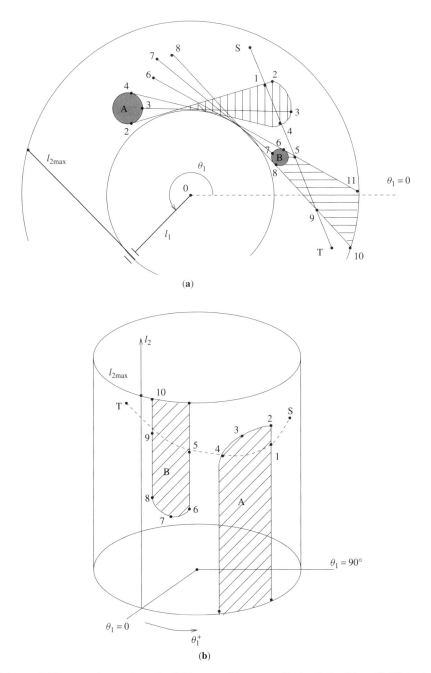

(a)

(b)

Figure 5.25 Revolute–prismatic (RP) arm with perpendicular links. Line (S, T) is the M-line. **(a)** W-space, with obstacles A and B. Similar to Figure 5.24, from the related obstacle shadows only the rear shadow of obstacle A and the front shadow of obstacle B are shown (shaded). **(b)** C-space images of the corresponding virtual obstacles.

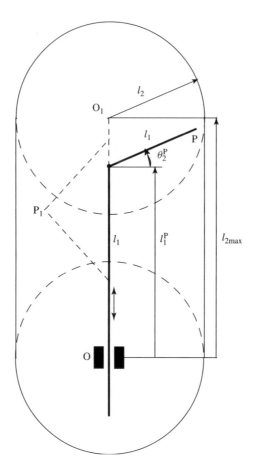

Figure 5.26 A PR (prismatic–revolute) arm. Any point in the workspace (such as point P) within the circles with centers O and O_1 and of radius l_2 has a single arm solution. Any point (such as P_1) outside those circles has two arm solutions—except on the border of W-space, where it also has a single solution.

lines are indicated in the figure by pairs of numbers, (1, 1), (2, 2), and so on. Note that the virtual line of obstacle A, which is outside the limit areas, is a closed curve: starting at some position of contact with A (say, point 1), the arm can pass around the obstacle while keeping in contact with it (points 2, 3, ..., 8, 1), returning eventually to the starting position. On the other hand, the virtual line of obstacle B is an open curve, whose endpoints 13 and 16 lie inside the W-space but on the boundary of a limit area. Finally, the virtual line of obstacle C is an open curve, whose endpoints 17 and 21 lie on the boundary of W-space (and of a limit area).

The C-space of our PR arm presents the surface of a cylinder whose height is equal to the upper limit of the first joint value, $l_{1\,max}$, and whose base circles

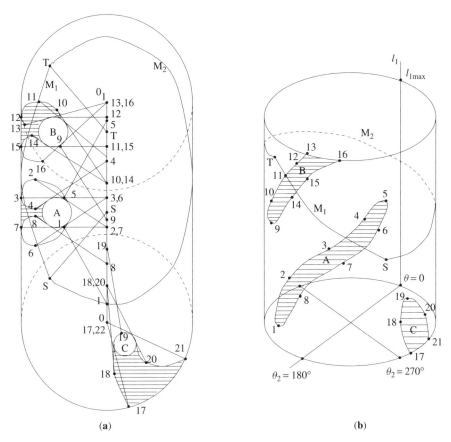

Figure 5.27 PR arm. (**a**) W-space. Shown are M-lines M_1 and M_2, along with some link positions during the arm's passing around obstacles A, B, and C. (**b**) The C-space images of the two M-lines, virtual obstacles, and those same positions.

correspond to the second joint value, θ_2. An obstacle virtual boundary presents in C-space a curve, which may or may not include as its part a segment of the base circle, depending on whether it corresponds to an inside, partially inside, or an outside obstacle (Figure 5.27b). No point of a base circle segment of a virtual boundary can be directly accessed by the arm.

Similar to the revolute–revolute (RR) arm studied in Section 5.2, it is easy to see that in case of an outside obstacle, (a) the corresponding virtual boundary in C-space presents a single simple closed curve—even if the obstacle virtual line in W-space has self-intersections or double points—and (b) this curve can be traced by the arm fully. Obstacle A in Figure 5.27 presents an example of this type.

Furthermore, similar to two revolute–prismatic (RP) arms considered in the two previous sections, it can be shown for the PR arm that in the case of inside and partially inside obstacles the corresponding virtual boundaries are simple curves. Each such curve includes two types of segments: (i) one or two open curve segments (and these can be followed by the arm) and (ii) one or two segments of the base circles of the C-space cylinder (and these cannot be accessed by the arm). Obstacles B and C in Figure 5.27 provide examples of this type.

That the curves are not closed in case (i) produces no algorithmic difficulties simply because the endpoints of the open curves always correspond to one or both limit values of link l_1 and hence are easily recognizable. Together with the simplicity of obstacle virtual boundaries, this fact ensures conditions necessary for convergence of the path planning procedure that we are about to develop.

The Choice of M-Line. A straight-line M-line in W-space would be a perfectly legitimate choice for M-line. From the practical standpoint, this choice is, however, not convenient because it may make it impossible to maintain continuous motion of both links. Try, for example, to follow a straight line between points S and T in Figure 5.27a. A discontinuity in the motion of one or both links will take place somewhere along the path. Similar to the RR arm case, it is preferable to define the M-line as a straight line in C-space.

A C-space straight M-line is a function $\theta_2 = p \cdot l_1 + q$ in the plane of variables l_1 and θ_2. Coefficients p and q are found from the coordinates of points S and T [97]. The image of this M-line, denoted as M_1-line in Figures 5.27a and 5.27b, is a geodesic line between points S and T on the surface of the C-space cylinder. If, because of the obstacles, point T cannot be reached from point S using this M-line, a complementary M-line, denoted as M_2-line in Figure 5.27, is used. M_2-line is defined similar to M_1-line, with the coordinate of point T being $(\theta_2 - 2\pi)$ instead of θ_2. In other words, M_1- and M_2-lines are mutually complementary over the joint angle θ_2, adding together to 2π. As one can see, the overall logic behind the choice of M-line here is very similar (though somewhat simpler because of two rather than four possible M-lines) to the logic in the case of RR-arm (Section 5.2).

Special Cases. To complete the study of interaction between the arm and the obstacles, and before we can formulate the motion planning algorithm proper, we will consider a few examples that show what kind of special cases can arise for the PR arm. For the sake of specificity, assume that every time the arm meets an obstacle and defines on it a hit point, it starts with the local direction "left."

In the example shown in Figure 5.28, the virtual obstacle A forms a swath along the C-space cylinder that extends all the way between the cylinder's two limit circles—that is, between both limits of l_1 values. That is, in this example the obstacle virtual boundary includes two separate open curves [passing through points (4, 1, 2, 3) and (5, 6, 7, 8), Figure 5.28b] and two segments of the cylinder limit bases [passing through points (4, 5) and (3, 8), accordingly, Figure 5.28b]. After leaving point S, following M-line and meeting obstacle A at point 1, the arm turns left along the virtual boundary and passes through points 2 and 3. Point

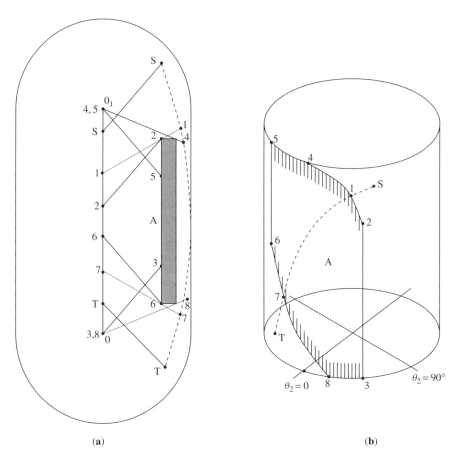

Figure 5.28   The PR arm. **(a)** A bar-like obstacle A in W-space forms **(b)** a swath-like virtual obstacle in C-space. Link positions shown are denoted in W-space and C-space with the same numbers.

3 is the first endpoint of the first open curve of the obstacle virtual boundary that the arm encounters. According to our algorithm, the arm will then attempt to complete the exploration of this curve, by returning to the hit point 1 and then trying to pass around the obstacle using the local direction "right." Along this path segment, point 4 will be reached, which is the other endpoint of the same open curve.

The difference between this example and, say, that of obstacle B in Figure 5.27 is that if in the latter both endpoints of the virtual boundary B correspond to the same (namely, upper) limit of the joint value l_1, the endpoints of the virtual boundary A in Figure 5.28 correspond to both limits of l_1. One consequence of this difference is that in the former case it is possible to pass around the obstacle, whereas in the latter case this is not possible. The fact that both endpoints of the open curve are located at the opposite base circles of the C-space cylinder

(Figure 5.28b) indicates that the arm is dealing with a Type II obstacle (see Section 5.2.1 for the definition of Type I and Type II obstacles).

As we know from the RR arm study, having explored one curve of a Type II obstacle virtual boundary brings in additional global knowledge: It tells the arm that somewhere there is another, second curve of the virtual boundary. It also tells the arm in which direction it can find this unknown curve. Now the arm knows it has to find and start exploring the second curve in order to draw conclusions about the target reachability. Because of the topology of a common cylinder, if one simple curve of an obstacle boundary connects to both base circles of C-space, in order for the virtual obstacle to be separated from the rest of the C-space cylinder, the two endpoints of the second curve of the boundary must lie at the two base circles as well. This means that the only way to reach the second curve is to go in the direction opposite to the current M-line. For that, a *complementary* M-line will be used. If, after reaching and following the second curve of the virtual boundary neither the M-line nor the target is met, this means the target T cannot be reached.

If an obstacle happen to interfere with the first link, as in the example in Figure 5.29a, the resulting virtual boundary forms a band around the C-space cylinder (Figure 5.29b). If the arm attempts to pass around the obstacle, starting, say, in the position 1, it will follow the whole virtual boundary, passing through points (1, 2, 3, 4, 5, 6, 1) and thus making a full circle. For the PR arm, this is the first example so far where the arm makes a full circle in θ_2. It suggests that, similar to the RR arm before, we may need a counter to indicate the fact of making a full circle in θ_2. Denote the counter C_2, to emphasize that it corresponds to our second joint variable, θ_2. (There will be no counter C_1.) The fact of completing a full circle will be detected by the counter C_2 as follows: Starting at the hit point, the counter integrates the angle θ_2, taking into account the sign. If a full circle is made without ever meeting the M-line (that is, the value of C_2 is 2π), then the target cannot be reached.

Continuing our observations of various special cases, if two (or more) obstacles happen to interfere with the first link the way it appears in Figure 5.30, two bands are formed on the surface of the C-space cylinder. In this example the distance between the obstacles along the line OO_1 is longer than the length of link l_2; this makes the two bands connect with each other in two places, forming two disconnected free areas in C-space. As a result, in Figure 5.30a position T cannot be reached from the arm position S.

Here is what will happen in this example under the algorithm: Starting at S, the arm follows the M-line, hits obstacle A at point 1 (Figure 5.30), goes through points (2, 3, 9, 8, 7) while trying to pass around the obstacle, and eventually returns to the hit point 1 without ever encountering the M-line. This trajectory of the arm endpoint in shown in Figure 5.30a. However, unlike in the example in Figure 5.29, the counter C_2 will now contain zero. Obviously, the target cannot be reached. [One may note that this outcome is quite different from the outcome in a similar situation with an RR arm, (Section 5.2).]

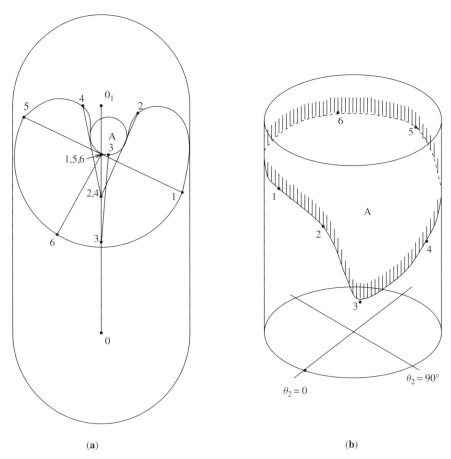

(a)   (b)

Figure 5.29 PR arm. (a) In W-space obstacle A interferes with the motion of link l_1. (b) In C-space this creates a band-like virtual obstacle. The arm can make a full circle around the obstacle, passing through points (1, 2, 3, 4, 5, 6, 1).

Local Cycles. As another similarity with the RR arm, the path planning procedure for the PR arm can create local cycles. Recall that a local cycle is a situation where the arm passes through some segment of its path more than once. (For definitions and further detail on local cycles, hit and leave points, and so on, see Section 5.2.1). In such a case, when the PR arm returns to the previously defined hit point, its counter C_2 will contain a value different from $|n| \cdot 2\pi, n = 0, 1, \ldots$. An example with a local cycle is shown in Figure 5.31, where the W-space scene contains four obstacles.

To see how a local cycle appears here, it is easier to follow the path first in C-space (Figure 5.31b) and then see what the path looks like in W-space. Assume that the chosen local direction is "left." The arm starts at point S and follows the M-line until it hits an obstacle, which happens to be obstacle D, and

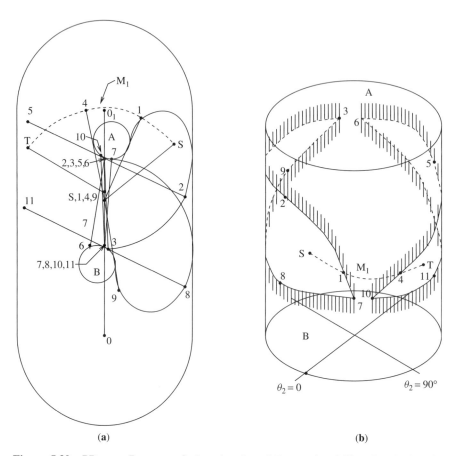

(a)   (b)

Figure 5.30 PR arm. Because of obstacles A and B, two band-like virtual obstacles appear in C-space and connect with each other in two spots. This creates two free areas of space disconnected from each other. As a result, point T cannot be reached from point S.

defines on it the first hit point, point 1 (Figure 5.31b). It then turns left (in the figure this corresponds to going up) and starts passing around obstacle D. On this path segment it passes point 2, and it meets the M-line again at point 3.

At point 3 the arm defines its first leave point; then it follows M-line until it hits obstacle B at point 4, which becomes the second hit point. Looking at Figure 5.31b, note that the "correct" way to proceed is so clear: One should turn right, go around obstacle C, meet there the M-line and follow it to T. So easy. Unfortunately, the arm has no information it would need to do this: It does not have the benefit of seeing the bird's-eye view of Figure 5.31 that we have. The only thing it knows is what comes to it from its sensors. Therefore, as required by the chosen local direction "left," the arm will turn left and attempt to pass around obstacle C. On this path segment, while passing through point 5, it will

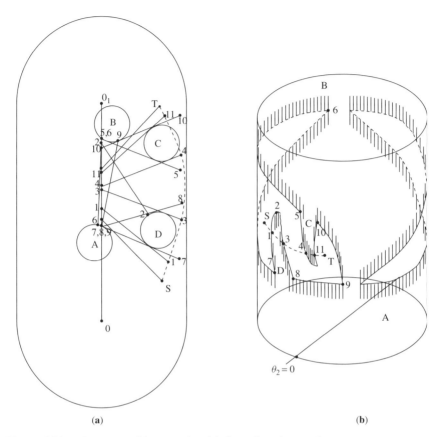

Figure 5.31 PR arm. In this example with four obstacles, as the arm attempts to move from S to T under the planning algorithm, a local cycle is created.

encounter obstacle B and and will try to pass it around. This will take the arm to point 6, where it will encounter obstacle A. As the arm starts following obstacle A, at point 7 it will meet obstacle D. While following the latter, it will arrive to point 1, the first hit point.

This moment will signal that a local cycle has been produced. (This should not be confused with completing a closed curve of a virtual boundary. No closed curve of a virtual boundary has been explored so far in this example). As the algorithm will prescribe the arm to look for a leave point that is closer to T than the lastly defined hit point, point 4, the arm will pass point 1 without even noticing it since the algorithm will not require remembering it. Nor will the counter C_2 contain anything of interest at this point: Since C_2 was last turned on at the second hit point, point 4, at the time of completion of the local cycle (point 1), its content will be some arbitrary number different from the prescribed $|n| \cdot 2\pi, n = 1, 2, \ldots$.

Hence the arm will pass point 1 and continue following the obstacle boundary. It will pass point 2 and then point 3. At point 8 the arm will encounter again obstacle A (albeit at this new point), switch to following it, encounter at point 9 obstacle B, follow it to point 10 where it meets obstacle D, and follow that obstacle to point 11, where it—finally!—meets the M-line again. Since point 11 is closer to T than the last hit point (point 4), point 11 is defined as the second leave point.

Notice that by the time the arm reaches point 1 for the second time, on its way to point 11 it will have encountered all four obstacles A, B, C, and D. Since during this motion the arm is passing from one obstacle to the other in an uninterrupted obstacle following, with no path segments in free space, the arm will perceive all four as one obstacle. [This is very clear from the C-space picture (Figure 5.31b).] Thereafter, starting at point 11, the arm will switch to following the M-line and eventually will arrive happily at the target point T.

The torturous exercise that we have just gone through is given only for those courageous souls who want to understand how a local cycle is formed and why it causes no difficulty to the algorithm. The arm will have no idea that it went through a local cycle. The theory developed here and in Section 5.2.1 guarantees that the cycle will not become an infinite loop. The different cases that we have considered thus far are necessary only to establish such a guarantee in the algorithm.

We emphasize again that the described path has been produced by the arm equipped with simple tactile sensing. As described in Section 5.2.5, a more advanced sensing will in general improve the performance—that is, shorten the path—quite markedly.

The Algorithm. We are now ready to formulate the sensor-based motion-planning algorithm for the prismatic-revolute (PR) arm. The following notation and parameters will be used in the procedure:

- Parameter F is used to handle an open curve of a Type II obstacle. It is set according to this rule:

    $F = +1$ when the arm, while following a virtual boundary, reaches the upper limit of joint value l_1.

    $F = -1$ when, while following a virtual boundary, the arm reaches the lower limit of joint value l_1.

    $F = 0$ at the first hit point of the virtual boundary.

- A flag is used to distinguish between the first and second open curves of the virtual boundary of a Type II obstacle.
- Counter C_2 is used to handle closed curves of virtual boundaries.
- Complementary M-lines, M_1 and M_2, are defined as before.
- Hit, H_j, and leave, L_j, points are defined as before (Section 5.2.1); $L_o = S$.

- Distances are Euclidean distances along M-line in the plane (l_1, θ_2).
- For specificity only, the local direction for passing around an obstacle is taken as "left" when the first hit point is defined.

The *PR-Arm Algorithm* consists of the following steps:

1. M_1-line is designated as the M-line. Set the flag down. Set $j = 1$. Go to Step 2.
2. Set counter C_2 to zero. Set $F = 0$. From point L_{j-1}, the arm moves along the M-line until one of the following occurs:
   (a) Target T is reached. The procedure stops.
   (b) An obstacle is encountered and a hit point, H_j, is defined. Choose the local direction "left". Turn on counter C_2. Go to Step 3.
3. The arm follows the virtual boundary until one of the following occurs:
   (a) Target is reached. The procedure stops.
   (b) M-line is met at a distance d from T such that $d < d(H_j, T)$; point L_j is defined. Set the flag down. Increment j. Go to Step 2.
   (c) The arm reaches one of the limits of link l_1 (this corresponds to an endpoint of one open curve of a virtual boundary) without ever meeting the M-line. Go to Step 4.
   (d) The arm returns to H_j (i.e., a closed curve along the virtual boundary has been completed) without ever encountering the M-line. The target cannot be reached. The procedure stops.
4. Depending on the value F, the flag condition, and the current arm position, one of the following occurs:
   (a) $F = 0$. Set F to $+1$ or -1 according to the rule above. Return to the last hit point. Choose the local direction "right." Go to Step 3.
   (b) Value F corresponds to the current curve endpoint (i.e., $+1$ for the upper limit and -1 for the lower limit of l_1); this means that the whole obstacle has been explored. The target cannot be reached. The procedure stops.
   (c) Value F does not correspond to the current curve endpoint, and the flag is down; this means the first open curve of a Type II obstacle has been explored. Set $j = 1$. Set the flag up. Designate M_2-line as the M-line. Return to S. Go to Step 2.
   (d) Value F does not correspond to the current curve endpoint, and the flag is up; this means that the second open curve of a Type II obstacle has been investigated. The target cannot be reached. The procedure stops.

5.8 TOPOLOGY OF ARM'S FREE CONFIGURATION SPACE

In the previous sections of this chapter we have considered an exhaustive list of five kinematic configurations of two-link arm manipulators (see Figure 5.1).

We observed that by studying the configuration space of each particular arm and making appropriate modifications in the basic sensor-based path planning procedure (which take into account topological properties of the arm workspace and configuration space), the basic algorithms developed in Chapter 3 for point mobile robots can be applied to planning the arm motion in space with arbitrary and previously unknown obstacles. Realization of these algorithms requires availability of sensing hardware that provides information about potential collision at every point of the arm body. We further learned in Section 5.2.5 that the developed algorithms can be extended to take advantage of more complex nontactile sensing, using the ideas of "algorithms with vision" developed in Section 3.6.

One may find it odd that each kinematic arm configuration in Figure 5.1 requires its own motion planning algorithm. While in general this is rather natural (after all, animals with different kinematics of their body have different gaits—compare a cat and a kangaroo), it would be indeed interesting to approach all these arms in a unified manner and attempt a unified motion planning strategy that would serve them all. In this section we will attempt a unified theory, and a unified motion planning algorithm, of planar arm manipulators.

Two comments on what follows:

- Looking ahead, we will see, in particular, that the topology of configuration space and planning algorithms for all arm configurations depicted in Figure 5.1 are special cases of the RR (revolute–revolute) arm. The consequence of this is that the sensor-based motion planning algorithm for RR Arm is the universal 2-link-Arm Algorithm. While, on the positive side, this means that one algorithm can serve all cases of 2D arm manipulators, it also means, on the negative side, that in cases of simpler arm configurations one would use a strategy that is more complex than the case in hand requires.

- The theory developed in this section is somewhat more complex than that presented elsewhere in this text (and it will not be used in the following chapters). It requires a prior exposure to topology. If this presents a problem to the reader, and/or if the comment above convinces the reader that a unified motion planning algorithm may be of a limited value, the reader can skip the rest of this section.

Let us recall some basics. The sought commonality of the five two-link arms of Figure 5.1 lies in the connectivity properties of their free space. Clearly, a path between two points in space exists if and only if both points lie in a connected area [or volume, in the three-dimensional (3D) case] of the free space. According to our model (Section 5.1.1), we deal with the case of highly incomplete information, with a situation when input information appears in real time and is of strictly local nature, as when coming from robot sensors. Since potential obstacles in the robot's environment are not known beforehand, the hope is that the robot can (a) infer some essential topological properties of the scene from a few incomplete encounters with obstacles and (b) use this knowledge to guarantee the solution to the motion planning task.

The sensor-based approach is thus a *topological approach*. The question being posed is as follows: Is there a solution to the robot motion planning problem based on topological (rather than geometrical and algebraic) characteristics of the space at hand—that is, a solution that does not require knowing shapes and dimensions of the objects involved, be it the robot itself or obstacles in its environment? Such procedures would allow robots to operate in a previously unknown environment filled with arbitrarily shaped obstacles. As we saw in the prior sections, a positive answer to this question carries significant advantages: It allows, first, to drop the computationally expensive requirement of algebraic representation and, second, drop the equally expensive requirement of complete information. The algorithms developed in earlier sections suggest that at least in some cases of arm kinematics the answer to the said question is "yes."

For the topological approach to work correctly, in the prior sections of this chapter it was vital that obstacle boundaries presented appropriate manifolds in the corresponding configuration space. In this section we will study the spatial relationships between the robot and obstacles and develop a set of conditions under which the obstacle boundaries present manifolds in the configuration space [107]. The analysis makes use of topology of the arm workspace and does not require algebraic representations.

Recall that kinematically a robot arm manipulator consists of connected rigid bodies, *links* and *joints*, which together possess some—say, d—*degrees of freedom (d-DOF)*, $d = 1, 2, \ldots$. The spatial arrangement (position and orientation) of the links and joints in the robot arm's workspace makes its kinematic *configuration*.

In all practical cases a robot configuration can be uniquely described by a finite number of parameters. Assuming that each DOF of the robot is implemented via the (most popular in practice) translational (prismatic and sliding are other terms used) or rotational (revolute) joint, each joint value represents such a parameter. For example, the three-dimensional robot in Figure 5.32 has nine degrees of freedom (9-DOF), and so its configuration can be described by the nine-tuple $(l_1, l_2, \theta_3, \theta_4, \theta_5, \theta_6, \theta_7, l_8, l_9)$. Here translational variables $l_1, l_2 \in \Re$ describe the Cartesian coordinates of the robot base unit; revolute variables $\theta_3 \ldots \theta_7 \in [0, 2\pi)$, respectively, parameterize the "waist," left and right "shoulders," and "elbows" of the robot arms; and translational variables l_8 and l_9 relate to the left- and right-hand effectors, respectively.

Two different sets of the nine-tuple parameters above would describe two distinct robot positions (configurations would be another term) in space. The collection of all possible robot configurations define the robot *configuration space (C-space)*. To emphasize the theoretical nature of this section, we will drop the terms W-space and C-space that we used above for the workspace and configuration space and we will use instead abbreviations *WS* and *CS*, accordingly.

Due to the presence of obstacles in *WS*, some regions in *CS* are not reachable; these regions collectively form the *configuration space obstacle*, denoted *CSO* or O_C. A reachable configuration is called a *free configuration (FC)*; the subspace that contains all free configurations is called the *free configuration space, FCS*. Points in *CS* represent robot configurations. A path in *CS* represents continuous

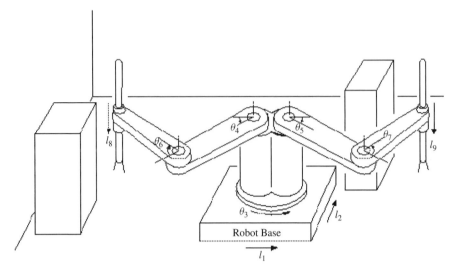

Figure 5.32 A 9-DOF robot with two arms attached to a common base.

motion of the robot in *WS*. By introducing *FCS*, the robot motion planning problem can be studied under a unified mathematical framework [108].

For sensor-based motion planning algorithms to work, it is essential that the *CSO* boundary presents manifolds. This topological property is not trivial and cannot be simply assumed. It has been shown in Ref. 109 that in general *CSO* boundaries are not manifolds. Consider, for example, the example shown in Figure 5.33a. The setting is such that the mobile robot *R* can barely squeeze into the opening in the obstacle *O*, while touching both opposite walls of *O* simultaneously. As a consequence, the *CSO* boundary, which consists of two rectangles and a straight-line segment that connects them (Figure 5.33b), is not a manifold.

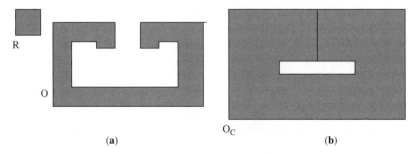

(a)   **(b)**

Figure 5.33 Interaction between a square-shaped mobile robot *R* and an obstacle *O*. **(a)** *WS*: The robot can barely squeeze into the opening of obstacle *O*. **(b)** *CS*: The corresponding *CSO* boundary consists of the inner and outer rectangles plus a straight-line segment that connects them.

Under what conditions will *CSO* boundaries present manifolds? We will show below that if a certain unrestrictive spatial relationship between the robot and obstacle is satisfied—for example, under no condition is the robot immobilized, nor does it need to squeeze between two obstacles as in Figure 5.33—then *FCS* is *uniformly locally connected (ULC)*. Although the ULC property is not sufficient to ensure manifold boundaries for the general case, we will show that for the two-dimensional (2D) case the ULC property guarantees that *FCS* is bounded by manifolds—in this case, simple closed curves.

We proceed as follows. First, a general robot with d translational and/or revolute degrees of freedom will be defined. *CS* is defined as a Euclidean space formed by the robot parameter variables. A physical obstacle is the interior of a connected compact point set in *WS*. We will show that *CSO* is a closed subset of *CS* (Corollary 5.8.1).

Then we will study the interaction between the robot and obstacles, and will define a set of conditions that correspond to certain undesirable degenerate situations (Conditions 5.8.1 to 5.8.4), such as when a part of or the whole robot is immobilized or when the robot can move between two obstacles only by simultaneously touching them both (Figure 5.33). We will show in Section 5.8.3 that after these situations are removed, *CSO* presents a uniformly locally connected subset of *CS*.

We will then show in Section 5.8.4 that ULC is a necessary condition for an open subset of a compact space to have manifold boundaries. This is also a sufficient condition for a 2D open subset of a compact space to have manifold boundaries—that is, simple closed curves. We will thus conclude that *FCS* of a 2-DOF robot is bounded by simple closed curves. More details pertinent to the material in this section can be found in [107].

5.8.1 Workspace; Configuration Space

As said above, kinematically a robot arm manipulator is an assembly of rigid links connected to each other by joints that permit the links' motion relative to each other [8]. Joints and links form *kinematic pairs*. As in prior sections, without loss of generality we limit the types of kinematic pairs to either translational (prismatic) or rotational (revolute). The degrees of freedom (DOF) of a robot are often referred to as its *mobility*, which is the number of independent variables that must be specified in order to locate all the links relative to each other.

The 9-DOF robot shown in Figure 5.32 has nine joints and nine links. The robot base is a link in which two translational joints l_1 and l_2 are implemented. The number of degrees of freedom of a robot is not necessarily equal to the number of links or the number of joints. Closed kinematic chains often have fewer DOF than the number of their links (joints). For example, a triangle-shaped planar closed kinematic chain with three links and three revolute joints has mobility zero.

We choose arbitrarily d independent joints, $J_1, \ldots, J_d$, to form a d-DOF robot, and we parameterize the robot configuration using the corresponding joint variables. With a reference system defined at its connecting joint, each kinematic pair can be specified by four scalar parameters. So, for the joint i, a_i is the link length, α_i is the link twist, θ_i is the angle between links L_i and L_{i-1}, and l_i is the distance between links L_i and L_{i-1}, $i = 1, \ldots, d$ [8].

Assume that a revolute joint has no joint limit, and a translational (prismatic) joint has its lower and upper limits. Let θ_i denote a revolute joint variable, let l_i be a translational joint variable, and let j_i be a (revolute or translational) joint variable when the exact type is not important; $i = 1, \ldots, d$. Assume for simplicity, and without loss of generality, that $l_i \in I^1$ and $\theta_i \in S^1$ (a 1-circle). The *configuration space (CS)* is defined as $C \overset{\triangle}{=} C_1 \times \cdots \times C_n$, where $C_i = I^1$ if the ith joint is translational and $C_i = S^1$ if it is revolute. In all combinations of cases, $C \cong I^{d_t} \times S^{d_r}$ for all d-DOF robots, where d_t and d_r are respectively the numbers of independent translational and revolute joints, $d_t + d_r = d$.

Lemma 5.8.1. *CS is compact and is of finite volume (area).*

Proof: The compactness is obvious since, by definition, *CS* is the cross product of a finite number of unit intervals (length 1) and circles (length 2π). The volume of *CS* is $(2\pi)^{d_r}$. **Q.E.D.**

For example, for a robot with two revolute joints, $C \cong S^1 \times S^1$ with area $2\pi \cdot 2\pi = 4\pi^2$; for a robot with two revolute joints and one prismatic joint, $C \cong S^1 \times S^1 \times I^1$ with volume $2\pi \cdot 2\pi \cdot 1 = 4\pi^2$.

We define the 3D robot *workspace* (denoted by *WS* or $\mathcal{W}$) as follows (its 2D counterpart can be defined accordingly by replacing $\mathfrak{R}^3$ with $\mathfrak{R}^2$):

Definition 5.8.1. *A robot link L_i, $i = 1, \ldots, d$, is defined as the interior of a connected and compact subset of $\mathfrak{R}^3$ homeomorphic to an open ball; for any point $x \in L_i$, let $x(j) \in \mathfrak{R}^3$ be the point that x would occupy in $\mathfrak{R}^3$ when the joint vector of the robot is $j \in C$. Let $L_i(j) = \bigcup_{x \in L_i} x(j)$. Then, $L_i(j) \subset \mathfrak{R}^3$ is a set of points the ith link occupies when the robot's joint vector is $j \in C$. Similarly, $L(j) \overset{\triangle}{=} \bigcup_i^n L_i(j) \subset \mathfrak{R}^3$ is a set of points the whole robot occupies when its joint vector is $j \in C$. The workspace is defined as*

$$\mathcal{W} \overset{\triangle}{=} \bigcup_{j \in C} \overline{L(j)}$$

where $\overline{L(j)}$ is the closure of $L(j)$.

We assume that L_i has a finite volume; thus, $\mathcal{W}$ is bounded.

The robot workspace may contain obstacles; each obstacle is a rigid body of an arbitrary shape. In the 2D case, an obstacle is of finite area and its boundary presents a simple closed curve. In the 3D case, an obstacle has a finite volume, its surface has a finite area, and it presents one or more orientable 2D manifolds. The assumption that *WS* has a finite volume (area) implies that the number of obstacles present in *WS* must be finite. Being rigid bodies, obstacles cannot intersect. We define 3D obstacles as follows (2D obstacles are defined accordingly):

Definition 5.8.2. *An obstacle, O_k, $k = 1, 2, \ldots, M$, is the interior of a connected and compact subset of $\Re^3$ satisfying*

$$\overline{O_{k_1}} \cap \overline{O_{k_2}} = \emptyset, \qquad k_1 \neq k_2 \qquad (5.6)$$

When the index k is not important, we use notation $O \overset{\triangle}{=} \bigcup_{k=1}^{M} O_i$ to represent the union of all obstacles in *WS*.

Definition 5.8.3. *The free workspace is*

$$\mathcal{W}_f \overset{\triangle}{=} \mathcal{W} - O$$

Lemma 5.8.2 follows from Definition 5.8.1.

Lemma 5.8.2. $\mathcal{W}_f$ *is a closed set.*

In *WS* the robot may simultaneously touch more than one obstacle. In such cases the obstacles involved effectively present one obstacle for the robot; in *CS* they present a single body.

Definition 5.8.4. *Configuration space obstacle (CSO) $O_C \subset \mathcal{C}$ is defined as*

$$O_C \overset{\triangle}{=} \{j \in \mathcal{C} : L(j) \cap O \neq \emptyset\}.$$

The free configuration space (FCS) is

$$\mathcal{C}_f \overset{\triangle}{=} \mathcal{C} - O_C$$

CSO may consist of many separate components. For convenience, we use the term "configuration space obstacle" to also refer to a component of O_C when the exact meaning is obvious from the context. A workspace obstacle can map into any large but finite number of disconnected *CSO* components (Theorem 5.8.2).

Theorem 5.8.1. O_C *is an open set in $\mathcal{C}$.*

Proof: Let $j^* \in O_C$. By Definition 5.8.4, there exists a point $x \in L$ such that $y = x(j^*) \in O$. Since O is an open set (Definition 5.8.2), there must exist an $\epsilon > 0$ such that the neighborhood $U(y, \epsilon) \subset O$. On the other hand, since $x(j)$ is a continuous function[7] from $\mathcal{C}$ to $\mathcal{W}$, there exists $\delta > 0$ such that for all $j \in U(j^*, \delta)$, $x(j) \in U(y, \epsilon) \subset O$; thus, $U(j^*, \delta) \subset O_C$, and O_C is an open set. **Q.E.D.**

The theorem gives rise to this statement:

Corollary 5.8.1. *FCS is a closed set.*

Being a closed set, $\mathcal{C}_f = \overline{\mathcal{C}_f}$. Thus, points on $\mathcal{C}_f$ boundary can be considered reachable by the robot.

---

[7]If $x \in L$ is a reference point on the robot, then $x(j)$ is the forward kinematics with respect to x and is thus continuous [8].

5.8.2 Interaction Between the Robot and Obstacles

Below we will need the property of space *uniform local connectedness (ULC)*. To derive it, we need to properly define the notion of a contact between the robot and an obstacle. To this end, four conditions will be stated (Conditions 5.8.1 to 5.8.4) that together define a contact. Mathematically, at position (joint vector) j^*, robot L is in contact with obstacle O if

$$L(j^*) \cap O = \emptyset \qquad \text{and} \qquad \overline{L(j^*)} \cap \overline{O} \neq \emptyset \tag{5.7}$$

The first relation in (5.7) states that $j^* \notin O_C$ (Definition 5.8.4), while the second relation states that $j^* \in \partial O_C$, where ∂O_C refers to the boundary of O_C. However, there may be situations where both relations of (5.7) hold but no obstacle exists in CS. Consider a robot manipulator with a fixed base, one link, and one revolute joint, along with a circular obstacle centered at the robot base O, as shown in Figure 5.34. Here relation (5.7) is satisfied at every robot configuration. Note, though, that the link can rotate freely in WS; this means that there are no obstacles, and hence no obstacle boundaries, in CS.

Therefore, robot configurations that satisfy Eq. (5.7) do not necessarily correspond to points on CSO boundaries. We modify the notion of contact by imposing additional conditions on the admissible robot and obstacle spatial relationships. As with any physical system, the term "contact" implies an existence of a force at the point of contact between the robot and the obstacle. In other words, for an object to present an obstacle for the robot, it must be possible for the robot to move in the direction of the force if the object were removed. With this definition of a contact, the robot shown in Figure 5.34 is not in contact with the obstacle at any position θ because at a point of "contact" it cannot exert a force upon

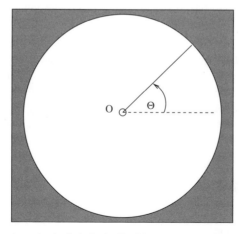

Figure 5.34 Shown is a single-link "robot" with a revolute joint at point O, along with a circular obstacle (shaded) also centered at O. With no obstacles in CS, the link can freely rotate about point O.

the obstacle. Mathematically, the removal of such "false contacts" translates into the following condition, which guarantees that each *CSO* component has at least one interior point:

Condition 5.8.1.   *Let $j^* \in C$ satisfy (5.7); that is, there exists $u \in \overline{L}$ such that $w = u(j^*) \in \overline{O}$. For given $\delta > 0$ and $\epsilon > 0$, define $\Delta O = \overline{O} \cap U(w, \delta)$, $\Delta L = \overline{L} \cap U(u, \delta)$, and $\Delta O_C = \{j \in U(j^*, \epsilon) : \Delta L(j) \cap \Delta O \neq \emptyset\}$. For any given $\gamma > 0$, there must exist $\epsilon \in (0, \gamma)$ and $\delta \in (0, \gamma)$ such that $\Delta O_C \neq \emptyset$.*

Theorem 5.8.2.   *An obstacle in WS can map into any large but finite number of CSO components in CS.*

Proof: We first design a simplified example showing that a simple obstacle in *WS* can map into two *CSO* components in *CS*. In Figure 5.35, the *WS* obstacle *O* produces two separate *CSO* components, each resulting from the interaction between *O* and each of the two vertical walls on the robot. Clearly, one can add additional vertical walls to the robot (and reduce the size of the obstacle if necessary) so that the number of *CSO* components will increase as well. This way one can create as many *CSO* components as one wishes.

On the other hand, by Condition 5.8.1, a *CSO* component must have an interior point. Also, by Theorem 5.8.1, *CSO* is an open set, and so its any interior point must have a neighborhood of positive radius r that is entirely enclosed in a *CSO* component. Thus the *CSO* component must occupy in *CS* a finite volume (area). By Lemma 5.8.1, C has a finite volume or area; hence the number of *CSO* components in *CS* must be finite. **Q.E.D.**

Figure 5.36a demonstrates another case of a "false contact," more complicated than the previous one. The corresponding *CSO* indeed has interior points, Figure 5.36b. By our definition of contact, at the configuration shown the robot is not in contact with the obstacle because it cannot exert any force upon the

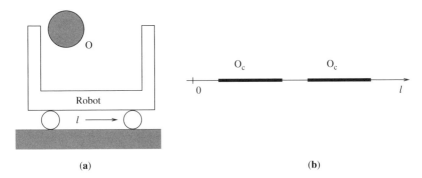

(a)   (b)

Figure 5.35   Illustration for Theorem 5.8.2. A single physical obstacle, O, can produce more than one CSO component. **(a)** WS: A simple robot with one translational joint. **(b)** CS: The corresponding two separate CSO components.

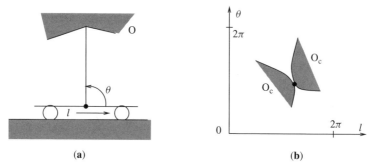

(a)   (b)

Figure 5.36   A 2-DOF robot with one translational joint and one revolute joint. **(a)** *WS*: The configuration shown is singular. **(b)** *CS*: The two resultant components of *CSO* almost touch each other—almost because at the corresponding point the robot is not in contact with obstacles, and hence the two *CSO* components are disjoint.

obstacle in the vertical direction. Thus the configuration corresponds not to a point on the *CSO* boundaries, but to an interior point of *FCO*.

Note that the robot configuration shown in Figure 5.36 is a *singular* configuration. That is, in this configuration a certain direction of motion in *WS*—here, along the vertical line—is impossible (or, theoretically, requires infinite joint velocities). In this configuration the robot cannot exert a force in the upward vertical direction. On the other hand, a small change in any joint variable will enable the robot to exert a force onto the obstacle. In *CS* this means that the two *CSO* components are very close but not touching each other (Figure 5.36b). We require that if the robot touches obstacles in two configurations that are arbitrarily close to each other, then the robot shall be able to move from one configuration to the other while maintaining the contact. This translates into the following condition:

Condition 5.8.2.   *Let ΔO, ΔL, and ΔO_C be as defined in Condition 5.8.1. $\forall c^1, c^2 \in \Delta O_C, c^1 \neq c^2$, there exists a continuous path $p_C : I \to \Delta O_C$ such that*

$$p_C(0) = c^1, \qquad p_C(1) = c^2, \qquad p_C(t) \in \Delta O_C \quad for\ t \in I.$$

We further restrict that only such interactions between the robot and an obstacle are allowed in which, for any robot configuration, at least some robot motion is possible:

Condition 5.8.3.   *For any $j^* \in C_f$ and any $\epsilon > 0$, there exists $c \in U(j^*, \epsilon) \subset C_f$, such that $c \neq j^*$ and c and j^* are connected within C_f.*

Condition 5.8.3 rules out the possibility that C_f contains isolated points. A degenerate case where the robot can barely squeeze between two or more obstacles while being in contact with both of them is not allowed. Removing this case simplifies the theory and is not restrictive for practice. Stated more precisely, this condition is as follows:

Condition 5.8.4. *Let $j^* \in C$ satisfy (5.7), and $\overline{L(j^*)} \cap \overline{O}$ contain at least two points. Let $u^1, u^2 \in \overline{L}$ be such that $w^i = u^i(j^*) \in \overline{O}$, $i = 1, 2$. For given δ^i and ϵ, define $\Delta O^i = \overline{O} \cap U(w^i, \delta^i)$, $\Delta L^i = \overline{L} \cap U(u^i, \delta^i)$, and $\Delta O_C^i = \{c \in U(j^*, \epsilon) : \Delta L^i(c) \cap \Delta O^i \neq \emptyset\}$, $i = 1, 2$. For any ϵ, $\delta^1, \delta^2 > 0$, $\Delta O_C^1 \cap \Delta O_C^2 \neq \emptyset$.*

We can now define the term "contact" in mathematical terms:

Definition 5.8.5. *The robot is in contact with an obstacle if and only if Eq. (5.7) and Conditions 5.8.1 to 5.8.4 are all satisfied.*

5.8.3   Uniform Local Connectedness

Together, Conditions 5.8.2 and 5.8.4 bring about an important topological property of CSO, the *uniform local connectedness (ULC)*. ULC guarantees that ∂O_C presents manifolds in the 2D case.

Definition 5.8.6. *Let E be a subset of a space X, and let x be any point of X. The set E is locally connected at x if, given any positive ϵ, there exists a positive δ such that any two points of $E \cap U(x, \delta)$ are joined by a connected set in $E \cap U(x, \epsilon)$.*

Note that x in this definition is not necessarily a point in E. However, if $x \notin \overline{E}$, then E is certainly locally connected at x, since for sufficiently small δ, there are no points in $E \cap U(x, \delta)$.

Definition 5.8.7. *A space or a set of points is uniformly locally connected if, given a positive ϵ, there exists a positive δ such that all pairs of points, x and y, of distance $\|x - y\| < \delta$ are joined by a connected subset of the space, of diameter less than ϵ.*

Theorem 5.8.3. *The open set CSO is uniformly locally connected.*

Proof: Since CS is compact and CSO is open and locally connected (Theorem 5.8.1), according to Ref. 110, VI.13.1, we only need to prove that CSO is locally connected at CSO boundary points.

Let $j^* \in C$ satisfy Eq. (5.7). If $\overline{L(j^*)} \cap \overline{O}$ contains only one single point, then Condition 5.8.2 guarantees that O_C is locally connected at j^*. Now assume $\overline{L(j^*)} \cap \overline{O}$ contains at least two points. Let $u^i \in \overline{L}$ satisfy $w^i = u^i(j^*) \in \overline{O}$; let ϵ^i and δ^i be such as to satisfy Condition 5.8.2 with respect to u^i, $i = 1, 2$. Let $\epsilon = \min(\epsilon^1, \epsilon^2)$, and let ΔO_C^i be as defined in Condition 5.8.4. According to Condition 5.8.4, there exists a point $c^+ \in \Delta O_C^1 \cap \Delta O_C^2$. According to Condition 5.8.2, every point c, $c \in \Delta O_C^i \subset U(j^*, \delta^i)$ is connected to $c^+ \in \Delta O_C^i \subset U(j^*, \delta^i)$ within ΔO_C^i. Thus, any two points $c^1, c^2 \in \Delta O_C^1 \cup \Delta O_C^2$ are connected in $U(j^*, \max(\epsilon^1, \epsilon^2))$, and so O_C is locally connected at j^*. **Q.E.D.**

5.8.4 The General Case of 2-DOF Arm Manipulators

We will now show that in 2D space ULC guarantees that *CSO* is bounded by simple closed curves. This result provides an effective tool for reducing the robot motion problem to the analysis of simple closed curves in *CS*.

Lemma 5.8.3. *CS of a 2-DOF robot presents one of the following: $C \cong I^1 \times I^1$, the unit square; $C \cong S^1 \times I^1$, a cylinder surface; or $C \cong S^1 \times S^1$, a torus surface.*

Proof: The lemma follows from the fact that $C_i \cong I^1$ for a translational joint, $C_i \cong S^1$ for a revolute joint, $i = 1, 2$, and $C = C_1 \times C_2$. **Q.E.D.**

Recall that the *CS* of a 2-DOF manipulator with two revolute joints (RR arm, Figure 5.37) presents the surface of a common torus. The *CS* of a PR or RP arm [arms (c), (d), and (e), Figure 5.1] is a cylinder, which is topologically a piece of the torus. A cylinder is obtained from the torus (Figure 5.5), for example, by making two vertical cuts in it. Finally, the *CS* of a PP arm [arm (b), Figure 5.1] is a rectangular piece of plane (formally, a unit square). It can be obtained from the torus by cutting a patch out of it. Figure 5.38 demonstrates how a cylinder and a patch are obtained from the torus.

The previous discussion and Lemma 5.8.3 suggest an important fact: The RR arm presents the most general case among all 2-DOF robot arms (see Figure 5.1). This means that the motion planning algorithm for the RR arm (Section 5.2.2), can be used to handle any two-link XX arm, X being X = (P or R). This will result in more calculations than the arm in question really needs, but it will work. For example, for the PR arm, the general (RR) algorithm may call for the third and fourth M-line, in which case the control will need to infer that for the PR arm those M-lines do not exist (see Figure 5.5).

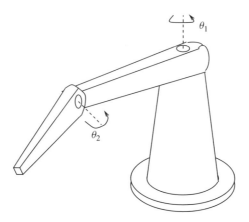

Figure 5.37 A 2-DOF robot with two revolute joints.

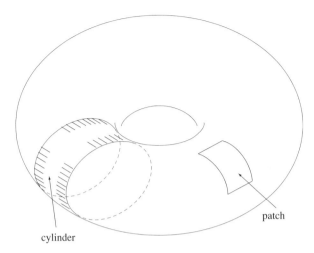

Figure 5.38   A cylinder (**left**) and a patch (**right**) can be cut out of a common torus.

Theorem 5.8.4.   *If a connected open set D in the torus surface T^1 is uniformly locally connected, then each component of its boundary is a simple closed curve, or a point, or null.*

The counterpart of this theorem in a closed plane is the so-called *Converse of Jordan's Theorem* [110, VI.16.2], which states that if a connected open set D in a closed plane $\overline{\Re^2}$ is uniformly locally connected, then each component of its boundary is a simple closed curve, or a point, or null.

It is of no surprise that the two theorems share the same necessary conditions. By definition, a simple closed curve is a continuum whose connectivity is destroyed by the removal of *two* points; this is a local property. The proof of Theorem 5.8.4 is analogous to its counterpart; due to its length, the proof appears in the Appendix to this chapter.

By Condition 5.8.3, the boundary of an obstacle cannot consist of isolated points. In addition, the boundary of the subset *CSO* on the torus is null if and only if *CSO* is either null, in which case there is no obstacle, or is the torus itself, which is a case of no interest. In summary, the following statement describes the *CSO* boundaries for a 2-DOF robot:

Corollary 5.8.2.   *For a 2-DOF robot with two revolute joints, if Conditions 5.8.1 to 5.8.4 are met, then the corresponding CSO is bounded by simple closed curves.*

Proof: The proof follows directly from Theorems 5.8.3 and 5.8.4. **Q.E.D.**

Theorem 5.8.5.   *For a 2-DOF robot, assuming Conditions 5.8.1 to 5.8.4 are met and joint limits, if any, are treated as obstacles, CSO is bounded by simple closed curves.*

Proof: CS of any 2-DOF robot can be considered as a closed subset of a torus, written as $C \subseteq T^1$. Let C^{-1} be the complement of C in T^1 and $O_C \subset C$ be *CSO*. Then, the set $O'_C \triangleq O_C \cup C^{-1}$ is open and locally connected in T^1; thus, according to Theorems 5.8.3 and 5.8.4, O'_C is bounded by simple closed curves in T^1. **Q.E.D.**

Now consider kinematic configurations of 2-DOF robots other than RR arms—that is, arms (b) to (e) in Figure 5.1. By Lemma 5.8.3, CS of each of these robot arms is homeomorphic to either the surface of a cylinder (RP or PR arms) or a disk (PP arm). In each of these cases, CS can be thought of as a closed subset of the torus. This also applies to 2-DOF arms with two revolute joints, one or both of which are constrained. The physical constraints on the joint range can be due to either the robot design or the obstacles in *WS*. This indicates that the constraints on joint limits can be treated as obstacles.

One might argue that, since the information about joint limits is known beforehand, there is no reason to treat them as unknown obstacles. This is true, especially if incorporating those limits is easy. Note, however, that the joint limits are not necessarily mutually independent. There are commercial robots in which the limit values of one joint depend on the values of other joints. This dependence is a function of the robot design and may be quite complex. Treating joint limits as obstacles is an elegant way to combine simplicity with universality.

To conclude, we have shown that if the spatial relationship between the robot arm and obstacles satisfies some reasonable and nonrestrictive for practice conditions, as defined by Conditions 5.8.1 to 5.8.4, then the corresponding configuration space obstacle (*CSO*) is uniformly locally connected. In particular, for the case of 2-DOF robot arms, this property guarantees that the free configuration space obstacle (*FCS*) is bounded by simple closed curves, which is an important feature upon which various sensor-based motion strategies developed above are based.

Both the simplicity and closedness of the boundary curves are important to these algorithms: It is these features that allow the algorithms to solve the motion planning problem with very little input information (local sensing only) about the robot environment. This is true for the simpler Bug family algorithms presented in Chapter 3, as well as for the more sophisticated algorithms developed in this chapter. The robot can correctly conclude that the target position is not reachable—a global property—by circumnavigating only parts of the obstacles involved.

5.9 APPENDIX

Proof of Theorem 5.2.1 (Section 5.2.1). Suppose the statement of the theorem does not hold, and the virtual boundary of some obstacle is formed in *C*-space by one or more simple (which is guaranteed by Lemma 5.2.1) but not closed curves. Take one such simple open curve and consider one of its two endpoints ("dead ends"). This endpoint corresponds to some arm position $P_* = (a_*, b_*)$. Along the curve, take a close, but distinct from P_*, arm position, $P_1 = (a_1, b_1)$. Apparently, once the arm moves from P_1 to P_*, the only way for it to continue its motion is to return to P_1.

Because this curve corresponds to the same virtual boundary, in both positions P_* and P_1 the arm is in contact with the obstacle. For P_* to be a dead end—that is, to be qualitatively different from P_1—there must be some other obstacle that affects the arm in the position P_* but does not affect it in position P_1. The idea of the proof is to show that for all possible occurrences of such additional obstacles at position P_*, there is a direction for passing around the obstacle different from the direction toward P_1. This means that the arm, after moving from P_1 to P_*, can then use this alternative direction. Hence, no dead ends are possible, and a virtual boundary must consist solely of simple closed curves.

Since the numbers a_i and b_i, $i = *, 1$, come in pairs, there are four possible combinations: (1) $a_* = a_1$, $b_* = b_1$; (2) $a_* \neq a_1$, $b_* = b_1$; (3) $a_* = a_1$, $b_* \neq b_1$; and (4) $a_* \neq a_1$, $b_* \neq b_1$. The first combination is of no interest because positions P_* and P_1 on the virtual boundary curve are assumed to be distinct. The second combination presents two arm solutions for a single point in W-space. For P_* and P_1 to be close to each other in C-space, this point must be located in the vicinity of the W-space boundary. If it is on the W-space boundary, then $a_* = a_1$, which brings us to the first combination. If the point is not on the W-space boundary, then consider a point P_2 on the curve such that it lies between, and is distinct from, points P_* and P_1. Because of the curve continuity, such a point exists.

It must be that $b_2 \neq b_*$ because otherwise P_2 would make the third distinct arm solution for the same point of W-space, and this is not possible. Using P_2 instead of P_1, we replace the second combination by either the third or the fourth combination. Therefore, out of the four possible combinations above, the ones to study are the third and the fourth. Now we consider the types of situations with the arm motion that lead to those combinations: Case 1 (for the third combination) and Cases 2 and 3 (for the fourth combination). (In Figures 5.39a, 5.39b, and 5.39c, known entities are shown in solid lines, and guessed entities are shown in dashed lines).

Case 1 (Figure 5.39a); $a_* = a_1$, $b_* \neq b_1$. For this case to occur, link l_1 must be constrained from one side by some obstacle, call it A. It cannot be constrained from both sides as it would be immobilized permanently. The only possibility for the position $P_* = (a_*, b_*)$ to be a dead end is if some other obstacle, B, constrains link l_2 as shown in Figure 5.39a. Clearly, it is possible to continue passing around the virtual obstacle while moving from P_* to some other position in its vicinity, $P_2 = (a_2, b_2)$, instead of P_1. Since P_2 does not lie between P_1 and P_* on the virtual boundary and since the virtual boundary is a simple curve, position P_* cannot be a dead end. The only other possibility to consider is having the obstacle A on the other side of the link l_1; this creates a symmetric situation.

Case 2 (Figure 5.39b); $a_* \neq a_1$, $b_* \neq b_1$. In this case, the segments l_2 in both positions P_1 and P_* do not intersect. This may occur only if in both positions the

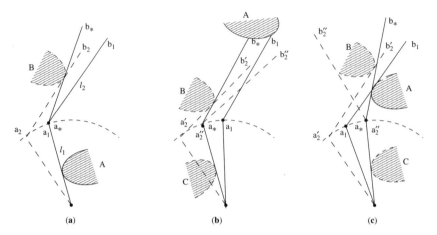

Figure 5.39 An illustration for the proof of Theorem 5.2.1.

arm endpoint (and not any other point of the arm body) is in contact with some obstacle, A. The dead end position may occur only if one or two other obstacles, B and C, appear as shown in Figure 5.39a. Clearly, it is always possible to move from P_* to a position distinct from P_1 (here, P_2' or P_2'', respectively). Thus, P_* cannot be a dead end position.

Case 3 (Figure 5.39c); $a_* \neq a_1$, $b_* \neq b_1$. In this case, the segments l_2 in both positions P_1 and P_* intersect each other. This may occur only if l_2 is "rolling" around some obstacle, A. Here, P_* may be a dead end only if one or two other obstacles, B and C, appear as shown in Figure 5.39c. Observe that positions P_2' and P_2'', respectively, are good alternatives to P_1. Therefore, P_* is not a dead-end position. This exhausts all possible cases and completes the proof. **Q.E.D.**

Proof of Lemma 5.2.2 (Section 5.2.1). Torus is a closed orientable manifold. The maximum number of closed curves needed to divide a given closed orientable manifold into two separate domains is determined by its connectivity numbers. The *first connectivity number* is known to define the maximum number of closed cuts that can be made on the surface without dividing it into separate domains. On the torus, the first connectivity number is equal to two [105]. The only arrangement for two closed cuts (two closed curves), a and b, that can be made on the torus without dividing it into separate domains is shown in Figure 5.40. According to Theorem 5.2.1, a virtual boundary consists of simple closed curves and thus cannot have self-intersections. Any other arrangement of two closed curves on the torus such that they do not touch or intersect each other produces at least two separate domains. Similarly, more than two simple closed curves produce more than two separate domains. Therefore, if some area on the torus is separated from the rest of it by simple closed curves then the boundary of this area consists of no more that two such curves. **Q.E.D.**

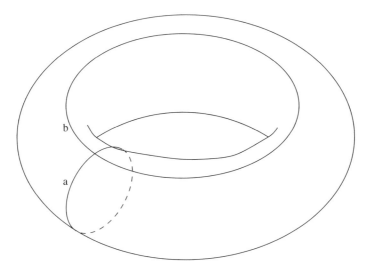

Figure 5.40   Closed cuts *a* and *b* do not divide the surface of the torus into separate areas but leave it as one area. However, addition of any other closed curve would cut it into two separate areas.

Separation Theorems on the Common Torus. (See Section 5.8.4). The purpose of this subsection is to prove Theorem 5.8.4. The proof for its planar counterpart can be found in Ref. 110, VI.16.2, which uses the concept of *regular grating* as the fundamental tool. We will use an analogous strategy to prove Theorem 5.8.4. Since the topology of a torus is different from that of a plane, we start with the modified definitions of *regular grating*, *k-chains*, and *k-cycles* in torus T^1, and proceed with the corresponding operations and properties (Theorems 5.9.1 to 5.9.7).

Several intermediate results are needed in order to prove Theorem 5.8.4. The proofs for some of these are the same as their planar counterparts, in which case we simply restate the statements and cite the source [110]. Proofs will be given to statements that are valid only for T^1.

Regular grating is a convenient tool for studying the connectivity of a subset of T^1. We show in Theorem 5.9.8 that a 1-cycle (a simple closed curve) does not necessarily separate a torus into two halves as it would in a plane. This major fact makes the proof of Theorem 5.8.4 different from its planar counterpart.

Finally, to prove Theorem 5.8.4 we need to show that if a region (a connected subset) D in T^1 is uniformly connected, then the connectivity of any of its boundary components is destroyed by the removal of two single points. This is done by drawing a *cross-cut L* connecting the same boundary component of D and showing that D-L has exactly two components (Theorem 5.9.12). The proof of Theorem 5.9.12 in turn requires the intermediate results of Theorems 5.9.9 to 5.9.11.

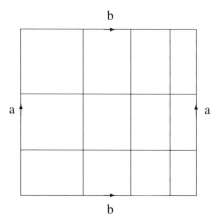

Figure 5.41   Rectangular grating of a torus surface.

Regular Grating. Let us represent T^1 as a unit square with its opposite sides, marked by a and b respectively, identified. A *regular grating* **G** in T^1 is formed by drawing a finite number of lines parallel to a or b. The lines become closed curves after the identified sides are glued together (see Figure 5.41). A grating must contain at least two lines, a and b, that cut the torus surface into a "rectangle."

The *2-cells* of **G** are the closures of the rectangular regions into which the unit square is divided. The *edges*, or *1-cells*, are the sides of the 2-cells, and the *vertices*, or *0-cells*, are their corners. All cells are bounded closed sets of points. When **G** contains a and b and at least another two lines that are parallel to a and b respectively, such as the case in Figure 5.41, then every edge of the grating evidently belongs to just two 2-cells, which lie on the opposite sides of it.

Theorem 5.9.1. *If F_1 and F_2 are nonintersecting closed sets in T^1, there exists a grating **G** no cells of which meet both F_1 and F_2.*

Proof: The nonintersecting compact sets F_1 and F_2 in T^1 are at positive distance δ apart. Let m be an integer exceeding $4/\delta$. A grating with the required property can evidently be formed by drawing m equidistant straight lines between, and parallel to, each pair of the opposite sides of the unit square. **Q.E.D.**

A *k-chain*, C^k $(k = 0, 1, 2)$, on grating **G** is any set of k-cells of **G**. The *sum* (modulo 2) of two k-chains, C_1^k and C_2^k, is the set of k-cells that belong to one and only one of C_1^k and C_2^k. It is denoted by $C_1^k + C_2^k$. The complement $(C^2)^{-1}$ of a 2-chain C^2 is the set of 2-cells of **G** not belonging to C^2. Thus, if Ω^2 denotes the 2-chain containing all the 2-cells of **G**, $(C^2)^{-1} = C^2 + \Omega^2$.

Addition of k-chains is commutative and associative, and

$$\sum_{1}^{q} C_i^k = C_1^k + C_2^k + \cdots + C_q^k \tag{5.8}$$

is the set of k-cells that belong to an odd number of the k-chains C_i^k. For any C^k, $C^k + C^k = 0$ (the "zero chain"), hence the equation $C_1^k + X^k = C_2^k$ is satisfied by $X^k = C_1^k + C_2^k$ and no other k-chain. Thus the k-chain on **G** form a commutative group under the operation of addition modulo 2.

The *boundary*, ∂C^k, of the k-chain C^k on **G** (for $k = 1, 2$) is the set of $(k - 1)$-cells of **G** that are contained in an odd number of k-cells of C^k. (The boundary of a 0-chain is not defined.)

Theorem 5.9.2. *[110, V.2.1]*

$$\partial(C_1^k + C_2^k) = \partial C_1^k + \partial C_2^k \qquad (k = 1, 2)$$

Since $\partial \Omega^2 = 0$, it follows from Theorem 5.9.2 that for any C^2,

$$\partial(C^2)^{-1} = \partial(C^2 + \Omega^2) = \partial C^2$$

A k-*cycle*, for $k = 1, 2$, is a k-chain whose boundary is zero; a 0-*cycle* is a 0-chain with an even number of 0-cells. The sum mod 2 of any set of k-cycle is a k-cycle (by Theorem 5.9.2, or directly from $k = 0$).

Theorem 5.9.3. *[110, V.2.2]. The boundary of any k-chain is a $(k - 1)$-cycle $(k = 1, 2)$.*

The k-chain C^k (a finite set whose members are k-cells) is to be distinguished from the union of its k-cells, a set of points denoted by $|C^k|$ and called the *locus of C^k*, or *the set of points covered by C^k*.

Clearly, $|C_1^k + C_2^k| \subset |C_1^k| \cup |C_2^k|$ in all cases, and $|C_1^k + C_2^k| = |C_1^k| \cup |C_2^k|$ if and only if the C_1^k and C_2^k have no common k-cells. Note that whereas $|C^2|^{-1}$ is an open set, $|(C^2)^{-1}|$ is a closed set, and in fact $|(C^2)^{-1}| = \overline{|C^2|^{-1}}$.

A k-chain C^k is, by definition, connected if its locus $|C^k|$ is connected. The maximal connected k-chains contained in any k-chain C^k are called the *components of C^k*. They have as loci the components of $|C^k|$.

Theorem 5.9.4. *[110, V.3.1] If K^k is a component of C^k, ∂K^k is the part of ∂C^k in K^k $(k > 0)$.*

Theorem 5.9.5. *If x and y form the boundary of a 1-chain C^1, they are connected in $|C^1|$.*

Proof: If this were not so, the component of C^1 containing x would have, by Theorem 5.9.4, the single vertex x as its boundary. This is contrary to Theorem 5.9.3. **Q.E.D.**

Theorem 5.9.6. *[110, V.3.3] For any C^2, $\partial|C^2| = |\partial C^2|$.*

It follows from Theorem 5.9.6 that Ω^2 and 0 are the only 2-cycles. Hence, if ∂C^2 is connected then so is C^2, because either $C^2 = \Omega^2$ or else every component of C^2 contains a non-null component of ∂C^2. On the other hand, if C^2 is connected, it does not necessarily mean ∂C^2 is connected.

It is sometimes necessary to pass from one grating, **G**, to another, **G***, by introducing additional lines. Such a new grating **G*** is called a *refinement* of the old. (It is convenient to agree that **G** is a refinement of itself.) A common refinement can be formed for any two gratings $\mathbf{G}_1$ and $\mathbf{G}_2$, by taking all the lines of $\mathbf{G}_1$ and $\mathbf{G}_2$ as cross lines.

To each k-chain, C^k on **G** corresponds to the *subdivided k-chain* C_*^k on **G***; C_*^k is the sum of the k-chains into which the k-chain C^k are subdivided. (0-chains are unaltered by subdivision: $C_*^0 = C^0$.) A subdivided chain has the same locus as its original: $|C_*^k| = |C^k|$.

Theorem 5.9.7. *[110, V.4.1]*

$$(C_1^k + C_2^k)_* = C_{1*}^k + C_{2*}^k, \qquad (\partial C^k)_* = \partial(C_*^k)$$

Corollary 5.9.1.   *C_*^k is a k-cycle if and only if C^k is a k-cycle.*

Separation Theorems. Let E be an open set of T^1, and let Γ^k be a k-cycle on a grating **G** defined in E. We say *the cycle Γ^k bounds in E*, denoted as $\Gamma^k \sim 0$, if there is, on some refinement **G*** of **G**, a $(k+1)$-chain C^{k+1} such that $|C^{k+1}| \subseteq E$ and $\partial C^{k+1} = \Gamma^k$ on **G***. We say Γ^k *is nonbounding* in E if $|\Gamma^k| \subseteq E$ but Γ^k does not bound in E. The notion of bounding can be used to study the connectivity of a subset of T^1: A simple closed curve (a 1-cycle) bounds if it separates a subset from the rest of the space; if two vertices (a 0-cycle) in **G** do not bound in E, then E is not connected.

By Jordan Curve Theorem [110, V.10.2], a simple closed curve always bounds in a plane (a sphere). The following statement indicates this is not necessarily true in a torus.

Theorem 5.9.8. *Every 1-cycle on a rectangular grating in T^1 is the boundary of either none 2-chain or just two 2-chains.*

Proof: The proof is by induction on the number of lines drawn across the unit square that represents T^1. On the grating consisting of a and b alone, the only 1-cycles are (a) the null sets, which bound two 2-chains (the zero chain and Ω^2), and (b) the cycles a, b, and $a + b$, each of which does not bound (i.e, does not separate a region from T^1) because there is only one rectangle in the grating, which is Ω and $\partial\Omega = 0$.

Assume the given grating, $\mathbf{G}_1$, is formed from a grating $\mathbf{G}_0$, for which the theorem holds true, by the addition of a line λ across the square; assume λ is parallel to a. Let Γ^1 be the given 1-cycle on $\mathbf{G}_1$. We denote by C^2 the sum of the 2-cells of $\mathbf{G}_1$ whose lower edges lie in the line λ and belong to Γ^1. The 1-cycle $\Gamma^1 + \partial C^2$ therefore contains no edge in λ. It follows that this 1-cycle is *the subdivided form of a 1-cycle* Γ_0^1 on $\mathbf{G}_0$, because $\Gamma^1 + \partial C^2$ contains no horizontal edge at a vertex x of λ and therefore, since it is a cycle, contains both or neither of the vertical edges at x, which together make up an edge of $\mathbf{G}_0$.

By hypothesis, if F_0^1 bounds, then there is a 2-chain C_0^2 on $\mathbf{G}_0$ such that $\Gamma_0^1 = \partial C_0^2$; and, if C_1^2 is the subdivided form of C_0^2 on $\mathbf{G}_1$, then $\partial C_1^2 = \Gamma^1 + \partial C^2$. Hence by Theorem 5.9.2

$$\partial(C_1^2 + C_2^2) = (\Gamma^1 + \partial C^2) + \partial C^2 = \Gamma^1$$

The 2-chain $(C_1^2 + C^2)^{-1}$ also has boundary Γ^1.

For every 1-cycle Γ_0^1 in $\mathbf{G}_0$ there is a 1-cycle Γ_1^1 in $\mathbf{G}_1$, which is the subdivided form, or a refinement, of Γ_0^1. Therefore, by Corollary 5.9.1, if every 1-cycle in $\mathbf{G}_1$ bounds, so will every 1-cycle in $\mathbf{G}_0$, which is impossible.

It has thus been shown that Γ^1 bounds none or two 2-chains. No 1-cycle bounds more than two 2-chains. For if $\partial C_a^2 = \partial C_b^2$, then $\partial(C_a^2 + C_b^2) = 0$ and therefore $C_a^2 + C_b^2$ is 0 or Ω, i.e. $C_b^2 = C_a^2$ or $(C_a^2)^{-1}$. **Q.E.D.**

Theorem 5.9.9. *[110, V.6.4] If 1-cycles Γ_1^1 and Γ_2^1 bound in T^1, and x and y do not lie on $|\Gamma_1^1|$ or $|\Gamma_2^1|$, at least one of the 1-cycles Γ_1^1, Γ_2^1, $\Gamma_1^1 + \Gamma_2^1$ bounds in $T^1 - (x \cup y)$.*

Theorem 5.9.10. *[110, V.9.1.2] Let F_1 and F_2 be closed sets in T^1. If the points x and y bound the 1-chain k_i not meeting F_i (for $i = 1, 2$) and if $k_1 + k_2 \sim 0$ in $T^1 - F_1 F_2$, then x and y are not separated by $F_1 \cup F_2$.*

Theorem 5.9.11. *[110, V.10.1] A simple arc in T^1 has a single complementary region.*

For a region D in T^1, a simple arc L, with one endpoint on ∂D and all its other points in D, is called an *end-cut*. If both endpoints are on ∂D and the rest in D, the arc is a *cross-cut*.

Theorem 5.9.12. *Given a region D of T^1, a cross-cut L can be drawn in D such that both endpoints of L in D are on the same component of ∂D, $(D-L)$ has two components, and L is contained in the boundary of both.*

Proof: First we prove that for any cross-cut L in D, $(D - L)$ has at most two components. Let u and v be the two endpoints of L. Suppose x, y, z are points of three different regions of $D - L$. By Theorem 5.9.11, there exists a 1-chain k_1 bounded by x and y, such that k_1 does not meet L; and by hypothesis, there exists a 1-chain k_2 bounded by x and y, such that k_1 does not meet D^{-1}, the complement of D in T^1. Since x and y are separated by $L \cup D^{-1}$, it follows from Theorem 5.9.10 that $k_1 + k_2$ is nonbounding in $(LD^{-1})^{-1}$, that is, in $T^1 - (u \cup v)$.

Similarly, $(y) + (z)$ is the common boundary of two 1-chains k_3 and k_4 that do not meet L and D^{-1} respectively, and $k_3 + k_4$ is nonbounding in $T^1 - (u \cup v)$. Therefore, by Theorem 5.9.9,

$$(k_1 + k_2) + (k_3 + k_4) \sim 0 \qquad \text{in} \qquad T^1 - (u \cup v) \qquad (5.9)$$

The 1-chain $k_1 + k_3$ does not meet L and is bounded by $(x + y) + (y + z)$—that is, by x and z. Similarly, $k_2 + k_4$ does not meet D^{-1} and is bounded by x and z. The sum

$$(k_1 + k_3) + (k_2 + k_4)$$

is identical to (5.9) and therefore bounds in $(LD^{-1})^{-1}$. Hence, x and z are not bounded by $L \cup D^{-1}$, contrary to the hypothesis.

Now we show that a cross-cut L can be drawn in D such that $(D - L)$ has at least two components. Consider one component, F, of ∂D and assume that it contains at least two points, u and v. Define a grating $\mathbf{G}$ on T^1 such that no 2-cells meet more than one component of D or D^{-1} (Theorem 5.9.1), and that both u and v are vertices of $\mathbf{G}$. Let K be the set of 2-cells in $\mathbf{G}$ that meet F. Then, ∂K bounds at least two regions; one contains F and the others do not. Let κ_1 be 1-chain such that $|\kappa_1| = |\partial K| \cap D$ and let κ_2 be 1-chain such that $|\kappa_2| = |\partial K| \cap D^{-1}$. Clearly, $|\kappa_1|$ and $|\kappa_2|$ are disjoint. Thus, $D - |\kappa_1|$ separates D into at least two components. Since $u \in F$ is a vertex in $\mathbf{G}$, one of the four 1-cells adjacent to u must have its other vertex in κ_1. Denote the 1-cell that connects u and κ_1 by C_1^1, denote the 1-cell that connects v and κ_1 by C_2^1, and let L be the locus of any 1-chain in $C_1^1 + C_2^1 + \kappa_1$ bounded by u and v. $(D - L)$ has at least two components. **Q.E.D.**

A point of ∂D is *accessible* from D if it is an endpoint of an end-cut in D.

Theorem 5.9.13. *[110, VI.14.4] If D is locally connected at u, a point of ∂D, then u is accessible from D.*

If u is an accessible point of ∂D and v a point of D, then there is an end-cut in D joining u and v. If L_u is an end-cut in D joining u and x, and L_v a segment arc joining x and v in D, let y be the first point of L_u, counting from u, that lies in L_v. Then, the arc segments uy of L_u and yv of L_b together form the required end-cut. Similarly, if u and v are accessible points of ∂D, there is a cross-cut uv in D. We are now ready to prove Theorem 5.8.4.

Proof of Theorem 5.8.4. The proof is analogous to that of Ref. 110, VI.16.2. Let F be a component of ∂D. Suppose F contains at least two points, u and v. Since D is uniformly locally connected, u and v are accessible (Theorem 5.9.13) and can be joined by a cross-cut, L, such that $(D - L)$ determines two regions, D_1 and D_2, in D (Theorem 5.9.12). No points of F except u and v can belong to the boundary of both of these regions, because if w were such a point, any arc in D joining two points x and y of D_1 and D_2 within $\frac{1}{2}\rho(w, L)$ of w would necessarily meet L and would therefore have diameter at least $\frac{1}{2}\rho(w, L)$. The region would not be locally connected at w. Thus, the sets

$$E_i = (F \cap \overline{D_i}) - (u \cup v), \qquad i = 1, 2$$

have no common points. Their union is $(F - (u \cup v))$, and since

$$E_i = (F - (u \cup v)) \cap \overline{D_i}$$

they are closed in $(F - (u \cup v))$. Thus, $(F - (u \cup v))$ is not connected because it has partition $E_1|E_2$, so F is a continuum whose connectivity is destroyed by the removal of any two points. F is therefore a simple closed curve. **Q.E.D.**

5.10 EXERCISES

1. Consider a planar RR (revolute–revolute) arm manipulator (see Figure 5.E.1). The arm's links l_1 and l_2 are line segments of the same length l. There is an

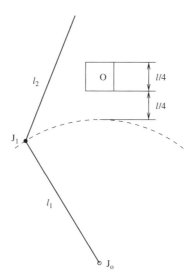

Figure 5.E.1

obstacle O that lies fully in the arm workspace, as shown. Obstacle O is a square with the side length $l/4$.

Recall that the method of motion planning with complete information (Piano Mover's model) [13] requires one to first approximate the configuration space (C-space) image of obstacle O by a polygon. Call the obstacle in C-space O_C and call the approximating polygon P_C. Let δ be the maximum linear deviation of P_C from O_C. Evaluate the minimum order (the number of sides) of the polygon P_C that will keep δ within 1% of the perimeter of P_C.

2. Consider a two-link planar robot arm with revolute joints, as shown in Figure 5.E.2. Joint values θ_1 and θ_2 of the joints J_0 and J_1 can change within the range $(0, 2\pi)$, with no mechanical stops. The arm has a sensing capability spread uniformly along its body so that any point of its body can sense surrounding objects within the distance r_v, called the arm's radius of vision. The relevant dimensions are: $l_1 = 30$, $l_2 = 20$, $R = 2$, $r_v = 3$.

   Develop *general equations* necessary to compute the boundary of the robot's sensing field (or fields if applicable) in the corresponding configuration space (C-space), as a function of the joint angles θ_1 and θ_2. The resulting boundary will consist of pieces of straight lines and curves. If you use any simplifications or approximations, acknowledge and justify them.

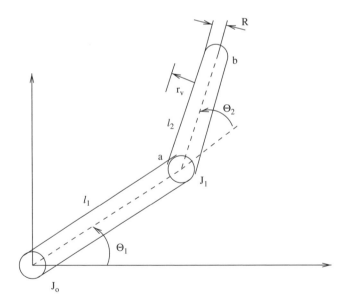

Figure 5.E.2

Draw the C-space as a flattened torus—that is, a square in the plane (θ_1, θ_2), with the coordinates of its corners, respectively, $(0, 0)$, $(0, 2\pi)$, $(2\pi, 2\pi)$, $(2\pi, 0)$. For the six positions of the arm characterized by the six pairs (θ_1, θ_2)

below, show in this plane the outline of the sensing field(s) for each point. The points (θ_1, θ_2) to check are, respectively:

$$(90°, 0°), \quad (90°, 45°), \quad (90°, 90°), \quad (30°, 150°), \quad (30°, 135°), \quad (30°, 90°)$$

3. Come up with an example of sensor-based motion planning for the prismatic–prismatic (PP) arm manipulator algorithm (Section 5.4). The scene in the example shall feature nonconvex shapes of the arm link l_2 and of obstacles in the workspace. Compare the results with the example in Figure 5.23.

4. Modify the sensor-based motion planning algorithm for the PP arm (Section 5.4), so as to incorporate in it a possibility of the rear end of link l_2 to collide with obstacles. Connect this modification with the test for target reachability, and show that the modified algorithm converges. Provide an example where the arm goes through a local cycle.

CHAPTER 6

Motion Planning for Three-Dimensional Arm Manipulators

> The robot is going to lose. Not by much. But when the final score is tallied, flesh and blood is going to beat the damn monster.
>
> — *Adam Smith, philosopher and economist, 1723–1790*

6.1 INTRODUCTION

We are continuing developing SIM (Sensing–Intelligence–Motion) algorithms for robot arm manipulators. The cases considered in Chapter 5 all deal with arm manipulators whose end effectors (hands) move along a two-dimensional (2D) surface. Although applications do exist that can make use of those algorithms—for example, assembly of microelectronics on a printed circuit board is largely limited to a 2D operation—most robot arm manipulators live and work in three-dimensional (3D) space. From this standpoint, our primary objective in Chapter 5 should be seen as preparing the necessary theoretical background and elucidating the relevant issues, before proceeding to the 3D case. Sensor-based motion planning algorithms should be able to handle 3D space and 3D arm manipulators. Developing such strategies is the objective of this chapter. As before, the arm manipulators that we consider are simple open kinematic chains.

Is there a fundamental difference between motion planning for two-dimensional (2D) and 3D arm manipulators? The short answer is yes, but the question is not that simple. Recall a similar discussion about mobile robots in Chapter 3. From the standpoint of motion planning, mobile robots differ from arm manipulators: They have more or less compact bodies, kinematics plays no decisive role in their motion planning, and their workspace is much larger compared to their dimensions. For mobile robots the difference between the 2D and 3D cases is absolute and dramatic: Unequivocally, if the 2D case has a definite and finite solution to the planning problem, the 3D case has no finite solution in general.

The argument goes as follows. Imagine a bug moving in the two-dimensional plane, and imagine that on its way the bug encounters an object (an obstacle).

Sensing, Intelligence, Motion, by Vladimir J. Lumelsky
Copyright © 2006 John Wiley & Sons, Inc.

Assume the bug's target location is somewhere on the other side of the obstacle. One way for it to continue its motion is to first pass around the obstacle. The bug has only two options: It can pass the obstacle from the left or it can pass it from the right, clockwise or counterclockwise. If neither option leads to success—let us assume it is a smart bug, with a reasonably good motion planning skills—the goal is not achievable. In this case, by slightly exaggerating the bug's stubbornness, we will note that eventually the bug will come back to the point on the obstacle where it started. It took only one simple going around, all in one direction, to explore the whole obstacle. That is the essence of the 2D case.

Imagine now a fly that is flying around a room—that is, in 3D space. Imagine that on its way the fly encounters a (3D) obstacle—say, a child's balloon hanging on a string. Now there is an infinite number of routes the fly can take to pass around the obstacle. The fly would need to make a great many loops around the obstacle in order to explore it completely. That's the fundamental difficulty of the 3D case; in theory it takes an infinitely long path to explore the whole obstacle, even if its dimensions and volume are finite and modest.[1] The point is that while in the 2D case a mobile robot has a theoretically guaranteed finite solution, no such solution can be guaranteed for a 3D mobile robot. *The 3D sensor-based motion planning problem is in general intractable.*

The situation is more complex, but also not as hopeless, for 3D arm manipulators. Try this little experiment. Fix your shoulder and try to move your hand around a long vertical pole. Unlike a fly that can make as many circles around the pole as it wishes, your hand will make about one circle around the pole and stop. What holds it from continuing moving in the same direction is the arm's kinematics and also the fact that the arm's base is "nailed down." The length of your arm links is finite, the links themselves are rigid, and the joints that connect the links allow only so much motion. These are natural constraints on your arm movement. The same is so for robot arm manipulators. In other words, the *kinematic constraints of an arm manipulator impose strong limitations on its motion.*

This fact makes the problem of sensor-based motion planning for 3D arm manipulators manageable. The hope is that the arm kinematics can be effectively exploited to make the problem tractable. Furthermore, those same constraints promise a constructive test of target reachability, similar to those we designed above for mobile robots and 2D arm manipulators.

As noted by Brooks [102], the motion planning problem for a manipulator with revolute joints is inherently difficult because (a) the problem is nondecomposable, (b) there may be difficulties associated with rotations, (c) the space representation and hence the time execution of the algorithm are exponential in the number of robot's degrees of freedom of the objects involved, and (d) humans are especially poor at the task when much reorientation is needed, which makes it difficult to

---

[1]One may argue that the fly can use its vision to space its loops far enough from each other, making the whole exercise quite doable. This may be true, but not so in general: The room may be dark, or the obstacle may be terribly wrinkled, with caves and overhangs and other hooks and crannies so that the fly's vision will be of little help.

develop efficient heuristics. This is all true, and indeed more true for arms with revolute joints—but these difficulties have been formulated for the motion planning problem with complete in formation. Notice that these difficulties above did not prevent us from designing rather elegant sensor-based planning algorithms for 2D arms with revolute joints, even in the workspace with arbitrarily complex obstacles. The question now is how far we can go with the 3D case.

It was said before that this is a difficult area of motion control and algorithm design. As we will see in Chapter 7, human intuition is of little help in designing reasonable heuristics and even in assessing proposed algorithms. Doing research requires expertise from different areas, from topology to sensing technology. There are still many unclear issues. Much of the exciting research is still waiting to be done. Jumping to the end of this chapter, today there are still no provable algorithms for the 3D kinematics with solely revolute joints. While this type of kinematics is just one mechanism among others in today's robotics, it certainly rules the nature.

As outlined in Section 5.1.1, we use the notion of a *separable* arm [103], which is an arm naturally divided into the *major linkage* responsible for the arm's *position planning* (or *gross motion*), and the *minor linkage* (the hand) responsible for the *orientation planning* of the arm's end effector. As a rule, existing arm manipulators are separable. Owing to the fact that three degrees of freedom (DOF) is the minimum necessary for reaching an arbitrary point in 3D space, and another three DOF are needed to provide an arbitrary orientation for the tool—six DOF in total as a minimum—many 3D arm manipulators' major linkages include three links and three joints, and so do typical robot hands. Our motion planning algorithms operate on the major linkage—that is, on handling gross motion and making sure that the hand is brought into the vicinity of the target position. The remaining "fine tuning" for orientation is usually a simpler task and is assumed to be done outside of the planning algorithm. For all but very few unusual applications, this is a plausible assumption.

While studying topological characteristics of the robot configuration space for a few 3D kinematics types, we will show that obstacle images in these configuration spaces exhibit a distinct property that we call *space monotonicity*: For any point on the surface of the obstacle image, there exists a direction along which all the remaining points of the configuration space belong to the obstacle. Furthermore, the sequential connection of the arm links results in the property called *space anisotropy* of the configuration space, whereby the obstacle monotonicity presents itself differently along different space axes.

The space monotonicity property provides a basis for selecting directions of arm motion that are more promising than others for reaching the target position. By exploiting the properties of space monotonicity and anisotropy, we will produce motion planning algorithms with proven convergence. No explicit or implicit beforehand calculations of the workspace or configuration space will be ever needed. All the necessary calculations will be carried out in real time in the arm workspace, based on its sensing information. No exhaustive search will ever take place.

As before with 2D arms, motion planning algorithms that we will design for 3D arms will depend heavily on the underlying arm kinematics. Each kinematics type will require its own algorithm. The extent of algorithm specialization due to arm kinematics will be even more pronounced in the 3D case than in the 2D case. Let us emphasize again that this is not a problem of depth of algorithmic research but is instead a fundamental constraint in the relationship between kinematics and motion. The same is true, of course, in nature: The way a four-legged cat walks is very different from the way a two-legged human walks. Among four-legged, the gaits of cats and turtles differ markedly. One factor here is the optimization process carried out by the evolution. Even if a "one fits all" motion control procedure is feasible, it will likely be cumbersome and inefficient compared to algorithms that exploit specific kinematic peculiarities. We observed this in the 2D case (Section 5.8.4): While we found a way to use the same sensor-based motion planning algorithm for different kinematics types, we also noted the price in inefficiency that this universality carried. Here we will attempt both approaches.

This is not to say that the general approach to motion planning will be changing from one arm to another; as we have already seen, the overall SIM approach is remarkably the same independent of the robot kinematics, from a sturdy mobile robot to a long-limbed arm manipulator.

As before, let letters P and R refer to prismatic and revolute joints, respectively. We will also use the letter X to represent either a P or a R joint, $X = [P, R]$. A three-joint robot arm manipulator (or the major linkage of an arm), XXX, can therefore be one of eight basic kinematic linkages: PPP, RPP, PRP, RRP, PPR, RPR, PRR, and RRR. As noted in Ref. 111, each basic linkage can be implemented with different geometries, which produces 36 linkages with joint axes that are either perpendicular or parallel to one another. Among these, nine degenerate into linkages with only one or two DOF; seven are planar. By also eliminating equivalent linkages, the remaining 20 possible spatial linkages are further reduced to 12, some of which are of only theoretical interest.

The above sequence XXX is written in such an order that the first four linkages in it are XXP arms; in each of them the outermost joint is a P joint. Those four are among the five major 3D linkages (see Figure 6.1) that are commonly seen in industry [111–113] and that together cover practically all today's commercial and special-purpose robot arm manipulators. It turns out that these four XXP arms are better amenable to sensor-based motion planning than the fifth one (Figure 6.1e) and than the remaining four arms in the XXX sequence. It is XXP arms that will be studied in this chapter.

While formally these four linkages—PPP, RPP, PRP, and RRP—cover a half of the full group XXX, they represent four out of five, or 80%, of the linkages in Figure 6.1. Many, though not all, robot arm applications are based on XXP major linkages. Welding, assembly, and pick-and-place robot arms are especially common in this group, one reason being that a prismatic joint makes it easy to produce a straight-line motion and to expand the operating space. The so-called SCARA arm (Selective Compliance Assembly Robot Arm), whose major linkage

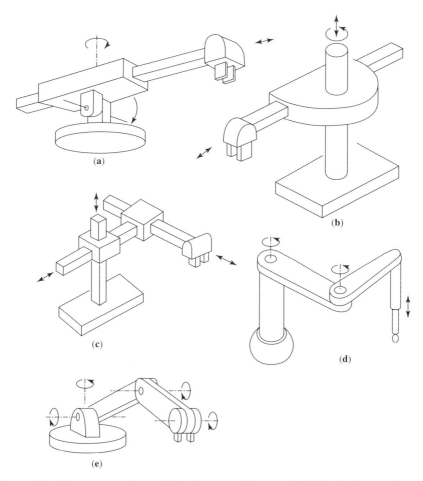

Figure 6.1 Five commonly used 3D robot arm linkages: **(a)** *RRP*, "polar coordinates" arm, with a spherical workspace; **(b)** *PRP*, "cylindrical coordinates" arm, with a cylindrical workspace; **(c)** *PPP*, Cartesian arm, with a "cubicle" workspace; **(d)** *RRP*, SCARA-type arm, with a cylindrical workspace; and **(e)** *RRR*, "articulate" arm, with a spherical workspace.

is, in our notation, an *RRP* arm, is especially popular among assembly oriented industrial arm manipulators.

Recall our showing in Chapter 5 that *RR*, the arm with two revolute joints, is the most general among *XX* arms, $X = [P, R]$. We will similarly show that the arm *RRP* is the most general case among *XXP* arms. The algorithm that works for the *RRP* arm will work for other *XXP* arms.

While arm configurations shown in Figure 6.1 have nice straight-line links that are mutually perpendicular or parallel, our SIM approach does not require such properties unless specified explicitly.

We will first analyze the *PPP* arm (Section 6.2), an arm with three slid-ing (prismatic) joints (it is often called the *Cartesian arm*), and will develop a sensor-based motion planning strategy for it. Similar to the 2D Cartesian arm, the SIM algorithm for a 3D Cartesian arm turns out to be the easiest to visu-alize and to design. After mastering in this case the issues of 3D algorithmic machinery, in Section 6.3 we will turn our attention to the general case of an *XXP* linkage. Similar to the material in Section 5.8, some theory developed in Section 6.2.4 and Sections 6.3 to 6.3.6 is somewhat more complex than most of other sections.

As before, we assume that the arm manipulator has enough sensing to sense nearby obstacles at any point of its body. A good model of such sensing mech-anism is a sensitive skin that covers the whole arm body, similar to the skin on the human body. Any other sensing mechanism will do as long as it guarantees not missing potential obstacles. Similar to the algorithm development for the 2D case, we will assume tactile sensing: As was shown in prior chapters, the algo-rithmic clarity that this assumption brings is helpful in the algorithm design. We have seen in Sections 3.6 and 5.2.5 that extending motion planning algorithms to more information-rich sensing is usually relatively straightforward. Regarding the issues of practical realization of such sensing, see Chapter 8.

6.2 THE CASE OF THE PPP (CARTESIAN) ARM

The model, definitions, and terminology that we will need are introduced in Section 6.2.1. The general idea of the motion planning approach is tackled in Section 6.2.2. Relevant analysis appears in Sections 6.2.3 and 6.2.4. We formu-late, in particular, an important necessary and sufficient condition that ties the question of existence of paths in the 3D space of this arm to existence of paths in the projection 2D space (Theorem 6.2.1). This condition helps to lay a foun-dation for "growing" 3D path planning algorithms from their 2D counterparts. The corresponding existential connection between 3D and 2D algorithms is for-mulated in Theorem 6.2.2. The resulting path planning algorithm is formulated in Section 6.2.5, and examples of its performance appear in Section 6.2.6.

6.2.1 Model, Definitions, and Terminology

For the sake of completeness, some of the material in this section may repeat the material from other chapters.

Robot Arm. The robot arm is an open kinematic chain consisting of three *links*, l_1, l_2, and l_3, and three *joints*, J_1, J_2, and J_3, of *prismatic* (sliding) type [8]. Joint axes are mutually perpendicular (Figure 6.2). For convenience, the arm endpoint P coincides with the upper end of link l_3. Point J_i, $i = 1, 2, 3$, also denotes the center point of joint J_i, defined as the intersection point between the axes of link l_i and its predecessor. Joint J_1 is attached to the robot base O and is the origin

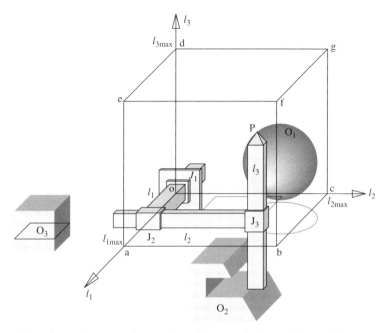

Figure 6.2 The work space of a 3D Cartesian arm: l_1, l_2, and l_3 are links; J_1, J_2, and J_3 are prismatic joints; P is the arm endpoint. Each link has the front and rear end; for example, J_3 is the front end of link l_2. O_1, O_2, and O_3 are three physical obstacles. Also shown in the plane (l_1, l_2) are obstacles' projections. The cube *abcodefg* indicates the volume whose any point can be reached by the arm endpoint.

of the fixed reference system. Value l_i also denotes the joint variable for link l_i; it changes in the range $l_i = [l_{i\,min}, l_{i\,max}]$. Assume for simplicity zero minimum values for all l_i, $l_i = [0, l_{i\,max}]$; all $l_{i\,max}$ are in general different.

Each link presents a *generalized cylinder* (briefly, a *cylinder*)—that is, a rigid body characterized by a straight-line axis coinciding with the corresponding joint axis, such that the link's cross section in the plane perpendicular to the axis does not change along the axis. A cross section of link l_i presents a simple closed curve; it may be, for example, a circle (then, the link is a common cylinder), a rectangle (as in Figure 6.2), an oval, or even a nonconvex curve. The link cross section may differ from link to link.[2]

The *front ends* of links l_1 and l_2 coincide with joints J_2 and J_3, respectively; the front end of link l_3 coincides with the arm *endpoint P* (Figure 6.2). The opposite end of link l_i, $i = 1, 2, 3$, is its *rear end*. Similarly, the *front (rear) part* of link l_i is the part of variable length between joint J_i and the front (rear) end of the link. When joint J_i is in contact with an obstacle, the contact is considered to be with link l_{i-1}.

---

[2]More precisely, we will see that only link l_3 has to be a generalized cylinder to satisfy the motion planning algorithm; links l_1 and l_2 can be of arbitrary shape.

For the sensing mechanism, we assume that the robot arm is equipped with a kind of "sensitive skin" that covers the surfaces of arm links and allows any point of the arm surface to detect a contact with an approaching obstacle. Other sensing mechanisms are equally acceptable as long as they provide information about potential obstacles at every point of the robot body. Depending on the nature of the sensor system, the contact can be either physical—as is the case with tactile sensors—or proximal. As said above, solely for presentation purposes we assume that the arm sensory system is based on tactile sensing.[3]

The Task. Given the start and target positions, S and T, with coordinates $S = (l_{1S}, l_{2S}, l_{3S})$ and $T = (l_{1T}, l_{2T}, l_{3T})$, respectively, the robot is required to generate a continuous collision-free *path* from S to T if one exists. This may require the arm to maneuver around obstacles. The act of *maneuvering around an obstacle* refers to a motion during which the arm is in constant contact with the obstacle. Position T may or may not be reachable from S; in the latter case the arm is expected to make this conclusion in finite time. We assume that the arm knows its own position in space and those of positions S and T at all times.

Environment and Obstacles. The 3D volume in which the arm operates is the *robot environment*. The environment may include a finite number of *obstacles*. Obstacle positions are fixed. Each obstacle is a 3D rigid body whose volume and outer surface are finite, such that any straight line may have only a finite number of intersections with obstacles in the workspace. Otherwise obstacles can be of arbitrary shape. At any position of the arm, at least some motion is possible. To avoid degeneracies, the special case where a link can barely squeeze between two obstacles is treated as follows: We assume that the clearance between the obstacles is either too small for the link to squeeze in between, or wide enough so that the link can cling to one obstacle, thus forming a clearance with the other obstacle. The number, locations, and geometry of obstacles in the robot environment are not known.

W-Space and W-Obstacles. The robot *workspace (W-space or W)* presents a subset of Cartesian space in which the robot arm operates. It includes the *effective workspace*, any point of which can be reached by the arm end effector (Figure 6.3a), and the outside volumes in which the rear ends of the links may also encounter obstacles and hence also need to be protected by the planning algorithm (Figure 6.3b). Therefore, W is the volume occupied by the robot arm when its joints take all possible values $l = (l_1, l_2, l_3)$, $l_i = [0, l_{i\,max}]$, $i = 1, 2, 3$. Denote the following:

- v_i is the set of points reachable by point J_i, $i = 1, 2, 3$;
- V_i is the set of points (the volume) reachable by any point of link l_i. Hence,

---

[3]On adaptation of "tactile" motion planning algorithms to more complex sensing, see Sections 3.6 and 5.2.5.

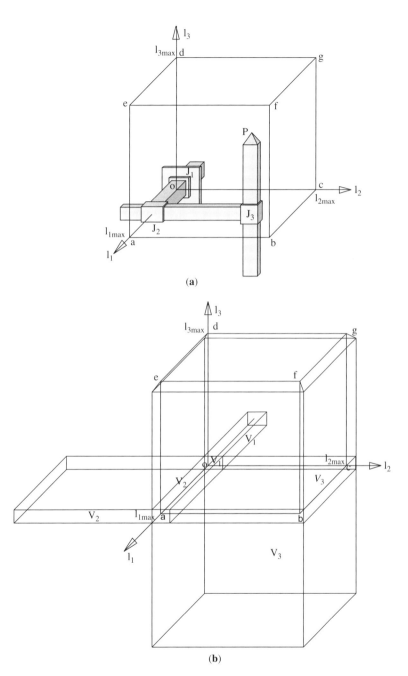

Figure 6.3 **(a)** The effective workspace of the 3D Cartesian arm—the volume that can be reached by the arm endpoint—is limited by the cubicle *abcodefg*. **(b)** Since the rear end of every link may also encounter obstacles, the workspace that has to be protected by the planning algorithm is larger than the effective workspace, as shown.

- v_1 is a single point, O;
- v_2 is a unit line segment, Oa;
- v_3 is a unit square, $Oabc$;
- V_1 is a cylinder whose (link) cross section is s_1 and whose length is $2l_{1\,max}$;
- V_2 is a slab of length $2l_{2\,max}$ formed by all possible motions of the front and rear ends of link l_2 within the joint limits of l_1 and l_2;
- V_3 is a "cubicle" of height $2l_{3\,max}$ formed by all possible motions of the front and rear ends of link l_3 within the joint limits of l_1, l_2, and l_3.

The total volume V_W of W-space is hence $V_W = V_1 \cup V_2 \cup V_3$. Out of this, the set $\{l\} = \{l \in [0, l_{max}]\}$, where $l_{max} = (l_{1\,max}, l_{2\,max}, l_{3\,max})$, represents points reachable by the arm end effector; $\{l\}$ is a closed set.

An obstacle in W-space, called W-*obstacle*, presents a set of points, none of which can be reached by any point of the robot body. This may include some areas of W-space which are actually free of obstacles but still not reachable by the arm because of interference with obstacles. Such areas are called the *shadows* of the corresponding obstacles. A W-obstacle is thus the sum of volumes of the corresponding physical obstacle and the shadows it produces. The word "interference" refers here only to the cases where the arm can apply a force to the obstacle at the point of contact. For example, if link l_1 in Figure 6.2 happens to be sliding along an obstacle (which is not so in this example), it cannot apply any force onto the obstacle, the contact would not preclude the link from the intended motion, and so it would not constitute an interference. W-obstacles that correspond to the three physical obstacles—O_1, O_2, and O_3—of Figure 6.2 are shown in Figure 6.4.

C-Space, C-Point, and C-Obstacle. The vector of joint variables $l = (l_1, l_2, l_3)$ forms the robot *configuration space* (*C-space* or C). In C-space, the arm is presented as a single point, called the *C-point*. The C-space of our Cartesian arm presents a parallelepiped, or generalized cubicle, and the mapping $W \rightarrow C$ is unique.[4] For the example of Figure 6.2, the corresponding C-space is shown in Figure 6.5. For brevity, we will refer to the sides of the C-space cubicle as its *floor* (in Figure 6.2 this is the side $Oabc$), its *ceiling* (side $edgf$), and its *walls*, the remaining four sides. *C-obstacle* is the mapping of a W-obstacle into C. In the algorithm, the planning decisions will be based solely on the fact of contact between the links and obstacles and will never require explicit computation of positions or geometry of W-obstacles or C-obstacles.

M-Line, M-Plane, and V-Plane. As before, a desired path, called the *main line* (*M-line*), is introduced as a simple curve connecting points S and T (start and target) in W-space. The M-line presents the path that the arm end effector would

---

[4]In general, the mapping $W \rightarrow C$ is not unique. In some types of kinematics, such as arm manipulators with revolute joints, a point in W may correspond to one, two, or even an infinite number of points in C [107].

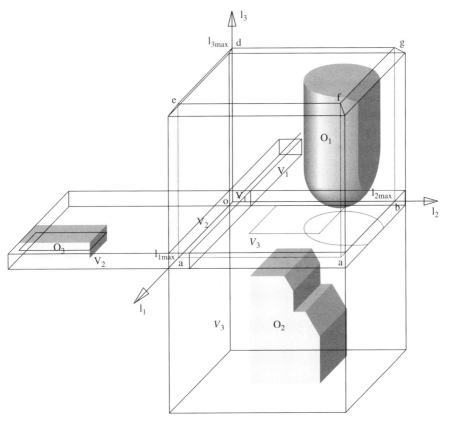

Figure 6.4 The W-obstacles produced by obstacles shown in Figure 6.2 consist of the parts of physical obstacles that intersect W-space plus their corresponding shadows.

follow if no obstacles interfered with the arm motion. Without loss of generality, we assume here that the M-line is a straight-line segment. We will also need two planes, *M-plane* and *V-plane*, that will be used in the motion planning algorithm when maneuvering around obstacles (see Figures 6.7 and 6.9):

- *M-plane* is a plane that contains an M-line and the straight line perpendicular to both the M-line and link l_3 axis. M-plane is thus undetermined only if the M-line is collinear with l_3 axis. This special case will present no difficulty: Here motion planning is trivial and amounts to changing only values l_3; hence we will disregard this case.
- *V-plane* contains the M-line and is parallel to link l_3 axis.

For our Cartesian arm, the M-line, M-plane, and V-plane map in C-space into a straight line and two planes, respectively.

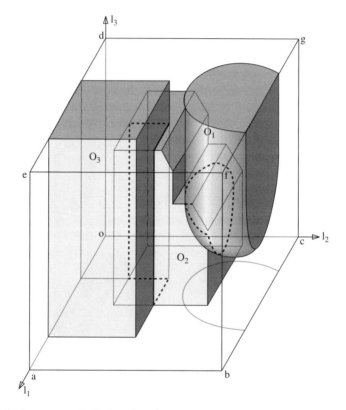

Figure 6.5 *C*-space and *C*-obstacles that correspond to *W*-space in Figures 6.2 and 6.4. Thicker dotted and solid lines show intersections between obstacles. Shown also are projections of the three obstacles on the plane l_1, l_2.

Local Directions. Similar to other algorithms in previous chapters, a common operation in the algorithm here will be the choice of a *local direction* for the next turn (say, left or right). This will be needed when, while moving along a curve, the *C*-point encounters a sort of T-intersection with another curve (which is here the horizontal part of "T"). Let us define the vector of current motion **p** and consider all possible cases.

1. The *C*-point moves along the M-line or along an intersection curve between the M-plane and an obstacle and is about to leave M-plane at the cross-point. Define the normal vector **m** of the M-plane [97]. Then the local direction **b** is *upward* if $\mathbf{b} \cdot \mathbf{m} > 0$ and *downward* if $\mathbf{b} \cdot \mathbf{m} \leq 0$.

2. The *C*-point moves along the M-line or along an intersection curve between the V-plane and an obstacle, and it is about to leave V-plane at the cross-point. Let $\vec{l_3}$ be the vector of l_3 axis. Then, local direction **b** is *left* if $\mathbf{b} \cdot (\mathbf{p} \times \vec{l_3}) > 0$ and *right* if $\mathbf{b} \cdot (\mathbf{p} \times \vec{l_3}) \leq 0$.

3. In a special case of motion along the M-line, the directions are $\vec{ST} = forward$ and $\vec{TS} = backward$.

Consider the motion of a C-point in the M-plane. When, while moving along the M-line, the C-point encounters an obstacle, it may define on it a *hit point*, H. Here it has two choices for following the intersection curve between the M-plane and the obstacle surface: Looking from S toward T, the direction of turn at H is either *left* or *right*. We will see that sometimes the algorithm may replace the current local direction by its opposite. When, while moving along the intersection curve in the M-plane, the C-point encounters the M-line again at a certain point, it defines here the *leave point*, L. Similarly, when the C-point moves along a V-plane, the local directions are defined as "upward" and "downward," where "upward" is associated with the positive and "downward"—with the negative direction of l_3 axis.

6.2.2 The Approach

Similar to other cases of sensor-based motion planning considered so far, conceptually we will treat the problem at hand as one of moving a point automaton in the corresponding C-space. (This does not mean at all, as we will see, that C-space needs to be computed explicitly.) Essential in this process will be sensing information about interaction between the arm and obstacles, if any. This information—namely, what link and what part (front or rear) of the link is currently in contact with an obstacle—is obviously available only in the workspace.

Our motion planning algorithm exploits some special topological characteristics of obstacles in C-space that are a function of the arm kinematics. Note that because links l_1, l_2, and l_3 are connected sequentially, the actual number of degrees of freedom available to them vary from link to link. For example, link l_1 has only one degree of freedom: If it encounters an obstacle at some value l_1', it simply cannot proceed any further. This means that the corresponding C-obstacle occupies all the volume of C-space that lies between the value l_1' and one of the joint limits of joint J_1. This C-obstacle thus has a simple structure: It allows the algorithm to make motion planning decisions based on the simple fact of a local contact and without resorting to any global information about the obstacle in question.

A similar analysis will show that C-obstacles formed by interaction between link l_2 and obstacles always extend in C-space in the direction of one semi-axis of link l_2 and both semi-axes of link l_3; it will also show that C-obstacles formed by interaction between link l_3 and obstacles present generalized cylindrical holes in C-space whose axes are parallel to the axis l_3. No such holes can appear, for example, along the axes l_1 or l_2. In other words, C-space exhibits an *anisotropy* property; some of its characteristics vary from one direction to the other. Furthermore, C-space possesses a certain property of *monotonicity* (see below), whose effect is that, no matter what the geometry of physical obstacles in W-space, no holes or cavities can appear in a C-obstacle.

From the standpoint of motion planning, the importance of these facts is in that the local information from the arm's contacts with obstacles allow one to infer some global characteristics of the corresponding C-obstacle that help avoid directions of motion leading to dead ends and thus avoid an exhaustive search.

Whereas the resulting path planning algorithm is used in the workspace, without computations of C-space, it can be conveniently sketched in terms of C-space, as follows. If the C-point meets no obstacles on its way, it will move along the M-line, and with no complications the robot will happily arrive at the target position T. If the C-point does encounter an obstacle, it will start moving along the intersection curve between the obstacle and one of the planes, M-plane or V-plane. The on-line computation of points along the intersection curve is easy: It uses the plane's equation and local information from the arm sensors.

If during this motion the C-point meets the M-line again at a point that satisfies some additional condition, it will resume its motion along the M-line. Otherwise, the C-point may arrive at an intersection between two obstacles, a position that corresponds to two links or both front and rear parts of the same link contacting obstacles. Here the C-point can choose either to move along the intersection curve between the plane and one of the obstacles, or move along the intersection curve between the two obstacles. The latter intersection curve may lead the C-point to a wall, a position that corresponds to one or more joint limits. In this case, depending on the information accumulated so far, the C-point will conclude (correctly) either that the target is not reachable or that the direction it had chosen to follow the intersection curve would lead to a dead end, in which case it will take a corrective action.

At any moment of the arm motion, the path of the C-point will be constrained to one of three types of curves, thus reducing the problem of three-dimensional motion planning to the much simpler linear planning:

- The M-line
- An intersection curve between a specially chosen plane and the surface of a C-obstacle
- An intersection curve between the surfaces of two C-obstacles

To ensure convergence, we will have to show that a finite combination of such path segments is sufficient for reaching the target position or concluding that the target cannot be reached. The resulting path presents a three-dimensional curve in C-space. No attempt will be made to reconstruct the whole or part of the space before or during the motion.

Since the path planning procedure is claimed to converge in finite time, this means that never, not even in the worst case, will the generated path amount to an exhaustive search.

An integral part of the algorithm is the basic procedure from the Bug family that we considered in Section 3.3 for two-dimensional motion planning for a point automaton. We will use, in particular, the Bug2 procedure, but any other convergent procedure can be used as well.

6.2.3 Topology of *W*-Obstacles and *C*-Obstacles

Monotonicity Property. Obstacles that intersect the W-space volume may interact with the arm during its motion. As mentioned above, one result of such interaction is the formation of obstacle shadows. Consider the spherical obstacle O_1 in Figure 6.2. Clearly, no points directly above O_1 can be reached by any point of the arm body. Similarly, no point of W-space below the obstacle O_2 or to the left of the cubical obstacle O_3 can be reached. Subsequently, the corresponding W-obstacles become as shown in Figure 6.4, and their C-space representation becomes as in Figure 6.5. This effect, studied in detail below, is caused by the constraints imposed by the arm kinematics on its interaction with obstacles. Anisotropic characteristics of W-space and C-space present themselves in a special topology of W- and C-obstacles best described by the notion of the (W- and C-) *obstacle monotonicity*:

Obstacle Monotonicity. In all cases of the arm interference with an obstacle, there is at least one direction corresponding to one of the axes l_i, $i = 1, 2, 3$, such that if a value l_i' of link l_i cannot be reached due to the interference with an obstacle, then no value $l_i'' > l_i'$ in case of contact with the link front part, or, inversely, $l_i'' < l_i'$ in case of contact with the link rear part, can be reached either.

In what follows, most of the analysis of obstacle characteristics is done in terms of C-space, although it applies to W-space as well. Comparing Figures 6.2 and 6.5, note that although physical obstacles occupy a relatively little part of the arm's workspace, their interference with the arm motion can reduce, often dramatically, the volume of points reachable by the arm end effector. The kinematic constraints are due to the arm joints, acting differently for different joint types, and to the fact that arm links are connected in series. As a result, the arm effectively has only one degree of freedom for control of motion of link l_1, two degrees of freedom for control of link l_2, and three degrees of freedom for control of link l_3. A simple example was mentioned above on how this can affect path planning: If during the arm motion along M-line the link l_1 hits an obstacle, then, clearly, the task cannot be accomplished.

The monotonicity property implies that C-obstacles, though not necessarily convex, have a very simple structure. This special topology of W- and C-obstacles will be factored into the algorithm; it allows us, based on a given local information about the arm interaction with the obstacle, to predict important properties of the (otherwise unknown) obstacle beyond the contact point. The monotonicity property can be expressed in terms more amenable to the path planning problem, as follows:

Corollary 6.2.1. *No holes or cavities are possible in a C-obstacle.*

W-obstacle monotonicity affects differently different links and even different parts—front or rear—of the same link. This brings about more specialized notions of l_i-*front and l_i-rear monotonicity* for every link, $i = 1, 2, 3$ (see more below). By treating links' interaction with obstacles individually and by making use of the information on what specific part—front or rear—of a given link is currently in contact with obstacles, the path planning algorithm takes advantage of the obstacle monotonicity property. Because this information is not available in C-space, the following holds:

Information Loss due to Space Transition. Information is lost in the space transition $W \rightarrow C$. Since some of this information—namely, the location of contact points between the robot arm and obstacles—is essential for the sensor-based planning algorithm, from time to time the algorithm may need to utilize some information specific to W-space only.

We will now consider some elemental planar interactions of arm links with obstacles, and we will show that if a path from start to target does exist, then a combination of elemental motions can produce such a path. Define the following:

- *Type I obstacle* corresponds to a W- or C-obstacle that results from the interaction of link l_1 with a physical obstacle.
- *Type II obstacle* corresponds to a W- or C-obstacle that results from the interaction of link l_2 with a physical obstacle.
- *Type III obstacle* corresponds to a W- or C-obstacle that results from the interaction of link l_3 with a physical obstacle.

We will use subscripts "+" and "−" to further distinguish between obstacles that interact with the front and rear part of a link, respectively. For example, a Type III_+ obstacle refers to a C-obstacle produced by interaction of the front part of link l_3 with some physical obstacle.

In the next section we will analyze separately the interaction of each link with obstacles. Each time, three cases are considered: when an obstacle interacts with the front part, the rear part, or simultaneously with both parts of the link in question. We will also consider the interaction of a combination of links with obstacles, setting the foundation for the algorithm design.

Interaction of Link l_1 with Obstacles — Type I Obstacles. Since, according to our model, sliding along an obstacle does not constitute an interference with the link l_1 motion, we need to consider only those cases where the link meets an obstacle head-on. When only the front end of link l_1 is in contact with an obstacle—say, at the joint value l_1'—a Type I_+ *obstacle* is produced, which extends from C-space floor to ceiling and side to side (see Figure 6.6) which effectively reduces the C-space cubicle by the volume $(l_{1\max} - l_1') \cdot l_{2\max} \cdot l_{3\max}$.

A similar effect appears when only the rear end of link l_1 interacts with an obstacle—say, at a joint value l_1'. Then the C-space is effectively decreased by the volume $l_1' \cdot l_{2\max} \cdot l_{3\max}$. Finally, a simultaneous contact of both front and rear ends with obstacles at a value l_1' corresponds to a degenerate case where no motion of link l_1 is possible; that is, the C-obstacle occupies the whole C-space. Formally the property of Type I obstacle monotonicity is expressed as follows:

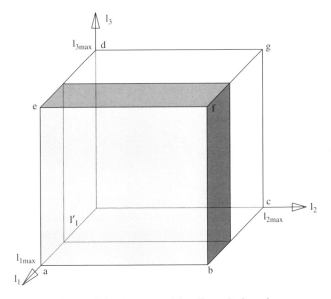

Figure 6.6 *C*-space with a Type *I* obstacle.

Type I Monotonicity. For any obstacle interacting with link l_1, there are three directions corresponding to the joint axes l_i, $i = 1, 2, 3$, respectively, along which the C-obstacle behaves monotonically, as follows: If a position (l'_1, l'_2, l'_3) cannot be reached by the arm due to an obstacle interference, then no position (l''_1, l''_2, l''_3), such that $l''_1 > l'_1$ in case of the (obstacle's) contact with the link's front part, or $l''_1 < l'_1$ in case of the contact with the link's rear part, and $l''_2 \in [0, l_{2\max}]$, $l''_3 \in [0, l_{3\max}]$, can be reached either.

Interaction of Link l_2 with Obstacles — Type II Obstacles

Front Part of Link l_2 — Type II_+ Obstacles. Consider the case when only the front part of link l_2 interferes with an obstacle (Figure 6.2). Because link l_2 effectively has two degrees of freedom, the corresponding Type II_+ obstacle will look in *C*-space as shown in Figure 6.7. The monotonicity property in this case is as follows:

Type II_+ Monotonicity. For any obstacle interacting with the front part of link l_2, there are two axes (directions), namely l_2 and l_3, along which the C-obstacle behaves monotonically, as follows: If a position (l'_1, l'_2, l'_3) cannot be reached by the arm due to an obstacle interference, then no position (l'_1, l''_2, l''_3), such that $l''_2 > l'_2$ and $l''_3 \in [0, l_{3\max}]$, can be reached either.

As a result, a Type II_+ collision, as at point H in Figure 6.7, indicates that any motion directly upward or downward from H along the obstacle will necessarily bring the *C*-point to one of the side walls of the *C*-space cubicle. This suggests

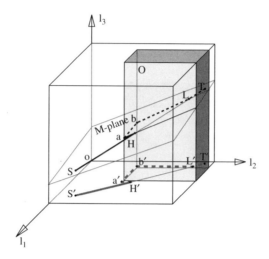

Figure 6.7   (a) W-space and (b) C-space with a Type II obstacle. (S, T) is the M-line; *HabL* is a part of the intersection curve between the obstacle O and M-plane.

that a plane can be chosen such that the exploration of the intersection curve between this plane and the Type II_+ obstacle will produce a more promising outcome that will result either in a success or in the correct conclusion that the target cannot be reached. In the algorithm, the M-plane will be used, which offers some technical advantages. In general, all three arm joints will participate in the corresponding motion.

For this case (front part of link l_2 interacting with an obstacle), the decision on which local direction, right or left, is to be taken at a hit point H in order to follow the intersection curve between an M-plane and a Type II_+ obstacle is made in the algorithm based on the following rule:

Rule 1:

If $l_{1H} > l_{1T}$, the current direction is "left."
If $l_{1H} < l_{1T}$, the current direction is "right."
If $l_{1H} = l_{1T}$, the target cannot be reached.

Rear Part of Link l_2 — Type II_- Obstacles.   Now consider the case when only the rear part of link l_2—that is, the link's part to the left of joint J_2—can interfere with obstacles (see obstacle O_3, Figure 6.2). This situation produces a C-space very similar to that in Figure 6.7. The direction of obstacle monotonicity along the axis l_2 will now reverse:

Type II_- Monotonicity.   *For any obstacle interacting with the rear part of link l_2, there are two axes (directions), namely l_2 and l_3, along which the C-obstacle behaves monotonically, as follows: If a position (l'_1, l'_2, l'_3) cannot be reached by* *the arm due to an obstacle interference, then no position* (l_1', l_2'', l_3''), *such that* $l_2'' < l_2'$ *and* $l_3'' \in [0, l_{3\,max}]$, *can be reached either.*

In terms of decision-making, this case is similar to the one above, except that the direction of obstacle monotonicity along l_2 axis reverses, and the choice of the current local direction at a hit point H obeys a slightly different rule:

Rule 2:

If $l_{1\,H} > l_{1\,T}$, the current direction is "right."

If $l_{1\,H} < l_{1\,T}$, the current direction is "left."

If $l_{1\,H} = l_{1\,T}$, the target cannot be reached.

Interaction of Both Parts of Link l_2 with Obstacles. Clearly, when both the front and the rear parts of link l_2 interact simultaneously with obstacles, the resulting Type II_+ and Type II_- obstacles fuse into a single C-obstacle that divides C-space into two separate volumes unreachable one from another (see Figure 6.8). If going from S to T requires the arm to cross that obstacle, the algorithm will conclude that the target position cannot be reached.

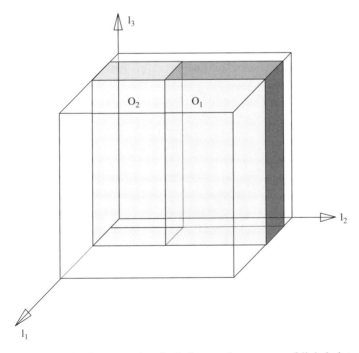

Figure 6.8 C-space in the case when both front and rear parts of link l_2 interact with obstacles, producing a single obstacle that is a combination of a Type II_+ and Type II_- obstacles.

Stalactites and Stalagmites: Type III Obstacles

Front Part of Link l_3 — Type III_+ Obstacles. Assume for a moment that only the front part of link l_3 can interfere with an obstacle (see, e.g., obstacle O_1, Figures 6.2 and 6.4). Consider the cross sections of the obstacle with two horizontal planes: one corresponding to the value l_3' and the other corresponding to the value l_3'', with $l_3' < l_3''$. Denote these cross sections a' and a'', respectively. Each cross section is a closed set limited by a simple closed curve; it may or may not include points on the C-space boundary. Because link l_3 is a generalized cylinder, the vertical projection of one cross section onto the other satisfies the relationship $a' \subseteq a''$. This is a direct result of the Type III_+ obstacle monotonicity property, which is formulated as follows:

Type III_+ Monotonicity. *For any obstacle interacting with the front part of link l_3, there is one axis (direction), namely l_3, along which the corresponding C-obstacle behaves monotonically, as follows: if a position (l_1', l_2', l_3') cannot be reached by the arm due to an obstacle interference, then no position (l_1', l_2', l_3'') such that $l_3'' > l_3'$ can be reached either.*

This property results in a special "stalactite" shape of Type III_+ obstacles. A typical property of icicles and of beautiful natural stalactites that hang down from the ceilings of many caves is that their horizontal cross section is continuously reduced (in theory at least) from its top to its bottom. Each Type III_+ obstacle behaves in a similar fashion. It forms a "stalactite" that hangs down from the ceiling of the C-space cubicle, and its horizontal cross section can only decrease, with its maximum horizontal cross section being at the ceiling level, $l_3 = l_{3\,max}$ (see cubicle $Oabcdefg$ and obstacle O_1, Figure 6.4). For any two horizontal cross sections of a Type III_+ obstacle, taken at levels l_3' and l_3'' such that $l_3'' > l_3'$, the projection of the first cross section (l_3' level) onto a horizontal plane contains no points that do not belong to the similar projection of the second cross section (l_3'' level). This behavior is the reflection of the monotonicity property.

Because of this topology of Type III_+ obstacles, the sufficient motion for maneuvering around any such obstacle—that is, motion sufficient to guarantee convergence—turns out to be motion along the intersection curves between the corresponding C-obstacle and either the M-plane or the V-plane (specifically, its part below M-plane), plus possibly some motion in the floor of the C-space cubicle (Figure 6.9).

Rear Part of Link l_3 — Type III_- Obstacles. A similar argument can be made for the case when only the rear end of link l_3 interacts with an obstacle (see, e.g., obstacle O_2, Figures 6.2, 6.4, and 6.5). In C-space the corresponding Type III_- obstacle becomes a "stalagmite" growing upward from the C-space floor. This shape is a direct result of the Type III_- obstacle monotonicity property, which is reversed compared to the above situation with the front part of link l_3, as follows:

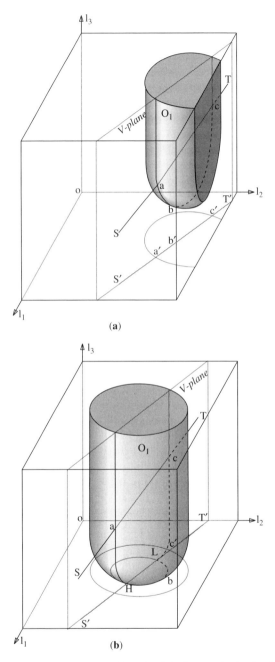

(a)

(b)

Figure 6.9 *C*-space with a Type *III* obstacle. **(a)** Curve *abc* is the intersection curve between the obstacle and V-plane that would be followed by the algorithm. **(b)** Here the Type *III* obstacle intersects the floor of *C*-space. Curve *aHbLc* is the intersection curve between the obstacle and V-plane and *C*-space floor that would be followed by the algorithm.

Type III_ Monotonicity. *For any obstacle interacting with the rear part of link l_3, there is one axis (direction), l_3, along which the corresponding C-obstacle behaves monotonically, as follows: if a position (l'_1, l'_2, l'_3) cannot be reached by the arm due to an obstacle interference, then no position (l'_1, l'_2, l''_3) such that $l''_3 < l'_3$ can be reached either.*

The motion sufficient for maneuvering around a Type III_- obstacle and for guaranteeing convergence is motion along the curves of intersection between the corresponding C-obstacle and either the M-plane or the V-plane (its part above M-plane), or the ceiling of the C-space cubicle.

Interaction of Both Parts of Link l_3 with Obstacles. This is the case when in C-space a "stalactite" obstacle meets a "stalagmite" obstacle and they form a single obstacle. (Again, similar shapes are found in some caves.) Then the best route around the obstacle is likely to be in the region of the "waist" of the new obstacle.

Let us consider this case in detail. For both parts of link l_3 to interact with obstacles, or with different pieces of the same obstacle, the obstacles must be of both types, Type III_+ and Type III_-. Consider an example with two such obstacles shown in Figure 6.10. True, these C-space obstacles don't exactly look like the stalactites and stalagmites that one sees in a natural cave, but they do have their major properties: One "grows" from the floor and the other grows from the ceiling, and they both satisfy the monotonicity property, which is how we think of natural stalactites and stalagmites.

Without loss of generality, assume that at first only one part of link l_3—say, the rear part—encounters an obstacle (see obstacle O_2, Figure 6.10). Then the arm will start maneuvering around the obstacle following the intersection curve between the V-plane and the obstacle (path segment aH, Figure 6.10). During this motion the front part of link l_3 contacts the other (or another part of the same) obstacle (here, obstacle O_1, Figure 6.10).

At this moment the C-point is still in the V-plane, and also at the intersection curve between both obstacles, one of Type III_+ and the other of Type III_- (point H, Figure 6.10; see also the intersection curve $H_2 cd L_2 fg$, Figure 6.12). As with any curve, there are two possible local directions for following this intersection curve. If both of them lead to walls, then the target is not reachable. In this example the arm will follow the intersection curve—which will depart from V-plane, curve $HbcL$—until it meets V-plane at point L, then continue in the V-plane, and so on.

Since for the intersection between Type III_+ and Type III_- obstacles the monotonicity property works in the opposite directions—hence the minimum area "waist" that they form—the following statement holds (it will be used below explicitly in the algorithm):

Corollary 6.2.2. *If there is a path around the union of a Type III_+ and a Type III_- obstacles, then there must be a path around them along their intersection curve.*

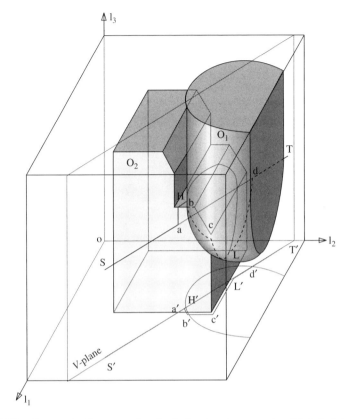

Figure 6.10 *C*-space in the case when both front and rear parts of link l_3 interact with obstacles, producing a single obstacle that is a combination of a Type III_+ and Type III_- obstacles.

Simultaneous Interaction of Combinations of Links with Obstacles.

Since Type *I* obstacles are trivial from the standpoint of motion planning—they can be simply treated as walls parallel to the sides of the *C*-space cubicle—we focus now on the combinations of Type *II* and Type *III* obstacles. When both links l_2 and l_3 are simultaneously in contact with obstacles, the *C*-point is at the intersection curve between Type *II* and Type *III* obstacles, which presents a simple closed curve. (Refer, for example, to the intersection of obstacles O_2 and O_3, Figure 6.11.) Observe that the Type *III* monotonicity property is preserved in the union of Type *II* and Type *III* obstacles. Hence,

Corollary 6.2.3. *If there is a path around the union of a Type II and a Type III obstacles, then there must be a path around them along their intersection curve.*

As in the case of intersection between the Type *III* obstacle and the V-plane (see above), one of the two possible local directions is clearly preferable to the

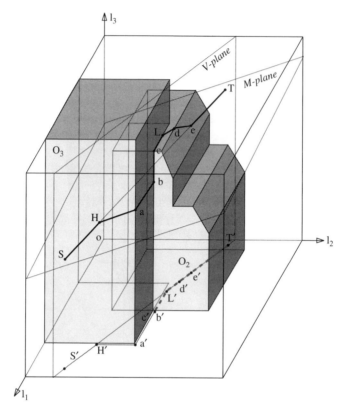

Figure 6.11   The path in C-space in the presence of obstacles O_2 and O_3 of Figure 6.2; $SHabcLdeT$ is the actual path of the arm endpoint, and curve $S'H'a' \dots T'$ is its projection onto the plane (l_1, l_2).

other. For example, when in Figure 6.11 the C-point reaches, at point b, the intersection curve between obstacles O_2 and O_3, it is clear from the monotonicity property that the C-point should choose the upward direction to follow the intersection curve. This is because the downward direction is known to lead to the base of the obstacle O_2 "stalagmite" and is thus less promising (though not necessarily hopeless) as the upward direction (see path segments bc and cL, Figure 6.11).

Let us stress that in spite of seeming multiplicity of cases and described elemental strategies, their logic is the same: All elemental strategies force the C-point to move either along the M-line, or along the intersection curves between a C-obstacle and a plane (M-plane, V-plane, or C-space side planes), or along the intersection curves between two Type *III* C-obstacles. Depending on the real workspace obstacles, various combinations of such path segments may occur. We will show in the next section that if in a given scene there exists a path to the target position, a combination of elemental strategies is sufficient to produce one.

6.2.4   Connectivity of $\mathcal{C}$

A space or manifold is *connected* relative to two points in it if there is a path that connects both points and that lies fully in the space (manifold). For a given path l, the corresponding *trajectory* $l(t)$ defines this path as a function of a scalar variable t; for example, t may indicate time. Denote the 2D Cartesian space formed by joint values l_1, l_2 as $\mathcal{C}_p$, $\mathcal{C}_p = [0, l_{1\,\max}] \times [0, l_{2\,\max}]$.

We intend to show here that for the 3D Cartesian arm the connectivity in $\mathcal{C}$ can be deduced from the connectivity in $\mathcal{C}_p$. Such a relationship will mean that the problem of path planning for the 3D Cartesian arm can be reduced to that for a point automaton in the plane, and hence the planar strategies of Chapter 3 can be utilized here, likely with some modifications.

Define the *conventional projection* $P_c(E)$ of a set of points $E = \{(l_1, l_2, l_3)\} \subseteq \mathcal{C}$ onto space $\mathcal{C}_p$ as $P_c(E) = \{(l_1, l_2) \mid \exists \, l_3^*, (l_1, l_2, l_3^*) \in E\}$. Thus, $P_c(S)$, $P_c(T)$, $P_c(\text{M-line})$, and $P_c(\{O\})$ are, respectively, the projections of points S and T, the M-line, and C-obstacles onto $\mathcal{C}_p$. See, for example, projections P_c of three obstacles, O_1, O_2, O_3 (Figure 6.12). It is easy to see that $P_c(O_1 \cap O_2) = P_c(O_1) \cap P_c(O_2)$.

Define the *minimal projection* $P_m(E)$ of a set of points $E = \{(l_1, l_2, l_3)\} \subseteq \mathcal{C}$ onto space $\mathcal{C}_p$ as $P_m(E) = \{(l_1, l_2) \mid \forall \, l_3, (l_1, l_2, l_3) \in E\}$. Thus, if a C-obstacle O stretches over the whole range of $l_3 \in [0, l_{3\,\max}]$, and E contains all the points in O, then $P_m(E)$ is the intersection between the (l_1, l_2)-space and the maximum cylinder that can be inscribed into O and whose axis is parallel to l_3. Note that if a set E is a cylinder whose axis is parallel to the l_3 axis, then $P_c(E) = P_m(E)$. Type I and Type II obstacles present such cylinders. In general, $P_m(S) = P_m(T) = \emptyset$.

Existence of Collision-Free Paths. We will now consider the relationship between a path in $\mathcal{C}$ and its projection in $\mathcal{C}_p$. The following statement comes directly from the definition of P_c and P_m:

Lemma 6.2.1. *For any C-obstacle O in $\mathcal{C}$ and any set E_p in $\mathcal{C}_p$, if $E_p \cap P_c(O) = \emptyset$, then $P_m^{-1}(E_p) \cap O = \emptyset$.*

If the hypothesis is not true, then $P_m^{-1}(E_p) \cap O \neq \emptyset$. We have $P_c(P_m^{-1}(E_p) \cap O) = P_c(P_m^{-1}(E_p)) \cap P_c(O) = E_p \cap P_c(O) \neq \emptyset$. Thus a contradiction.

The next statement provides a sufficient condition for the existence of a path in C-space:

Lemma 6.2.2. *Given a set of obstacles $\{O\}$ in $\mathcal{C}$ and the corresponding projections $P_c(\{O\})$, if there exists a path between $P_c(S)$ and $P_c(T)$ in $\mathcal{C}_p$, then there must exist a path between S and T in $\mathcal{C}$.*

Let $l_p(t) = \{l_1(t), l_2(t)\}$ be a trajectory of $P_c(C\text{-point})$ between $P_c(S)$ and $P_c(T)$ in $\mathcal{C}_p$. From Lemma 6.2.1, $P_m^{-1}(l_p(t)) \cap \{O\} = \emptyset$ in $\mathcal{C}$. Hence, for example, the path $l(t) = \{(l_p(t), (1-t)l_{3S} + t \cdot l_{3T})\} \in P_c^{-1}(l_p(t))$ connects S and T in $\mathcal{C}$.

To find the necessary condition, we will use the notion of a minimal projection. The following statement asserts that a zero overlap between two sets in C implies a zero overlap between their minimal projections in C_p:

Lemma 6.2.3. *For any set E and any C-obstacle O in C, if $O \cap E = \emptyset$, then $\mathcal{P}_m(E) \cap \mathcal{P}_m(O) = \emptyset$.*

By definition, $\mathcal{P}_m^{-1}(E_1 \cap E_2) = \mathcal{P}_m^{-1}(E_1) \cap \mathcal{P}_m^{-1}(E_2)$ and $\mathcal{P}_m^{-1}(\mathcal{P}_m(O)) \subset O$. Thus, if $\mathcal{P}_m(E) \cap \mathcal{P}_m(O) = \emptyset$, then $\emptyset \neq \mathcal{P}_m^{-1}(\mathcal{P}_m(E) \cap \mathcal{P}_m(O)) = \mathcal{P}_m^{-1}(\mathcal{P}_m(E) \cap \mathcal{P}_m^{-1}(\mathcal{P}_m(O))) \subset O \cap E$.

To use this lemma in the algorithm design, we need to describe minimal projections for different obstacle types. For any Type I or Type II obstacle O, $\mathcal{P}_c(O) = \mathcal{P}_m(O)$. For a Type III obstacle we consider three cases, using, as an example, a Type III_+ obstacle; denote it O_+.

- O_+ intersects the floor F of C. Because of the monotonicity property, $\mathcal{P}_m(O_+) = O_+ \cap F$. In other words, the minimal projection of O_+ is exactly the intersection area of O_+ with the floor F.
- O_+ intersects with a Type III_- obstacle, O_-. Then, $B(\mathcal{P}_m(O_+ \cup O_-)) = \mathcal{P}_c(B(O_+) \cap B(O_-))$, where $B(O)$ refers to the boundary of O. That is, the boundary curve of the combined minimal projection of O_+ and O_- is the conventional projection of the intersection curve between the boundary surfaces of O_+ and O_-.
- Neither of the above cases apply. Then $\mathcal{P}_m(O_+) = \emptyset$.

A similar argument can be carried out for a Type III_- obstacle.

We now turn to establishing a necessary and sufficient condition that ties the existence of paths in the plane C_p with that in C. This condition will provide a base for generalizing, in the next section, a planar path planning algorithm to the 3D space. Assume that points S and T lie outside of obstacles.

Theorem 6.2.1. *Given points S, T and a set of obstacles $\{O\}$ in C, a path exists between S and T in C if and only if there exists a path in C_p between points $\mathcal{P}_c(S)$ and $\mathcal{P}_c(T)$ among the obstacles $\mathcal{P}_m(\{O\})$.*

First, we prove the necessity. Let $l(t)$, $t \in [0, 1]$, be a trajectory in C. From Lemma 6.2.3, $\mathcal{P}_m(l(t)) \cap \mathcal{P}_m(\{O\}) = \emptyset$. Hence, the path $\mathcal{P}_m(l(t))$ connects $\mathcal{P}_c(S)$ and $\mathcal{P}_c(T)$ in C_p.

To show the sufficiency, let $l_p(t)$, $t \in [0, 1]$, be a trajectory in C_p and let $l_p(\cdot)$ be the corresponding path. Then $\mathcal{P}_m^{-1}(l_p(\cdot))$ presents a manifold in C. Define $E = \mathcal{P}_m^{-1}(l_p(\cdot)) \cap \{O\}$ and let E^c be the complement of E in $\mathcal{P}_m^{-1}(l_p(\cdot))$. We need to show that E^c consists of one connected component. Assume that this is not true. For any $t_* \in [0, 1]$, since $l_p(t_*) \cap \mathcal{P}_m(\{O\}) = \emptyset$, there exists l_{3*} such that point $(l_p(t_*), l_{3*}) \in E^c$. The only possibility for E^c to consist of two or more disconnected components is when there exists t_* and a set $(l_{3*}, l'_{3*}, l''_{3*})$, $l'_{3*} > l_{3*} > l''_{3*}$, such that $(l_p(t_*), l_{3*}) \in E$ while $(l_p(t_*), l'_{3*}) \in E^c$ and $(l_p(t_*), l''_{3*}) \in E^c$. However, this cannot happen because of the monotonicity property of obstacles. Hence E^c must be connected, and since points S and T lie outside of obstacles, then $S, T \in E^c$. **Q.E.D.**

Lifting 2D Algorithms into 3D Space. Theorem 6.2.1 establishes the relationship between collision-free paths in $\mathcal{C}$ and collision-free paths in $\mathcal{C}_p$. We now want to develop a similar relationship between motion planning algorithms for $\mathcal{C}$ and those for $\mathcal{C}_p$. We address, in particular, the following question: Given an algorithm A_p for $\mathcal{C}_p$, can one construct an algorithm A for $\mathcal{C}$, such that any trajectory (path) $l(t)$ produced by A in $\mathcal{C}$ in the presence of obstacles $\{O\}$ maps by $\mathcal{P}_m$ into the trajectory $l_p(t)$ produced by A_p in $\mathcal{C}_p$ in the presence of obstacles $\mathcal{P}_m(\{O\})$?

We first define the class of algorithms from which algorithms A_p are chosen. A planar algorithm A_p is said to belong to class $\mathcal{A}_p$ if and only if its operation is based on local information, such as from tactile sensors; the paths it produces are confined to the M-line, obstacle boundaries, and W-space boundaries; and it guarantees convergence. In other words, class $\mathcal{A}_p$ comprises only sensor-based motion planning algorithms that satisfy our usual model. In addition, we assume that all decisions about the direction to proceed along the M-line or along the obstacle boundary are made at the intersection points between M-line and obstacle boundaries.

Theorem 6.2.1 says that if there exists a path in $\mathcal{C}_p$ (between projections of points S and T), then there exists at least one path in $\mathcal{C}$. Our goal is to dynamically construct the path in $\mathcal{C}$ while A_p, the given algorithm, generates its path in $\mathcal{C}_p$. To this end, we will analyze five types of elemental motions that appear in $\mathcal{C}$, called Motion I, Motion II, Motion III, Motion IV, and Motion V, each corresponding to the $\mathcal{P}_c(C\text{-point})$ motion either along the $\mathcal{P}_c(\text{M-line})$ or along the obstacle boundaries $\mathcal{P}_c(\{O\})$. Based on this analysis, we will augment the decision-making mechanism of A_p to produce the algorithm A for $\mathcal{C}$.

Out of the three types of obstacle monotonicity property identified above, only Type *III* monotonicity is used in this section. One will see later that other monotonicity types can also be used, resulting in more efficient algorithms. Below, Type *I* and *II* obstacles are treated as C-space side walls; the C-space ceiling is treated as a Type III_+ obstacle; and the C-space floor is treated as a Type III_- obstacle. Note that when the arm comes in contact simultaneously with what it perceives as two Type *III* obstacles, only those of the opposite "signs" have to be distinguished—that is, a Type III_+ and a Type III_-. Obstacles of the same sign will be perceived as one. Below, encountering "another Type *III* obstacle" refers to an obstacle of the opposite sign. Then the projection $\mathcal{P}_m$ of the union of the obstacles is not zero, $\mathcal{P}_m(\cdot) \neq \emptyset$.

Among the six local directions defined in Section 6.2.1—*forward, backward, left, right, upward,* and *downward*—the first four can be used in a 2D motion planning algorithm. Our purpose is to design a general scheme such that, given any planar algorithm A_p, a 3D algorithm A can be developed that lifts the decision-making mechanism of A_p into 3D space by complementing the set of local directions by elements *upward* and *downward*. We now turn to the study of five fundamental motions in 3D, which will be later incorporated into the 3D algorithm.

Motion I — Along the M-Line. Starting at point S, the C-point moves along the M-line, as in Figure 6.7, segment SH; this corresponds to $\mathcal{P}_c(C$-point) moving along the $\mathcal{P}_c($M-line) (segment $S'H'$, Figure 6.7). Unless algorithm A_p calls for terminating the procedure, one of these two events can take place:

1. A wall is met; this corresponds to $\mathcal{P}_c(C$-point) encountering an obstacle. Algorithm A_p now has to decide whether $\mathcal{P}_c(C$-point) will move along $\mathcal{P}_c($M-line), or turn *left* or *right* to go around the obstacle. Accordingly, the C-point will choose to reverse its local direction along the M-line, or to turn *left* or *right* to go around the wall. In the latter case we choose a path along the intersection curve between the wall and the M-plane, which combines two advantages: (i) While not true in the worst case, due to obstacle monotonicity the M-plane typically contains one of the shortest paths around the obstacle; and (ii) after passing around the obstacle, the C-point will meet the M-line exactly at the point of its intersection with the obstacle (point L, Figure 6.7), and so the path will be simpler. In general, all three joints participate in this motion.

2. A Type III_+ or III_- obstacle is met. The C-point cannot proceed along the M-line any longer. The local objective of the arm here is to maneuver around the obstacle so as to meet the M-line again at a point that is closer to T than the encounter point. Among various ways to pass around the obstacle, we choose here motion in the V-plane. The intersection curve between the Type III obstacle and the V-plane is a simple planar curve. It follows from the monotonicity property of Type III obstacles that when the front (rear) part of link l_3 hits an obstacle, then any motion upward (accordingly, downward) along the obstacle will necessarily bring the C-point to the ceiling (floor) of the C-space. Therefore, a local contact information is sufficient here for a global planning inference—that the local direction *downward* (*upward*) along the intersection curve between the V-plane and the obstacle is a promising direction. In the example in Figure 6.9a, the resulting motion produces the curve abc.

Motion II — Along the Intersection Curve Between the M-Plane and a Wall. In $\mathcal{C}_p$, this motion corresponds to $\mathcal{P}_c(C$-point) moving around the obstacle boundary curve in the chosen direction (see Figure 6.12, segments H_1aL_1 and $H_1'a'L_1'$). One of these two events can take place:

1. The M-line is encountered, as at point L_1, Figure 6.12; in $\mathcal{C}_p$ this means $\mathcal{P}_c($M-line) is encountered. At this point, algorithm A_p will decide whether

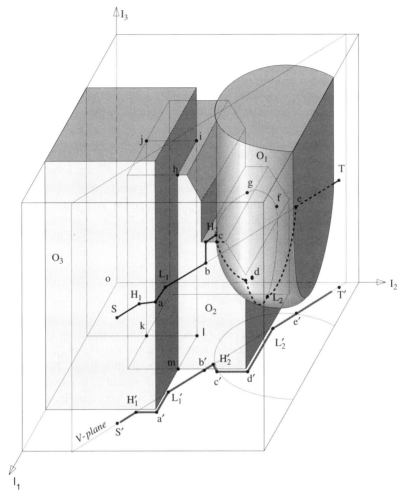

Figure 6.12 The path in C-space in the presence of obstacles O_1, O_2 and O_3 of Figure 6.2. Trajectory $SH_1aL_1bH_2cdL_2eT$ is the actual path of the arm endpoint; $S'H_1'a'b'H_2'c'd'L_2'e'T'$ is its projection onto the plane (l_1, l_2).

$\mathcal{P}_c(C\text{-point})$ should start moving along the $\mathcal{P}_c(\text{M-line})$ or continue moving along the intersection curve between the M-plane and the obstacle boundary. Accordingly, the C-point will either resume its motion along the M-line, as in Motion I, or will keep moving along the M-plane/obstacle intersection curve.

2. A Type III_+ or III_- obstacle is met. In $\mathcal{C}_p$, the $\mathcal{P}_c(C\text{-point})$ keeps moving in the same local direction along the obstacle boundary (path segments bcL and $b'L'$, Figure 6.11). As for C-point, there are two possible directions for it to follow the intersection curve between the Type III obstacle

and the wall. Since the Type III_+ (or III_-) monotonicity property is pre-served here, the only promising local direction for passing around the Type III_+ (III_-) obstacle is *downward* (*upward*). This possibility corresponds to Motion IV below.

Motion III — Along the Intersection Curve Between the V-Plane and a Type III_+ or III_- Obstacle. This corresponds to moving along the P_c(M-line) in C_p. One of the following can happen:

1. The M-line is met. C-point resumes its motion along the M-line as in Motion I; see path segments bc, cT and $b'c'$, $c'T'$, Figure 6.9a.
2. A wall is met. This corresponds to the P_m(C-point) encountering an obsta-cle. According to the algorithm A_p, P_c(C-point) will either reverse its local direction to move along P_c(M-line) or will make a turn to follow the obstacle. Accordingly, the C-point will either (a) reverse its local direction to follow the intersection curve between the V-plane and the (Type III) obstacle or (b) try to go around the union of the wall and the obstacle. For the latter motion we choose a path along the intersection curve between the wall and the Type III obstacle.
3. Another Type III_+ or III_- obstacle is met. Since the P_m projection of both Type III obstacles onto C_p is not zero, this corresponds to the P_c(C-point) encountering an obstacle, which presents the P_m projection of the intersection curve between both obstacles. According to A_p algorithm, the P_c(C-point) will either (a) reverse its local direction to move along P_c(M-line) or (b) make a turn to follow the obstacle. Accordingly, the C-point will either (a) reverse its local direction to follow the intersection curve between V-plane and the Type III obstacle or (b) try to go around the union of two Type III obstacles. For the latter motion, we choose a path along the intersection curve between the two Type III obstacles in the local direction *left* or *right* decided by A_p.

Motion IV — Along the Intersection Curve Between a Type III Obstacle and a Wall. In C_p, this corresponds to the P_c(C-point) moving along the bound-ary of $P_m(\{O\})$; see segments bcL and $b'L'$, Figure 6.11. One of the following can occur:

1. The C-point returns to the M-plane. Then C-point resumes its motion along the intersection curve between the M-plane and the wall, similar to Motion II.
2. The V-plane is encountered (see point L, Figure 6.11). In C_p this means that P_c(M-line) is encountered. At this point, algorithm A_p will decide whether the P_c(C-point) should start moving along the P_c(M-line) or should con-tinue moving along the obstacle boundary. Accordingly, the C-point will either (a) continue moving along the intersection curve between the Type III obstacle and the wall or (b) move along the intersection curve between

V-plane and the Type *III* obstacle in the only possible local direction, as in Motion III.

3. Another Type *III* obstacle is encountered. Then there will be a nonzero projection of the intersection curve between two Type *III* obstacles onto C_p; the $\mathcal{P}_c$(C-point) will continue following the obstacle boundary. Accordingly, the C-point will follow the intersection curve between the two Type *III* obstacles in the only possible local direction, see Motion V.

Motion V — Along the Intersection Curve Between Two Type III Obstacles. In C_p this corresponds to the $\mathcal{P}_c$(C-point) moving along the boundary of $\mathcal{P}_m(\{O\})$; see segments H_2cdL_2 and $H_2'c'd'L_2'$, Figure 6.12. One of the following two events can occur:

1. The V-plane is encountered (point L_2, Figure 6.12). In C_p this means that $\mathcal{P}_c$(M-line) is encountered. At this point, algorithm A_p will decide whether the $\mathcal{P}_c$(C-point) should start moving along the $\mathcal{P}_c$(M-line) or should continue moving along the obstacle boundary in one of the two possible directions. Accordingly, the C-point will either (a) move along the intersection curve between the V-plane and the Type *III* obstacle that is known to lead to the M-plane (as in Motion III.3 above) or (b) keep moving along the intersection curve between two Type *III* obstacles.

2. A wall is encountered. In C_p this corresponds to continuous motion of the $\mathcal{P}_c$(C-point) along the obstacle boundary. Accordingly, the C-point starts moving along the intersection curve between the newly encountered wall and one of the two Type *III* obstacles—the one that is known to lead to the M-line (as in Motion IV).

To summarize, the above analysis shows that the five motions that exhaust all distinct possible motions in C can be mapped uniquely into two categories of possible motions in C_p—along the $\mathcal{P}_c$(M-line) and along $\mathcal{P}_m(\{O\})$—that constitute the trajectory of the $\mathcal{P}_c$(C-point) in C_p under algorithm A_p. Furthermore, we have shown how, based on additional information on obstacle types that appear in C, any decision by algorithm A_p in C_p can be transformed uniquely into the corresponding decision in C. This results in a path in C that has the same convergence characteristics as its counterpart in C_p. Hence we have the following theorem:

Theorem 6.2.2. *Given a planar algorithm $A_p \in \mathcal{A}_p$, a 3D algorithm A can be constructed such that any trajectory produced by A in the presence of obstacles $\{O\}$ in C maps by $\mathcal{P}_c$ into the trajectory that A_p produces in the presence of obstacles $\mathcal{P}_m(\{O\})$ in C_p.*

6.2.5 Algorithm

Theorem 6.2.2 states that an algorithm for sensor motion planning for the 2D Cartesian robot arm can be extended to a 3D algorithm, while preserving the algorithm's convergence. This kind of extension will not, however, necessarily produce efficient paths, since it may or may not utilize all the information available in the arm (3D) workspace. For example, the obstacle monotonicity property suggests that some directions for passing around an obstacle are more promising than others. Following the analysis in Section 6.2.4, a more efficient algorithm, which preserves the convergence properties discussed in Section 6.2.4 and also takes into account the said additional information, is presented below. The algorithm is based on a variation of the Bug2 procedure (see Section 3.3.2).

The following notation is used in the algorithm's procedure:

- Flag F is used to check whether or not both local directions have been tried for a given curve.
- Variables $curr\_dir$ and $curr\_loc$, respectively, store the current local direction and current robot position.
- Function $dist(x, y)$ gives the Cartesian distance between points x and y.
- Rules 1 and 2 are as defined in Section 6.2.3.
- When the C-point encounters a curve on the surface of an obstacle that it has to follow, the directions *right* and *left* are defined similar to those for the M-plane above.
- In order to choose the local direction *upward* or *downward*, function $AboveMLine(curr\_loc)$ is used to determine if the current location is above or below the M-line.
- Variable $local\_dir1$ refers to the local directions *forward, backward, left* and *right*.
- Variable $local\_dir2$ refers to the local directions *upward* and *downward*.

The algorithm proceeds as follows.

Step 0 Initialization. Set $j = 0$. Start at S. Go to Step 1.

Step 1 Motion Along the M-Line. Move along the M-line toward T until one of the following occurs:

(a) T is reached. The procedure terminates.

(b) A wall or a Type I obstacle is encountered. T cannot be reached—the procedure terminates.

(c) A Type II_+ obstacle is encountered. Set $j = j + 1$ and define $H_j = curr\_loc$; set $F = 1$ and set $local\_dir1$ according to Rule 1. Go to Step 2.

(d) A Type II_+ obstacle is encountered. Set $j = j + 1$ and define $H_j = curr\_loc$; set $F = 1$ and set $local\_dir1$ according to Rule 2. Go to Step 2.

(e) A Type III obstacle is encountered. If $l_{3S} = l_{3T}$, T cannot be reached—the procedure terminates. Otherwise, set $local\_dir2 = downward$ or *upward* if the obstacle is of Type III_+ (or III_-); go to Step 3.

Step 2 Motion Along the Intersection Curve Between a Type II Obstacle and an M-Plane. Move along the intersection curve until one of the following occurs:

(a) The M-line is encountered. If $dist(\mathcal{P}_c(curr\_loc), \mathcal{P}_c(T)) > dist(\mathcal{P}_c(H_j), \mathcal{P}_c(T))$, go[5] to Step 2. Otherwise, define $L_j = curr\_loc$; go to Step 1.

(b) A wall or a Type *I* obstacle is encountered. *T* cannot be reached—the procedure terminates.

(c) Another Type *II* obstacle is encountered. *T* cannot be reached—the procedure terminates.

(d) A Type *III* obstacle is encountered. Set *local_dir2* = *downward* or *upward* if the obstacle is of Type *III₊* (or *III₋*). Go to Step 4.

Step 3 Motion Along the Intersection Curve Between a Type III Obstacle and a V-Plane. Move along the intersection curve until one of the following occurs:

(a) The M-line is met. Go to Step 1.

(b) A wall or a Type *I* obstacle is encountered. *T* cannot be reached—the procedure terminates.

(c) A Type *II₊* obstacle is encountered. Set $j = j + 1$ and define $H_j = curr\_loc$; set $F = 1$ and set *local_dir1* according to Rule 1. Go to Step 4.

(d) A Type *II₋* obstacle is encountered. Set $j = j + 1$ and define $H_j = curr\_loc$; set $F = 1$ and set *local_dir1* according to Rule 2. Go to Step 4.

(e) Another Type *III* obstacle is encountered. Set $j = j + 1$ and define $H_j = curr\_loc$; set $F = 2$ and set *local_dir1* = *right*. Go to Step 5.

Step 4 Motion along the intersection curve between a Type II obstacle and a Type III obstacle. Move along the intersection curve until one of the following occurs:

(a) A wall or a Type *I* obstacle is encountered. Target *T* cannot be reached—the procedure terminates.

(b) The M-plane is encountered. Go to Step 2.

(c) The V-plane is encountered. If $dist(\mathcal{P}_c(curr\_loc), \mathcal{P}_c(T)) > dist(\mathcal{P}_c(H_j), \mathcal{P}_c(T))$, then go to Step 2. Otherwise, define $L_j = curr\_loc$; go to Step 3.

(d) Another Type *II* obstacle is encountered. Target *T* cannot be reached—the procedure terminates.

(e) Another Type *III* obstacle is encountered. Go to Step 5.

---

[5]If $l_{3S} = l_{3T}$, this comparison will never be executed because the procedure will have terminated at Step 1d.

(f) The V-plane is encountered. If $dist(\mathcal{P}_c(curr\_loc),\ \mathcal{P}_c(T)) > dist(\mathcal{P}_c(H_j),\ \mathcal{P}_c(T))$, then repeat Step 4. Otherwise, define $L_j = curr\_loc$; go to Step 3.

Step 5 Motion Along the Intersection Curve Between Two Type III Obstacles. Move along the intersection curve until one of the following occurs:

(a) A wall or a Type *I* obstacle is encountered. $F = F - 1$. If $F = 0$, T cannot be reached—the procedure terminates. Otherwise, set *local_dir1* to its opposite; retrace to H_j; repeat Step 5.

(b) The V-plane is encountered. If $dist(\mathcal{P}_c(curr\_loc), \mathcal{P}_c(T)) > dist(\mathcal{P}_c(H_j), \mathcal{P}_c(T))$, then repeat Step 5. Otherwise, define $L_j = curr\_loc$; if $AboveMPlane(curr\_loc) = true$, then follow the intersection curve between the V-plane and the Type *III_* obstacle. Otherwise, follow the intersection curve between the V-plane and the Type *III_+* obstacle. Go to Step 3.

(c) A Type *II* obstacle is encountered. If $AboveMPlane(curr\_loc) = true$, then follow the intersection curve between the Type *II* obstacle and the Type *III_* obstacle. Otherwise, follow the intersection curve between the Type *II* obstacle and the Type *III_+* obstacle; go to Step 4.

6.2.6 Examples

Two examples considered here demonstrate performance of the motion planning algorithm presented in the previous section. Both examples make use of examples considered above in the course of the algorithm construction. To simplify the visualization of algorithm performance and not to overcrowd pictures with the arm links and joints, only the resulting paths are presented in Figures 6.11 and 6.12. Since for the Cartesian arm the C-space presentation of a path is the same as the path of the arm end effector in W-space, the paths are shown in C-space only.

Example 1. The arm's workspace contains only two obstacles, O_2 and O_3, of those three shown in Figure 6.2. Shown in Figure 6.11 are the corresponding C-obstacles, the start and target points S and T, the path ($SHabcLdeT$) of the arm end effector, and, for better visualization, the path's projection ($S'H' \ldots T'$) onto the plane (l_1, l_2). Between points S and H the end effector moves in free space along the M-line. At point H the rear part of link l_2 contacts obstacle O_3, and the arm starts maneuvering around this (Type *II*) obstacle, producing path segments Ha and ab. At point b the rear part of link l_3 contacts the (Type *III*) obstacle O_2. The next two path segments, bc and cL, correspond to the motion when the arm is simultaneously in contact with both obstacles. At point L the C-point encounters the V-plane, and the next two path segments, Ld and de, correspond to the motion in the V-plane; here the arm is in contact with only one obstacle, O_2. Finally, at point e the C-point encounters the M-line and the arm proceeds in free space along the M-line toward point T.

Example 2. The arm workspace here contains all three obstacles, O_1, O_2, and O_3, of Figure 6.2. The corresponding C-space and the resulting path are shown in Figure 6.12. Up until the arm encounters obstacle O_1 the path is the same as in Example 1: The robot moves along the M-line from S toward T until the rear part of link l_2 contacts obstacle O_3 and a hit point H_1 is defined. Between points H_1 and L_1, the arm maneuvers around obstacle O_3 by moving along the intersection curve between the M-plane and O_3 and producing path segments H_1a and aL_1. At point L_1 the M-line is encountered, and the arm moves along the M-line toward point T until the rear part of link l_3 contacts obstacle O_2 at point b. Between points b and H_2 the arm moves along the intersection curve between the V-plane and obstacle O_2 in the direction *upward*. During this motion the front part of link l_3 encounters obstacle O_1 at the hit point H_2. Now the C-point leaves the V-plane and starts moving along the intersection curve between obstacles O_1 and O_2 in the local direction *right*, producing path segments H_2c, cd, and dL_2. At point L_2 the arm returns to the V-plane and resumes its motion in it; this produces the path segment L_2e. Finally, at point e the arm encounters the M-line again and continues its unimpeded motion along the M-line toward point T.

6.3 THREE-LINK *XXP* ARM MANIPULATORS

In Section 6.2 we studied the problem of sensor-based motion planning for a specific type, *PPP*, of a three-dimensional three-link arm manipulator. The arm is one case of kinematics from the complete class *XXX* of arms, where each joint X is either P or R, prismatic or revolute. All three joints of arm *PPP* are of prismatic (sliding) type. The theory and the algorithm that we developed fits well this specific kinematic linkage, taking into account its various topological peculiarities—but it applies solely to this type. We now want to attempt a more universal strategy, for the whole group *XXP* of 3D manipulators, where X is, again, either P or R. As mentioned above, while this group covers only a half of the exhaustive list of *XXX* arms, it accounts for most of the arm types used in industry (see Figure 6.1).

As before, we specify the robot arm configuration in workspace (W-space, denoted also by $\mathcal{W}$) by its joint variable vector $j = (j_1, j_2, j_3)$, where j_i is either linear extension l_i for a prismatic joint, or an angle θ_i for a revolute joint, $i = 1, 2, 3$. The space formed by the joint variable vector is the arm's *joint space* or *J-space*, denoted also by $\mathcal{J}$. Clearly, $\mathcal{J}$ is 3D.

Define *free J-space* as the set of points in J-space that correspond to the collision-free robot arm configurations. We will show that free J-space of any *XXP* arm has a 2D subspace, called its *deformation retract*, that preserves the connectivity of the free J-space. This will allow us to reduce the problem's dimensionality. We will further show that a *connectivity graph* can be defined in this 2D subspace such that the existing algorithms for moving a point robot in a 2D metric space (Chapter 3) can be "lifted" into the 3D J-space to solve the motion planning problem for *XXP* robot arms.

In particular, in Section 6.3.1 we define the arm *configuration* $L_j = L(j)$ as an image of a continuous mapping from J-space to 3D Cartesian space $\Re^3$, which is the connected compact set of points the arm would occupy in 3D space when its joints take the value j. A real-world obstacle is the interior of a connected compact point set in $\Re^3$. Joint space obstacles are thus defined as sets of points in joint space whose corresponding arm configurations have nonempty intersections with real-world obstacles. The task is to generate a continuous collision-free motion between two given start and target configurations, denoted L_s and L_t.

Analysis of a J-space in Section 6.3.2 will show that a J-space exhibits distinct topological characteristics that allow one to predict global characteristics of obstacles in J based on the arm local contacts with (that is, sensing of) obstacles in the workspace. Furthermore, similar to the Cartesian arm case in Section 6.2, for all *XXP* arms the obstacles in J exhibit a property called the *monotonicity* property, as follows: For any point on the surface of the obstacle image, there exists one or more directions along which all the remaining points of J belong to the obstacle. The geometric representation of this property will differ from arm to arm, but it will be there and topologically will be the same property. These topological properties bring about an important result formulated in Section 6.3.3: *The free J-space, J_f, is topologically equivalent to a generalized cylinder.* This result will be essential for building our motion planning algorithm.

Deformation retracts $\mathcal{D}$ of J and $\mathcal{D}_f$ of J_f, respectively, are defined in Section 6.3.4. By definition, $\mathcal{D}_f$ is a 2D surface that preserves the connectivity of J_f. That is to say, for any two points $j_s, j_t \in J_f$, if there exists a path $p_J \subset J_f$ connecting j_s and j_t, then there must exist a path $p_D \subset \mathcal{D}_f$ connecting j_s and j_t, and p_D is topologically equivalent to p_J in J_f. Thus the dimensionality of the planning problem can be reduced.

When one or two X joints in *XXP* are revolute joints, $X = R$, J is somewhat less representative of $\mathcal{W}$, only because the mapping from J to $\mathcal{W}$ is not unique. That is, it may happen that $L(j) = L(j')$ for $j \neq j'$. Let $S_J = \{j \in J | L(j) = L_s\}$ and $T_J = \{j \in J | L(j) = L_t\}$. The task in J-space is to find a path between any pair of points $j_s \in S_J$ and $j_t \in T_J$. We define in Section 6.3.5 a *configuration space* or *C-space*, denoted by $\mathcal{C}$, as the quotient space of J over an equivalent relation that identifies all J-space points that correspond to the same robot arm configuration. It is then shown that $\mathcal{B}$ and $\mathcal{B}_f$, the quotient spaces of $\mathcal{D}$ and $\mathcal{D}_f$ over the same equivalent relation, are, respectively, deformation retracts of $\mathcal{C}$ and $\mathcal{C}_f$. Therefore, the connectivity between two given robot configurations in $\mathcal{C}$ can be determined in $\mathcal{C}_f$.

A *connectivity graph* $\mathcal{G}$ will be then defined in Section 6.3.6, and it will be shown that $\mathcal{G}$ preserves the connectivity of $\mathcal{D}_f$ and J_f. We will conclude that the workspace information available to the 3D robot is sufficient for it to identify and search the graph, and therefore the problem of 3D arm motion planning can be reduced to a graph search—something akin to the maze-searching problem in Chapter 3. Finally, in Section 6.3.7 we will develop a systematic approach, which, given a 2D algorithm, builds its 3D counterpart that preserves convergence.

The following notation is used throughout this section:

- $X, Y \subset \mathfrak{R}^3$ are point sets.
- ∂X denotes the boundary of X.
- $X \cong Y$ means X is *homeomorphic* to Y.
- $\overline{X}$ includes the closure of X, $\overline{X} \triangleq X \cup \partial X$.
- For convenience, define the closure of $\overline{R}^1 \triangleq R^1 \cup \{-\infty, +\infty\}$ and $\overline{R}^n \triangleq \overline{R}^1 \times \cdots \times \overline{R}^1$.
- It is obvious that $\overline{R}^n \cong I^n$.

6.3.1   Robot Arm Representation Spaces

To recap some notations introduced earlier, a *three-joint XXP robot arm manipulator* is an open kinematic chain consisting of three *links*, L_i, and three *joints*, J_i, $i = 1, 2, 3$; J_i also denotes the center point of joint J_i, defined as the intersection point between the axes of joints J_{i-1} and J_i. Joints J_1 and J_2 can be of either prismatic (sliding) or revolute type, while joint J_3 is of prismatic type. Joint J_1 is attached to the base O and is the origin of the fixed reference system. Figures 6.1a–d depict *XXP* arm configurations. Let p denote the arm end point; θ_i, a revolute joint variable, l_i, a prismatic joint variable, and j_i, either one of them, a revolute or a prismatic joint variable; $i = 1, 2, 3$. Figure 6.13 depicts the so-called SCARA type arm manipulator, which is of *RRP* type; it is arm (d) in Figure 6.1. We will later learn that from the standpoint of sensor-based motion planning the *RRP* arm presents the most general case among the *XXP* kinematic linkages.

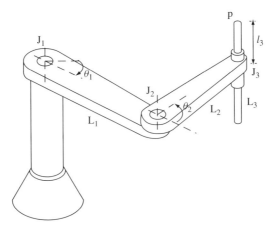

Figure 6.13   An *RRP* robot arm manipulator: p is the arm end point; J_i and L_i are, respectively, the ith joint and link, $i = 1, 2, 3$; θ_1, θ_2, and l_3 are the joint variables.

As before, assume the robot arm has enough sensing to (a) detect a contact with an approaching obstacle at any point of the robot body and (b) identify the location of that point(s) of contact on that body. The act of *maneuvering around an obstacle* refers to a motion during which the arm is in constant contact with the obstacle.

Without loss of generality, assume for simplicity a unit-length limit for joints, $l_i \in I^1$ and $\theta_i \in \overline{R}^1$, $i = 1, 2, 3$. Points at infinity are included for convenience. The *joint space* $\mathcal{J}$ is defined as $\mathcal{J} \triangleq \mathcal{J}_1 \times \mathcal{J}_2 \times \mathcal{J}_3$, where $\mathcal{J}_i = I^1$ if the ith joint is prismatic, and $\mathcal{J}_i = \overline{R}^1$ if the ith joint is revolute. In all combinations of cases, $\mathcal{J} \cong I^3$. Thus, by including points at infinity in $\mathcal{J}_i$, it is possible to treat all *XXP* arms largely within the same analytical framework.

Definition 6.3.1. *Let L_k be the set of points representing the kth robot link, $k = 1, 2, 3$; for any point $x \in L_k$, let $x(j) \in \mathfrak{R}^3$ be the point that x would occupy in $\mathfrak{R}^3$ when the arm joint vector is $j \in \mathcal{J}$. Let $L_k(j) = \bigcup_{x \in L_k} x(j)$. Then, $L_k(j) \subset \mathfrak{R}^3$ is a set of points the kth robot link occupies when the arm is at $j \in \mathcal{J}$. Similarly, $L(j) \triangleq L_1(j) \cup L_2(j) \cup L_3(j) \subset \mathfrak{R}^3$ is a set of points the whole robot arm occupies at $j \in \mathcal{J}$. The workspace (or W-space, denoted $\mathcal{W}$) is defined as*

$$\mathcal{W} \triangleq \bigcup_{j \in \mathcal{J}} \overline{L(j)} \tag{6.1}$$

We assume that L_i has a finite volume; thus $\mathcal{W}$ is bounded.

Arm links L_1 and L_2 can be of arbitrary shape and dimensions. Link L_3 is assumed to present a *generalized cylinder* —that is, a rigid body characterized by a straight-line axis coinciding with the corresponding joint axis, such that the link cross section in the plane perpendicular to the axis does not change along the axis. There are no restrictions on the shape of the cross section itself, except the physical-world assumption that it presents a simple closed curve—it can be, for example, a circle (then the link is a common cylinder), a rectangle, an oval, or any nonconvex curve.

We distinguish between the *front end* and *rear end* of link L_3. The *front end* coincides with the arm endpoint p (see Figure 6.13). The opposite end of link L_3 is its *rear end*. Similarly, the *front (rear) part* of link L_3 is the part of variable length between joint J_3 and the front (rear) end of link L_3. Formally, we have the following definition:

Definition 6.3.2. *At any given position $j = (j_1, j_2, l_3) \in \mathcal{J}$, the front part $L_{3+}(j)$ of link L_3 is defined as the point set*

$$L_{3_+}(j) = L_3(j) - L_3((j_1, j_2, 0))$$

Similarly, the rear part $L_{3_-}(j)$ of link L_3 is defined as the point set

$$L_{3_-}(j) = L_3(j) - L_3((j_1, j_2, 1))$$

The purpose of distinguishing between the front and rear parts of a prismatic (sliding) link as follows: When the front (respectively, rear) part of link L_3 approaches an obstacle, the only reasonable local direction for maneuvering around the obstacle is by decreasing (respectively, increasing) the joint variable l_3. This makes it easy to decide on the direction of motion based on the local contact only. Since the point set is $L_3((j_1, j_2, 0)) \cap L_3((j_1, j_2, 1)) \subset L_3((j_1, j_2, l_3))$ for any $l_3 \in I$, and it is independent of the value of joint variable l_3, the set is considered a part of L_2. These definitions are easy to follow on the example of the *PPP* (Cartesian) arm that we considered in great detail in Section 6.2: See the arm's effective workspace in Figure 6.3a and its complete workspace in Figure 6.3b.

The robot workspace may contain obstacles. We define an obstacle as a rigid body of an arbitrary shape. Each obstacle is of finite volume (in 2D of finite area), and its surface is of finite area. Since the arm workspace is of finite volume (area), these assumptions imply that the number of obstacles present in the workspace must be finite. Being rigid bodies, obstacles cannot intersect. Formally, we have the following definition:

Definition 6.3.3. *In the 2D (3D) case, an obstacle O_k, $k = 1, 2, \ldots$, is the interior of a connected and compact subset of $\mathfrak{R}^2$ ($\mathfrak{R}^3$) satisfying*

$$\overline{O_{k_1}} \cap \overline{O_{k_2}} = \emptyset, \qquad k_1 \neq k_2 \tag{6.2}$$

We use notation $O \triangleq \bigcup_{k=1}^{M} O_i$ to represent a general obstacle, where M is the number of obstacles in $\mathcal{W}$.

Definition 6.3.4. *The free W-space is*

$$\mathcal{W}_f \triangleq \mathcal{W} - O.$$

Lemma 6.3.1 follows from Definition 6.3.1.

Lemma 6.3.1. $\mathcal{W}_f$ *is a closed set.*

The robot arm can simultaneously touch more than one obstacle in the workspace. In this case the obstacles being touched effectively present one obstacle for the arm. They will present a single obstacle in the joint space.

Definition 6.3.5. *An obstacle in J-space (J-obstacle) $O_J \subset \mathcal{J}$ is defined as*

$$O_J \triangleq \{ j \in \mathcal{J} : L(j) \cap O \neq \emptyset \}.$$

Theorem 6.3.1. O_J *is an open set in $\mathcal{J}$.*

Let $j^* \in O_J$. By Definition 6.3.5, there exists a point $x \in L$ such that $y = x(j^*) \in O$. Since O is an open set (Definition 6.3.3), there must exist an $\epsilon > 0$ such that the neighborhood $U(y, \epsilon) \subset O$. On the other hand, since $x(j)$ is a continuous function[6] from J to W, there exists a $\delta > 0$ such that for all $j \in U(j^*, \delta)$, we have $x(j) \in U(y, \epsilon) \subset O$; thus, $U(j^*, \delta) \subset O_J$, and O_J is an open set.

The *free J-space* is $\mathcal{J}_f \overset{\triangle}{=} \mathcal{J} - O_J$. Theorem 6.3.1 gives rise to this corollary:

Corollary 6.3.1. *$\mathcal{J}_f$ is a closed set.*

Being a closed set, $\mathcal{J}_f = \overline{\mathcal{J}_f}$. Thus, a collision-free path can pass through $\partial \mathcal{J}_f$.

When the arm kinematics contains a revolute joint, due to the 2π repetition in the joint position it may happen that $L(j) = L(j')$ for $j \neq j'$. For an *RR* arm, for example, given two robot arm configurations, L_s and L_t, in W, $L(j_{s_{k_1,k_2}}) = L_{s_{0,0}} = L_s$ for $j_{s_{k_1,k_2}} = (2k_1\pi + \theta_{1s}, 2k_2\pi + \theta_{2s}) \in \mathcal{J}, k_1, k_2 = 0, \pm 1, \pm 2, \dots$. Similarly, $L(j_{t_{k_3,k_4}}) = L_{t_{0,0}} = L_t$ for $j_{t_{k_3,k_4}} = (2k_3\pi + \theta_{1t}, 2k_4\pi + \theta_{2t}) \in \mathcal{J}, k_3, k_4 = 0, \pm 1, \pm 2, \dots$. This relationship reflects the simple fact that in W every link comes to exactly the same position with the periodicity 2π. In physical space this is the same position, but in J-space these are different points.[7] The result is the tiling of space by tiles of size 2π. Figure 6.14 illustrates this situation in the plane. We can therefore state the motion planning task in J-space as follows:

Given two robot arm configurations in W, L_s and L_t, let sets $S_s = \{j \in \mathcal{J} : L(j) = L_s\}$ and $S_t = \{j \in \mathcal{J} : L(j) = L_t\}$ contain all the J-space points that correspond to L_s and L_t respectively. The task of motion planning is to generate a path $p_J \subset \mathcal{J}_f$ between j_s and j_t for any $j_s \in S_s$ and any $j_t \in S_t$ or, otherwise, conclude that no path exists if such is the case.

The motion planning problem has thus been reduced to one of moving a point in J-space. Consider the two-dimensional RR arm shown in Figure 6.14a; shown also is an obstacle O_1 in the robot workspace, along with the arm starting and target positions, S and T. Because of obstacle O_1, no motion from position S to position T is possible in the "usual" sector of angles $[0, 2\pi]$. In J-space (Figure 6.14b), this motion would correspond to the straight line between points $s_{0,0}$ and $t_{0,0}$ in the square $[0, 2\pi]$; obstacle O_1 appears as multiple vertical columns with the periodicity $(0, 2\pi)$.

However, if no path can be found between a specific pair of positions j_s and j_t in J-space, it does not mean that no paths between S and T exist. There may be paths between other pairs, such as between positions $j_{s_{0,0}}$ and $j_{t_{1,0}}$ (Figure 6.14b). On the other hand, finding a collision-free path by considering all pairs of j_s and

---

[6]If $x \in L$ is the arm endpoint, then $x(j)$ is the forward kinematics and is thus continuous.

[7]In fact, in those real-world arm manipulators that allow unlimited movement of their revolute joints, going over the 2π angle may sometimes be essential for collision avoidance.

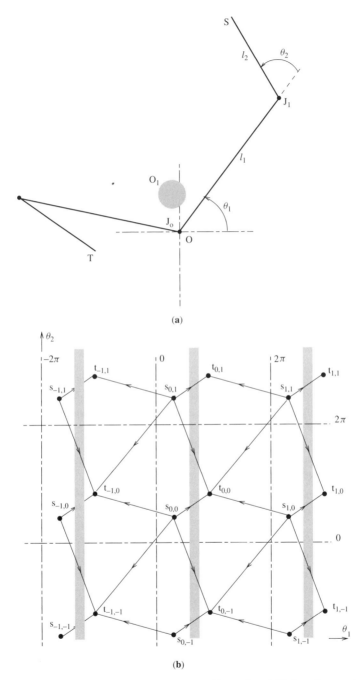

(a)

(b)

Figure 6.14 The L_s and L_t configurations (positions S and T) in robot workspace in part **(a)** produce an infinite number of S-T pairs of points in the corresponding J-space, part **(b)**; vertical shaded columns in J-space are J-obstacles that correspond to obstacle O_1 in the robot workspace.

j_t drawn from S_s and S_t, respectively, is not practical because S_s and S_t are likely to contain an infinite number of points. To simplify the problem further, in Section 6.3.5 we will introduce the notion of *configuration space*.

6.3.2   Monotonicity of Joint Space

Let $L_i(j) \subset \Re^3$ be the set of points that link L_i occupies when the manipulator joint value vector is $j \in \mathcal{J}$, $i = 1, 2, 3$ (Definition 6.3.1). Define joint space obstacles resulting from the interaction of L_i with obstacle O as *Type i obstacles*. For link L_3, let $L_{3_+}(j)$ and $L_{3_-}(j) \subset \Re^3$ be, respectively, the set of points that the *front part* and *rear part* of link L_3 occupy when the joint value vector is $j \in \mathcal{J}$ (Definition 6.3.2). Define *Type 3_+* and *Type 3_-* J-obstacles respectively resulting from the interaction of L_{3_+} and L_{3_-} with an obstacle O. More precisely:

Definition 6.3.6. *The Type i J-obstacle, $i = 1, 2, 3$, is defined as*

$$O_{Ji} \overset{\triangle}{=} \{j \in \mathcal{J} | L_i(j) \cap O \neq \emptyset\} \tag{6.3}$$

Similarly, the Type 3_+ and Type 3_- J-obstacles are defined as

$$O_{J3_+} \overset{\triangle}{=} \{j \in \mathcal{J} | L_{3_+}(j) \cap O \neq \emptyset\} \quad and \quad O_{J3_-} \overset{\triangle}{=} \{j \in \mathcal{J} | L_{3_-}(j) \cap O \neq \emptyset\} \tag{6.4}$$

Note that $O_{J3} = O_{J3_+} \cup O_{J3_-}$ and $O_J = O_{J1} \cup O_{J2} \cup O_{J3}$. We will also need notation for the intersection of Type 3_+ and Type 3_- obstacles: $O_{J3\cap} \overset{\triangle}{=} O_{J3_+} \cap O_{J3_+}$.

We now show that the underlying kinematics of the *XXP* robot arm results in a special topological properties of J-space, which is best described by the notion of *J-space monotonicity*:

J-**Space Monotonicity.** *In all cases of arm interference with obstacles, there is at least one of the two directions along the l_3 axis, such that if a value l_3' of link L_3 cannot be reached because of the interference with an obstacle, then no value $l_3'' > l_3'$ (in case of contact with the front part of link L_3) or, inversely, $l_3'' < l_3'$ (in case of contact with the rear part of link L_3) or $l_3'' \in I^1$ (in case of contact with link L_1 or L_2) can be reached either.*

J-space monotonicity results from the fact that link L_3 of the arm manipulator presents a generalized cylinder. Because links are chained together successively, the number of degrees of freedom that a link has differs from one link to another. As a result, a specific link, or even a specific part—front or rear—of the same link can produce J-space monotonicity in one or more directions. A more detailed analysis appears further.

Lemma 6.3.2. *If $j = (j_1, j_2, l_3) \in O_{J1} \cup O_{J2}$, then $j' = (j_1, j_2, l_3') \in O_{J1} \cup O_{J2}$ for all $l_3 \in I^1$.*

Consider Figure 6.13. If $j \in O_{J1}$, then $L_1(j) \cap O \neq \emptyset$. Since $L_1(j)$ is independent of l_3, then $L_1(j') \cap O \neq \emptyset$ for all $j' = (j_1, j_2, l_3')$. Similarly, if $j \in O_{J2}$, then $L_2(j) \cap O \neq \emptyset$. Since $L_2(j)$ is independent of l_3, then $L_2(j') \cap O \neq \emptyset$ for $j' = (j_1, j_2, l_3')$ with any $l_3' \in I$.

Lemma 6.3.3. *If $j = (j_1, j_2, l_3) \in O_{J3_+}$, then $j' = (j_1, j_2, l_3') \in O_{J3_+}$ for all $l_3' > l_3$. If $j = (j_1, j_2, l_3) \in O_{J3_-}$, then $j' = (j_1, j_2, l_3') \in O_{J3_-}$ for all $l_3' < l_3$.*

Using again an example in Figure 6.13, if $j \in O_{J3}$, then $L_3(j) \cap O \neq \emptyset$. Because of the linearity and the (generalized cylinder) shape of link L_3, $L_3(j') \cap O \neq \emptyset$ for all $j' = (j_1, j_2, l_3')$ and $l_3' > l_3$. A similar argument can be made for the second half of the lemma.

Let us call the planes $\{l_3 = 0\}$ and $\{l_3 = 1\}$ the *floor* and *ceiling* of the joint space. A corollary of Lemma 6.3.3 is that if $O_{3_+} \neq \emptyset$, then its intersection with the ceiling is not empty. Similarly, if $O_{3_-} \neq \emptyset$, then its intersection with the floor is nonempty. We are now ready to state the following theorem, whose proof follows from Lemmas 6.3.2 and 6.3.3.

Theorem 6.3.2. *J-obstacles exhibit the monotonicity property along the l_3 axis.*

This statement applies to all configurations of *XXP* arms. Depending on the specific configuration, though, *J*-space monotonicity may or may not be limited to the l_3 direction. In fact, for the Cartesian arm of Figure 6.15 the monotonicity property appears along all three axes: Namely, the three physical obstacles O_1, O_2, and O_3 shown in Figure 6.15a produce the robot workspace shown in Figure 6.15b and produce the configuration space shown in Figure 6.15c. Notice that the Type 3 obstacles O_1 and O_2 exhibit the monotonicity property only along the axis l_3, whereas the Type 2 obstacle O_3 exhibits the monotonicity property along two axes, l_1 and l_2. A Type 1 obstacle (not shown in the figure) exhibits the monotonicity property along all three axes (see Figure 6.6 and the related text).

6.3.3 Connectivity of $\mathcal{J}_f$

We will now show that for *XXP* arms the connectivity of $\mathcal{J}_f$ can be deduced from the connectivity of some planar projections of $\mathcal{J}_f$. From the robotics standpoint, this is a powerful result: It means that the problem of path planning for a three-joint *XXP* arm can be reduced to the path planning for a point robot in the plane, and hence the planar strategies such as those described in Chapters 3 and 5 can be utilized, with proper modifications, for 3D planning.

Let $\mathcal{J}_p$ be the floor $\{l_3 = 0\}$. Clearly, $\mathcal{J}_f \cong \mathcal{J}_1 \times \mathcal{J}_2$. Since the third coordinate of a point in $\mathcal{J}_f$ is constant zero, we omit it for convenience.

Definition 6.3.7. *Given a set $E = \{j_1, j_2, l_3\} \subset \mathcal{J}$, define the conventional projection $P_c(E)$ of E onto space $\mathcal{J}_p$ as $P_c(E) = \{(j_1, j_2) \mid \exists l_3^*, (j_1, j_2, l_3^*) \in E\}$.*

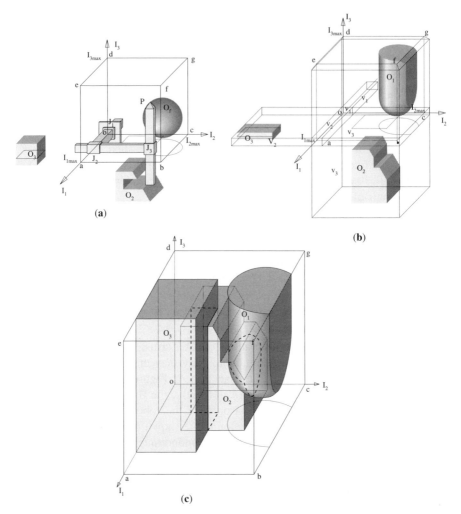

Figure 6.15 The physical obstacles O_1, O_2, and O_3 shown in figure **(a)** will be effectively perceived by the arm as those in figure **(b)**, and they will produce the rather crowded C-space shown in **(c)**.

Thus, for a joint space obstacle O_J, given start and target configurations j_s, j_t and any path p_J between j_s and j_t in $\mathcal{J}$, $P_c(O_J)$, $P_c(j_s)$, $P_c(j_t)$, and $P_c(p_J)$ are respectively the conventional projections of O_J, j_s, j_t, and p_J onto $\mathcal{J}_p$. See, for example, the conventional projections of three obstacles, O_1, O_2, and O_3, Figure 6.15b. It is easy to see that for any nonempty sets E_1, $E_2 \subset \mathcal{J}$, we have $P_c(E_1 \cap E_2) = P_c(E_1) \cap P_c(E_2)$.

Definition 6.3.8. *Define the minimal projection $P_m(E)$ of a set of points $E = \{(j_1, j_2, l_3)\} \subseteq \mathcal{J}$ onto space $\mathcal{J}_p$ as $P_m(E) = \{(j_1, j_2) \mid \forall l_3, \ (j_1, j_2, l_3) \in E\}$. For* any set $E_p = \{(j_1, j_2)\} \subset \mathcal{J}_p$, the inverse minimal projection is $P_m^{-1}(E_p) = (j_1, j_2, l_3) \mid (j_1, j_2) \in E_p$ and $l_3 \in I\}$.

The minimal projection of a single point is empty. Hence $P_m(\{j_s\}) = P_m(\{j_t\}) = \emptyset$. If $E \subset \mathcal{J}$ is homeomorphic to a sphere and stretches over the whole range of $l_3 \in I$, then $P_m(E)$ is the intersection between $\mathcal{J}_p$ and the maximum cylinder that can be inscribed into O_J and whose axis is parallel to l_3. If a set $E \subset \mathcal{J}$ presents a cylinder whose axis is parallel to the axis l_3, then $P_c(E) = P_m(E)$.

In general, O_J is not homeomorphic to a sphere and may be composed of many components. We extend the notion of a cylinder as follows:

Definition 6.3.9. *A subset $E \subset \mathcal{J}$ presents a generalized cylinder if and only if $P_c(E) = P_m(E)$.*

Type 1 and Type 2 obstacles present such generalized cylinders. It is easy to see that for any $E_p \subset \mathcal{J}_p$, $P_m^{-1}(E_p)$ is a generalized cylinder, and $P_c(P_m^{-1}(E_p)) = E_p$.

We will now consider the relationship between a path p_J in $\mathcal{J}$ and its projection $p_{J_p} \overset{\triangle}{=} P_c(p_J)$ in $\mathcal{J}_p$.

Lemma 6.3.4. *For any set $E \subset \mathcal{J}$ and any set $E_p \subset \mathcal{J}_p$, if $E_p \cap P_c(E) = \emptyset$, then $P_m^{-1}(E_p) \cap E = \emptyset$.*

If $P_m^{-1}(E_p) \cap E \neq \emptyset$, then we have $P_c(P_m^{-1}(E_p) \cap E) = P_c(P_m^{-1}(E_p)) \cap P_c(E) = E_p \cap P_c(E) \neq \emptyset$. Thus a contradiction.

The next statement provides a sufficient condition for the existence of a path in joint space:

Lemma 6.3.5. *For a given joint space obstacle O_J in $\mathcal{J}$ and the corresponding projection $P_c(O_J)$, if there exists a path between $P_c(j_s)$ and $P_c(j_t)$ in $\mathcal{J}_p$ that avoids obstacle $P_c(O_J)$, then there must exist a path between j_s and j_t in $\mathcal{J}$ that avoids obstacle O_J.*

Let $p_{J_p}(t) = \{(j_1(t), j_2(t))\}$ be a path between $P_c(j_s)$ and $P_c(j_t)$ in $\mathcal{J}_p$ avoiding obstacle $P_c(O_J)$. From Lemma 6.3.4, $P_m^{-1}(p_{J_p}(t)) \cap O_J = \emptyset$ in $\mathcal{J}$. Hence, for example, the path $p_J(t) = \{(p_{J_p}(t), (1-t)l_{3s} + t \cdot l_{3t})\} \in P_m^{-1}(\{p_{J_p}(t)\})$ connects positions j_s and j_t in $\mathcal{J}$ and avoids obstacle O_J.

To find the necessary condition, we use the notion of a minimal projection. The next statement asserts that a zero overlap between two sets in $\mathcal{J}$ implies a zero overlap between their minimal projections in $\mathcal{J}_p$:

Lemma 6.3.6. *For any sets $E_1, E_2 \subset \mathcal{J}$, if $E_1 \cap E_2 = \emptyset$, then $P_m(E_1) \cap P_m(E_2) = \emptyset$.*

By definition, $P_m^{-1}(E_{p1} \cap E_{p2}) = P_m^{-1}(E_{p1}) \cap P_m^{-1}(E_{p2})$, and $P_m^{-1}(P_m(E)) \subset E$. Thus, if $P_m(E_1) \cap P_m(E_2) \neq \emptyset$, then $\emptyset \neq P_m^{-1}(P_m(E_1) \cap P_m(E_2)) = P_m^{-1}(P_m(E_1)) \cap P_m^{-1}(P_m(E_2)) \subset E_1 \cap E_2$.

To use this lemma for designing a sensor-based motion planning algorithm, we need to describe minimal projections for different obstacle types. For a Type 1 or Type 2 obstacle O, we have $P_c(O) = P_m(O)$. For a Type 3 obstacle, we consider three events that cover all possible cases, using as an example a Type 3_+ obstacle; denote it O_+.

- O_+ intersects the floor $\mathcal{J}_f$. Because of the monotonicity property, $P_m(O_+) = O_+ \cap \mathcal{J}_f$. In other words, the minimal projection of O_+ is exactly the intersection area of O_+ with the floor $\mathcal{J}_f$.
- O_+ intersects with a Type 3_- obstacle, O_-. Then, $P_m(O_+ \cup O_-) = P_c(\partial O_+ \cap \partial O_-)$. That is, the combined minimal projection of O_+ and O_- is the conventional projection of the intersection curve between O_+ and O_-.
- Neither of the above cases apply. Then $P_m(O_+) = \emptyset$.

A similar argument can be carried out for a Type 3_- obstacle.

Define $\mathcal{J}_{pf} \stackrel{\triangle}{=} \mathcal{J}_p - P_m(O_J)$ and $\mathcal{J}_f \stackrel{\triangle}{=} P_m^{-1}(\mathcal{J}_{pf})$. It is easy to see that $\mathcal{J}_{pf} = P_c(\mathcal{J}_f)$. Therefore, $\mathcal{J}_f = P_m^{-1}(\mathcal{J}_{pf}) = P_m^{-1}(P_c(\mathcal{J}_f))$.

Theorem 6.3.3. $\mathcal{J}_f \cong \mathcal{J}'_f$; that is, $\mathcal{J}_f$ is topologically equivalent to a generalized cylinder.

Define $O'_J \stackrel{\triangle}{=} O_J - P_m^{-1}(P_m(O_J))$. Clearly, $\mathcal{J}'_f = \mathcal{J}_f \cup O'_J$ and $P_m(O'_J) = \emptyset$. By Theorem 6.3.2, each component of O'_J can be deformed either to the *floor* $\{l_3 = 0\}$ or to the *ceiling* $\{l_3 = 1\}$ and thus does not affect the topology of $\mathcal{J}_f$. Thus, $\mathcal{J}_f \cong \mathcal{J}'_f$ and, by definition, $\mathcal{J}'_f$ presents a generalized cylinder in $\mathcal{J}$.

From the motion planning standpoint, Theorem 6.3.3 indicates that the third dimension, l_3, of $\mathcal{J}_f$ is not easier to handle than the other two because $\mathcal{J}_f$ possesses the monotonicity property along l_3 axis. It also implies that as a direct result of the monotonicity property of joint space obstacles, the connectivity of $\mathcal{J}_f$ can be decided via an analysis of 2D surfaces.

We now turn to establishing a necessary and sufficient condition that ties the existence of paths in the plane $\mathcal{J}_p$ with that in 3D joint space $\mathcal{J}$. This condition will provide a base for generalizing planar motion planning algorithms to 3D space. Assume that points (arm positions) j_s and j_t lie outside of obstacles.

Theorem 6.3.4. *Given points* j_s, $j_t \in \mathcal{J}_f$ *and a joint space obstacles* $O_J \subset \mathcal{J}$, *a path exists between* j_s *and* j_t *in* $\mathcal{J}_f$ *if and only if there exists a path in* $\mathcal{J}_{pf}$ *between points* $P_c(j_s)$ *and* $P_c(j_t)$.

To prove the necessary condition, let $p_J(t)$, $t \in [0, 1]$, be a path in $\mathcal{J}_f$. From Lemma 6.3.6, $P_m(p_J(t)) \cap P_m(O_J) = \emptyset$. Hence the path $P_m(p_J(t))$ connects $P_c(j_s)$ and $P_c(j_t)$ in $\mathcal{J}_{pf}$.

To show the sufficiency, let $p_{J_p}(t)$, $t \in [0, 1]$, be a path in $\mathcal{J}_{pf}$. Then $P_m^{-1}(p_{J_p}(t))$ presents a "wall" in $\mathcal{J}$. Define $E = P_m^{-1}(p_{J_p}(t)) \cap O_J$ and let E^{-1} be the complement of E in $P_m^{-1}(p_{J_p}(t))$. We need to show that E^{-1} consists of one connected component. Assume that this is not true. For any $t_* \in [0, 1]$, since $p_{J_p}(t_*) \notin P_m(O_J)$, there exists l_{3*} such that point $(p_{J_p}(t_*), l_{3*}) \in E^{-1}$. The only possibility for E^{-1} to consist of two or more disconnected components is when there exists t_* and a set $(l_{3*}, l'_{3*}, l''_{3*})$, $l'_{3*} > l_{3*} > l''_{3*}$, such that $(p_{J_p}(t_*), l_{3*}) \in E^{-1}$ while $(p_{J_p}(t_*), l'_{3*}) \in E$ and $(p_{J_p}(t_*), l''_{3*}) \in E$. However, this cannot happen because of the monotonicity property of obstacles. Hence E^{-1} must be connected.

6.3.4 Retraction of $\mathcal{J}_f$

Theorem 6.3.4 indicates that the connectivity of space $\mathcal{J}_f$ can indeed be captured via a space of lower dimension, $\mathcal{J}_{pf}$. However, space $\mathcal{J}_{pf}$ cannot be used for motion planning because, by definition, it may happen that $\mathcal{J}_{pf} \cap O_J \neq \emptyset$; that is, some portion of $\mathcal{J}_{pf}$ is not obstacle-free. In this section we define a 2D space $\mathcal{D}_f \subset \mathcal{J}_f$ that is entirely obstacle-free and, like $\mathcal{J}_{pf}$, captures the connectivity of $\mathcal{J}_f$.

Definition 6.3.10. *[57]. A subset $\mathcal{A}$ of a topological space $\mathcal{X}$ is called a retract of $\mathcal{X}$ if there exists a continuous map $r : \mathcal{X} \longrightarrow \mathcal{A}$, called a retraction, such that $r(a) = a$ for any $a \in \mathcal{A}$.*

Definition 6.3.11. *[57]. A subset $\mathcal{A}$ of space $\mathcal{X}$ is a deformation retract of $\mathcal{X}$ if there exists a retraction r and a continuous map*

$$f : \mathcal{X} \times I \to \mathcal{X} \tag{6.5}$$

such that

$$\left. \begin{array}{l} f(x, 0) = x \\ f(x, 1) = r(x) \end{array} \right\} \quad x \in \mathcal{X} \\ f(a, t) = a, \qquad a \in \mathcal{A} \ \text{and} \ t \in I \tag{6.6}$$

In other words, set $\mathcal{A} \subset \mathcal{X}$ is a deformation retract of $\mathcal{X}$ if $\mathcal{X}$ can be continuously deformed into $\mathcal{A}$. We show below that $\mathcal{D}_f$ is a deformation retract of $\mathcal{J}_f$. Let $\mathcal{J}_p$, $\mathcal{J}_{pf}$, and $\mathcal{J}_f$ be as defined in the previous section; then we have the following lemma.

Lemma 6.3.7. *$\mathcal{J}_p$ is a deformation retract of $\mathcal{J}$, and $\mathcal{J}_{pf}$ is a deformation retract of $\mathcal{J}_f$.*

Define $r(j_1, j_2, l_3) = (j_1, j_2, 0)$. It follows from Lemma 6.3.2 that r is a retraction. Since for Type 1 and 2 obstacles $P_m^{-1}(P_m(O_J)) = O_J$, then, if $\mathcal{J}$ contains only Type 1 and Type 2 obstacles—that is, $\mathcal{J}_f = \mathcal{J} - (O_{J1} \cup O_{J2})$—it follows that $\mathcal{J}_f = \mathcal{J}'_f$ and $\mathcal{J}_{pf}$ is a deformation retract of $\mathcal{J}_f$. In the general case, all obstacle types (including Type 3) can be present, and $\mathcal{J}_{pf}$ is not necessarily a deformation retract of $\mathcal{J}_f$.

Theorem 6.3.5. *Let* $\mathcal{Q}_1 \stackrel{\triangle}{=} \partial O_{J3_-} \cap \mathcal{J}_f$, *and* $\mathcal{Q}_2 \stackrel{\triangle}{=} \mathcal{J}_p \cap \mathcal{J}_f$. *Then,*

$$\mathcal{D}_f \stackrel{\triangle}{=} \mathcal{Q}_1 \cup \mathcal{Q}_2$$

is a deformation retract of $\mathcal{J}_f$.

$\mathcal{Q}_1$ and $\mathcal{Q}_2$, are respectively, the obstacle-free portion of ∂O_{J3_-} and $\mathcal{J}_p$. It is easy to see that $\mathcal{D}_f \cong \mathcal{J}_{pf}$. Since $\mathcal{J}_f \cong \mathcal{J}'_f$ (Theorem 6.3.3) and $\mathcal{J}_{pf}$ is a deformation retract of $\mathcal{J}'_f$ (Lemma 6.3.7), then $\mathcal{D}_f$ is a deformation retract of $\mathcal{J}_f$.

Let $\mathcal{D}$ denote the 2D space obtained by patching all the "holes" in $\mathcal{D}_f$ so that $\mathcal{D} \cong \mathcal{J}_p$. It is obvious that $\mathcal{D}$ is a deformation retract of $\mathcal{J}$.

Theorem 6.3.6. *Given two points* $j'_s, j'_t \in \mathcal{D}_f$, *if there exists a path* $p_J \subset \mathcal{J}_f$ *connecting* j'_s *and* j'_t, *then there exists a path* $p_D \subset \mathcal{D}_f$, *such that* $p_D \sim p_J$.

From Theorem 6.3.5, $\mathcal{D}_f$ is a deformation retract of $\mathcal{J}_f$. Let r be the retraction as in Ref. 57, II.4; then $p' = r(p)$ must be an equivalent path in $\mathcal{D}_f$.

On the other hand, if j'_s and j'_t are not connected in $\mathcal{J}_f$, then by definition j'_s and j'_t are not connected in $\mathcal{D}_f$ either. Hence the connectivity of j'_s and j'_t can be determined completely by studying $\mathcal{D}_f$, which is simpler than $\mathcal{J}_f$ because the dimensionality of $\mathcal{D}_f$ is lower than that of $\mathcal{J}_f$. Furthermore, a path between two given points $j_s = (j_{1s}, j_{2s}, l_{3s})$, $j_t = (j_{1t}, j_{2t}, l_{3t}) \in \mathcal{J}_f$ can be obtained by finding the path between the two points $j'_s = (j_{1s}, j_{2s}, l'_{3s})$, $j'_s = (j_{1t}, j_{2t}, l'_{3t}) \in \mathcal{D}_f$. Because of the monotonicity property (Theorem 6.3.2), j'_s and j'_t always exist and they can be respectively connected within $\mathcal{J}_f$ with j_s and j_t via vertical line segments. Hence the following statement:

Corollary 6.3.2. *The problem of finding a path in the 3D subset* $\mathcal{J}_f$ *between points* $j_s, j_t \in \mathcal{J}_f$ *can be reduced to one of finding a path in its 2D subset* $\mathcal{D}_f$.

6.3.5   Configuration Space and Its Retract

Our motion planning problem has thus been reduced to one of moving a point in a 2D subset of J-space, $\mathcal{J}$. However, as pointed out in Section 6.3.1, the joint space $\mathcal{J}$ is not very representative when revolute joints are involved, because the mapping from $\mathcal{J}$ to workspace $\mathcal{W}$ is not unique. Instead, we define *configuration space* $\mathcal{C}$:

Definition 6.3.12. *Define an equivalence relation* $\mathcal{F}$ *in* $\mathcal{J}$ *as follows: for* $j = (j_1, j_2, j_w)$, $j' = (j'_1, j'_2, j'_3) \in \mathcal{J}$, $j\mathcal{F}j'$ *if and only if* $(j_i - j'_i)\%2\pi = 0$, *for* $i =$

1, 2, 3, *where % is the modular operation. The quotient space* $\mathcal{C} \stackrel{\triangle}{=} \mathcal{J}/\mathcal{F}$ *is called the configuration space (C-space), with normal quotient space topology assigned, see Ref. 57, A.1. Let* $c = \mathcal{F}j$ *represent an equivalence class; then the project* $f : \mathcal{J} \to \mathcal{C}$ *is given by* $f(j) = \mathcal{F}j$.

Theorem 6.3.7. *The configuration space* $\mathcal{C}$ *is compact and of finite volume (area).*

By definition, $\mathcal{J} = \mathcal{J}_1 \times \mathcal{J}_2 \times \mathcal{J}_3$. Define equivalence relations $\mathcal{F}_i$ in $\mathcal{J}_i$ such that $j_i \mathcal{F}_i j_i'$ if and only if $(j_i - j_i')\%2\pi = 0$. Define $\mathcal{C}_i \stackrel{\triangle}{=} \mathcal{J}_i/\mathcal{F}_i$ and the project $f_i : \mathcal{J}_i \to \mathcal{C}_i$ given by $f_i(j) = \mathcal{F}_i j$. Apparently, $\mathcal{C}_i \cong S^1$ with length $v_i = 2\pi$ if $\mathcal{J}_i = \mathfrak{R}$, and $\mathcal{C}_i \cong I^1$ with length $v_i = 1$ if $\mathcal{J}_i = I^1$. Because f is the product of f_i's, f_i's are both open and closed, and the product topology and the quotient topology on $\mathcal{C}_1 \times \mathcal{C}_2 \times \mathcal{C}_3$ are the same (see Ref. 57, Proposition A.3.1); therefore, $\mathcal{C} \cong \mathcal{C}_1 \times \mathcal{C}_2 \times \mathcal{C}_3$ is of finite volume $v_1 \cdot v_2 \cdot v_n$.

For an *RR* arm, for example, $\mathcal{C} \cong S^1 \times S^1$ with area $2\pi \cdot 2\pi = 4\pi^2$; for an *RRP* arm, $\mathcal{C} \cong S^1 \times S^1 \times I^1$ with volume $2\pi \cdot 2\pi \cdot 1 = 4\pi^2$.

For $c \in \mathcal{C}$, we define $L(c) \stackrel{\triangle}{=} L(j)$, where $j \in f^{-1}(c)$, to be the area the robot arm occupies in $\mathcal{W}$ when its joint vector is j.

Definition 6.3.13. *The configuration space obstacle (C-obstacle) is defined as*

$$O_C \stackrel{\triangle}{=} \{c \in \mathcal{C} : L(c) \cap O \neq \emptyset\}$$

The free C-space is $\mathcal{C}_f \stackrel{\triangle}{=} \mathcal{C} - O_C$.

The proof for the following theorem and its corollary are analogous to those for Theorem 6.3.1.

Theorem 6.3.8. *A C-obstacle is an open set.*

Corollary 6.3.3. *The free C-space* $\mathcal{C}_f$ *is a closed set.*

The configuration space obstacle O_C may have more than one component. For convenience, we may call each component an obstacle.

Theorem 6.3.9. *Let* $L(c_s) = L_s$ *and* $L(c_t) = L_t$. *If there exists a collision-free path (motion) between* L_s *and* L_t *in* $\mathcal{W}$, *then there is a path* $p_C \subset \mathcal{C}_f$ *connecting* c_s *and* c_t, *and vice versa.*

If there exists a motion between L_s and L_t in $\mathcal{W}$, then there must be a path $p_J \subset \mathcal{J}_f$ between two points $j_s, j_t \in \mathcal{J}_f$ such that $L(j_s) = L_s$ and $L(j_t) = L_t$. Then, $p_C = f(p_J) \subset \mathcal{C}_f$ is a path between $c_s = f(j_s)$ and $c_t = f(j_t)$. The other half of the theorem follows directly from the definition of $\mathcal{C}_f$.

We have thus reduced the motion planning problem in the arm workspace to the one of moving a point from start to target position in C-space.

The following characteristics of the C-space topology of *XXP* arms are direct results of Theorem 6.3.7:

- For a *PPP* arm, $C \cong I^1 \times I^1 \times I^1$, the unit cube.
- For a *PRP* or *RPP* arm, $C \cong S^1 \times I^1 \times I^1$, a pipe.
- For an *RRP* arm, $C \cong S^1 \times S^1 \times I^1$, a solid torus.

Figure 6.16 shows the C-space of an *RRP* arm, which can be viewed either as a cube with its front and back, left and right sides pairwise identified, or as a solid torus.

The obstacle monotonicity property is preserved in configuration space. This is simply because the equivalent relation that defines C and C_f from $\mathcal{J}$ and $\mathcal{J}_f$ has no effect on the third joint axis, l_3. Thus we have the following statement:

Theorem 6.3.10. *The configuration space obstacle O_C possesses the monotonicity property along l_3 axis.*

As with the subset $\mathcal{J}_f$, $C_p \subset C$ can be defined as the set $\{l_3 = 0\}$; O_{C1}, O_{C2}, O_{C3}, O_{C3_+}, O_{C3_-}, P_c, P_m, C_f, C_{pf}, and C_f' can be defined accordingly.

Theorem 6.3.11. *Let $Q_1 \overset{\Delta}{=} \partial O_{C3_-} \cap C_f$ and $Q_2 \overset{\Delta}{=} C_p \cap C_f$. Then,*

$$B_f \overset{\Delta}{=} Q_1 \cup Q_2$$

is a deformation retract of C_f.

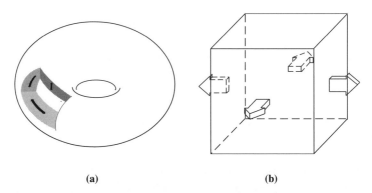

(a)   (b)

Figure 6.16   Two views of C-space of an *RRP* arm manipulator: **(a)** As a unit cube with its front and back, left and right sides pairwise identified; and **(b)** as a solid torus.

The proof is analogous to that of Theorem 6.3.5. Let $\mathcal{B}$ denote the 2D space obtained by patching all the "holes" in $\mathcal{B}_f$ so that $\mathcal{B} \cong \mathcal{C}_p$. It is obvious that $\mathcal{B} \cong \mathcal{C}_p \cong \mathcal{T}$ is a deformation retract of $\mathcal{C}$. We obtain the following statement parallel to Theorem 6.3.6.

Theorem 6.3.12. *Given two points j'_s, $j'_t \in \mathcal{B}_f$, if there exists a path $p_C \subset \mathcal{C}_f$ connecting j'_s and j'_t, then there must exist a path $p_B \subset \mathcal{B}_f$, such that $p_B \sim p_J$.*

A path between two given points $j_s = (j_{1s}, j_{2s}, l_{3s})$, $j_t = (j_{1t}, j_{2t}, l_{3t}) \in \mathcal{C}_f$ can be obtained by finding the path between the two points $j'_s = (j_{1s}, j_{2s}, l'_{3s})$, $j'_s = (j_{1t}, j_{2t}, l'_{3t}) \in \mathcal{B}_f$. Because of the monotonicity property (Theorem 6.3.10), j'_s and j'_t always exist and can be respectively connected within $\mathcal{C}_f$ with j_s and j_t via vertical line segments. Hence the following statement:

Corollary 6.3.4. *The problem of finding a path in $\mathcal{C}_f$ between points j_s, $j_t \in \mathcal{C}_f$ can be reduced to that of finding a path in its subset $\mathcal{B}_f$.*

6.3.6 Connectivity Graph

At this point we have reduced the problem of motion planning for an *XXP* arm in 3D space to the study of a 2D subspace $\mathcal{B}$ that is homeomorphic to a common torus.

Consider the problem of moving a point on a torus whose surface is populated with obstacles, each bounded by simple closed curves. The torus can be represented as a square with its opposite sides identified in pairs (see Figure 6.17a). Note that the four corners are identified as a single point. Without loss of generality, let the start and target points S and T be respectively in the center and the corners of the square. This produces four straight lines connecting S and T, each connecting the center of the square with one of its corners. We call each line a *generic path* and denote it by g_i, $i = 1, 2, 3, 4$.

Define a *connectivity graph* $\mathcal{G}$ on the torus by the obstacle-free portions of any three of the four generic paths and the obstacle boundary curves. We have the following statement:

Theorem 6.3.13. *On a torus, if there exists an obstacle-free path connecting two points S and T, then there must exist such a path in the connectivity graph $\mathcal{G}$.*

Without loss of generality, let g_1, g_2, and g_3 be the complete set of generic paths, as shown in Figure 6.17a, where the torus is represented as a unit square with its opposite sides identified.

The generic paths g_1, g_2, and g_3 cut the unit square into three triangular pieces. Rearrange the placements of the three pieces by identifying the opposite sides of the square in pairs along edges a and b, respectively (see Figures 6.17b and 6.17c). We thus obtain a six-gon (hexagon), with three pairs of S and T as its vertices and generic paths g_1, g_2, and g_3 as its edges. The hexagon presentation is called the *generic form* of a torus.

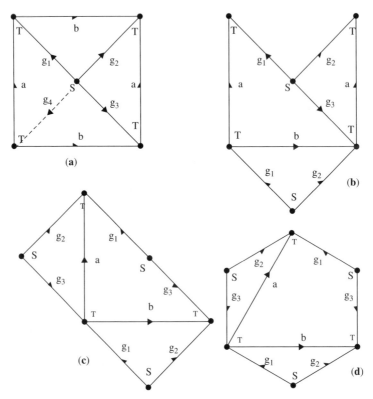

Figure 6.17 Paths g_1, g_2, and g_3 constitute a complete set of generic paths. A hexagon is obtained by **(a)** cutting the square along g_1, g_2, and g_3, **(b)** pasting along b, and **(c)** pasting along a. **(d)** The resulting hexagon.

Now consider the effects of the above operation on obstacles (see Figure 6.18a). Obstacle boundaries and the generic paths partition our hexagon into *occupied areas* (shaded areas in Figure 6.18b) and *free areas* (numbered *I*, *II*, *III*, *IV* and *V* in Figure 6.18b). Each free area is not necessarily simple, but it must be homeomorphic to a disc, possibly with one or more smaller discs removed (e.g., area *I* of Figure 6.18b has the disc that corresponds to obstacle O_2 removed). The free area's inner boundaries are formed by obstacle boundaries; its outer boundary consists of segments of obstacle boundaries and segments of the generic paths.

Any arbitrary obstacle-free path p that connects points S and T consists of one or more segments, $p_1, p_2, \ldots, p_n$, in the hexagon. Let x_i, y_i be the end points of segment p_i, where $x_1 = S$, $x_{i+1} = y_i$ for $i = 1, 2, \ldots, n - 1$, and $y_n = T$. Since p is obstacle-free, x_i and y_i must be on the outer boundary of the free area that contains p_i. Therefore, x_i and y_i can be connected by a path segment p_i' of the outer boundary of the free area. The path $p' \overset{\triangle}{=} p_1' p_2' \ldots p_n'$ that connects S and T and consists of segments of the generic paths and segments of obstacle boundaries is therefore entirely on the connectivity graph $\mathcal{G}$.

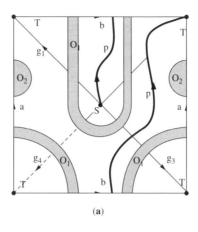

(a)

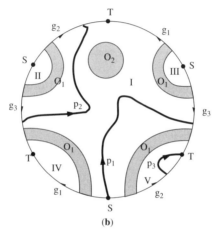

(b)

Figure 6.18 Illustration for Theorem 6.3.13. Shown are two obstacles O_1, O_2 (shaded areas) and path p (thicker line). The torus is represented, respectively, as **(a)** a unit square with its opposite sides a and b identified in pairs and **(b)** as a hexagon, with generic paths as its sides. Segments p_1, p_2 and p_3 in (b) are connected; they together correspond to the path p in (a).

Figure 6.18a presents a torus shown as a unit square, with its opposite sides a and b identified in pairs. O_1 and O_2 are two obstacles. Note that the three pieces of obstacle O_1 in the figure are actually connected. Segments g_1, g_2 and g_3 are (any) three of the four generic paths.

For an *XXP* arm, we now define *generic paths* and the *connectivity graph* in $\mathcal{B}$, which is homeomorphic to a torus.

Definition 6.3.14. *For any two points $a, b \in \mathcal{J}$, let $\overline{ab}$ be the straight line segment connecting a and b. A vertical plane is defined as*

$$\mathcal{V}_{ab} \stackrel{\triangle}{=} P_m^{-1}(P_c(\overline{ab}))$$

where P_c and P_m are respectively the conventional and minimal projections as in Definition 6.3.7 and Definition 6.3.8.

In other words, $\mathcal{V}_{ab}$ contains both a and b and is parallel to the l_3 axis. The degenerate case where $\overline{ab}$ is parallel to the l_3 axis is simple to handle and is not considered.

Definition 6.3.15. *Let L_s and L_t be the given start and target configurations of the arm, and let $S \overset{\triangle}{=} \{j \in \mathcal{J} | L(j) = L_s\} \subset \mathcal{J}$ and $T \overset{\triangle}{=} \{j \in \mathcal{J} | L(j) = L_t\} \subset \mathcal{J}$, respectively, be the sets of points corresponding to L_s and L_t. Let $f : \mathcal{J} \to \mathcal{C}$ be the projection as in Definition 6.3.12. Then the vertical surface $\mathcal{V} \subset \mathcal{C}$ is defined as*

$$\mathcal{V} \overset{\triangle}{=} \{f(j) \in \mathcal{C} | j \in \mathcal{V}_{st} \quad for\ all \quad s \in \mathcal{S} \quad and \quad t \in T\}$$

For the *RRP* arm, which is the most general case among *XXP* arms, $\mathcal{V}$ consists of four components $\mathcal{V}_i$, $i = 1, 2, 3, 4$. Each $\mathcal{V}_i$ represents a class of vertical planes in $\mathcal{J}$ and can be determined by the first two coordinates of a pair of points drawn respectively from S and T. If $j_s = (\theta_1^s, \theta_2^s, l_3^s)$ and $j_t = (\theta_1^t, \theta_2^t, l_3^t)$ are the robot's start and target configurations, the four components of the vertical surface $\mathcal{V}$ can be represented as follows:

$$\mathcal{V}_1 : (\theta_1^s, \theta_2^s)\ (\theta_1^t, \theta_2^t)$$
$$\mathcal{V}_2 : (\theta_1^s, \theta_2^s)\ (\theta_1^t, \theta_2^t - 2\pi \times \text{sign}(\theta_2^t - \theta_2^s)) \qquad (6.7)$$
$$\mathcal{V}_3 : (\theta_1^s, \theta_2^s))\ (\theta_1^t - 2\pi \times \text{sign}(\theta_1^t - \theta_1^s), \theta_2^t)$$
$$\mathcal{V}_4 : (\theta_1^s, \theta_2^s)\ (\theta_1^t - 2\pi \times \text{sign}(\theta_1^t - \theta_1^s), \theta_2^t - 2\pi \times \text{sign}(\theta_2^t - \theta_2^s))$$

where sign() takes the values $+1$ or -1 depending on its argument. Each of the components of V-surface determines a *generic path*, as follows:

$$g_i \overset{\triangle}{=} \mathcal{V}_i \cap \mathcal{B}, \qquad i = 1, 2, 3, 4$$

Since $\mathcal{B}$ is homeomorphic to a torus, any three of the four generic paths can be used to form a connectivity graph. Without loss of generality, let $g \overset{\triangle}{=} \bigcup_{i=1}^{3} g_i$ and denote $g' \overset{\triangle}{=} \mathcal{B}_f \cap g$. A *connectivity graph* can be defined as $\mathcal{G} \overset{\triangle}{=} g' \cup \partial \mathcal{B}_f$. If there exists a path in $\mathcal{C}_f$, then at least one such path can be found in the connectivity graph $\mathcal{G}$.

Now we give a physical interpretation of the connectivity graph $\mathcal{G}$; $\mathcal{G}$ consists of the following curves:

- $\partial \mathcal{C}_p$—the boundary curve of the floor, identified by the fact that the third link of the robot reaches its lower joint limit ($l_3 = 0$) and, simultaneously, one or both of the other two links reach their joint limits.

- $\mathcal{C}_p \cap \partial(O_{C1} \cup O_{C2})$—the intersection curve between the floor and the Type 1 or Type 2 obstacle boundary, identified by the fact that the third link of the robot reaches its lower joint limit ($l_3 = 0$) and simultaneously, one or both of the other two links contact some obstacles.

- $\partial O_{C3_-} \cap \partial \mathcal{J}$—the intersection curve between the Type 3_- obstacle boundary and *C*-space boundary, identified by the fact that the robot's third link touches obstacles with its rear part and, simultaneously, one or more links reach their joint limits; this case includes the intersection curve between a Type 3_- obstacle boundary and the ceiling.

- $\partial O_{C3_-} \cap \partial(O_{C1} \cup O_{C2})$—the intersection curve between the Type 3_- obstacle boundary and the Type 1 or Type 2 obstacle boundary, identified by the fact that the third link of the robot touches obstacles with its rear part and that, simultaneously, one or both of the other two links contact some obstacles.

- $\partial O_{C3_-} \cap \partial O_{C3_+}$—the intersection curve between the Type 3_+ obstacle boundary and the Type 3_- obstacle boundaries, identified by the fact that both front and rear parts of the third link contact obstacles.

- $g \cap \mathcal{J}_{pf}$—the obstacle-free portion of the generic path, identified by the fact that the robot is on the V-surface and that the third joint reaches its lower limit ($l_3 = 0$).

- $\mathcal{V} \cap \partial O_{C3_-}$—the intersection curve between V-surface and the Type 3_- obstacle boundaries, identified by the fact that the robot is on V-surface and simultaneously touches the obstacle with the rear part of its third link.

Note that the sensing capability of the robot arm manipulator allows it to easily identify the fact of being at any of the curves above.

6.3.7 Lifting 2D Algorithms into 3D

We have reduced the problem of motion planning for an *XXP* arm to the search on a connectivity graph. The search itself can be done using any existing graph search algorithm. The efficiency of a graph search algorithm is often in general—and, as we discussed above, in the Piano Mover's approach in particular—measured by the total number of nodes visited and the number of times each node is visited, regardless of the number of times that each edge is visited. In robotics, however, what is important is the total number of edge traverses, because physically each edge presents a part of the path whose length the robot may have to pass. For this reason, typical graph algorithms—for example, the *width-first search* algorithm [114]—would be inefficient from the standpoint of robot motion planning. On the other hand, *depth-first search* algorithms may work better; Tarry's rule [42] and Fraenkel's rule [45], which we discussed in Section 2.9.1, are two versions of such search algorithms. More efficient algorithms can be obtained by taking into account specifics of the connectivity graph topology [59].

In this section we intend to show how a 2D motion planning algorithm—we will call it A_p—can be carried out in $\mathcal{B}_f$. We assume that A_p operates according to our usual model—that is, using local information from the robot tactile sensors, the paths it produces are confined to generic paths, obstacle boundaries, and space boundaries, if any—and that it guarantees convergence.[8] As before, it is also assumed that all decisions as to what directions the robot will follow along the generic path or along an obstacle boundary are made at the corresponding intersection points.

Without loss of generality, side walls of the C-space, if any, are simply treated below as Type I and Type II obstacles, the C-space ceiling is treated as a Type III$_+$ obstacle, and the C-space floor is treated as a Type III$_-$ obstacle.

Since the robot start and target positions are not necessarily in $\mathcal{B}_f$, our first step is to bring the robot to $\mathcal{B}_f$. This is easily achieved by moving from j_s downward until a Type III$_-$ obstacle is encountered; that is, the link L_3 of the robot either reaches its joint limit or touches an obstacle with its rear end. Then, the robot switches to A_p, which searches for point j_t' directly above or below j_t, with the following identification of path elements:

- Generic path—the intersection curve of $\mathcal{V}$ and ∂O_{3_-}.
- Obstacle boundary—the intersection curve of ∂O_{3_-} and $\partial(O_1 \cup O_2 \cup O_{3_+})$.

If A_p terminates without reaching j_t', then the target j_t is not reachable. On the other hand, if j_t' is reached, then the robot moves directly toward j_t. Along this path segment the robot will either reach j_t or encounter an obstacle, in which case the target is not reachable. This shows how a motion planning algorithm for a compact 2D surface can be "lifted" into 3D to solve the motion planning problem of an *XXP* arm.

6.3.8 Step Planning

Similar to 2D algorithms in Chapter 5, realization of the above 3D motion planning algorithms requires a complementary lower-level control piece for step calculation. The required procedure for step calculation is similar to the one sketched in Section 5.2.3 for a 2D arm, except here the tangent to the C-obstacle at the local point of contact is three-dimensional. Since, according to the motion planning algorithm, the direction of motion is always unique—the curve along which the arm moves is either an M-line, or an intersection curve between an obstacle and one of the planes M-plane or V-plane, or an intersection curve between obstacles—the tangent to the C-obstacle at the contact point is unique as well. More detail on the step calculation procedure can be found in Refs. 106 and 115.

---

[8]The question of taking advantage of a sensing medium that is richer than tactile sensing (vision, etc.) has been covered in great detail in Section 3.6 and also in Section 5.2.5; hence we do not dwell on it in this chapter.

6.3.9   Discussion

As demonstrated in this section, the kinematic constraints of any *XXP* arm major linkage result in a certain property—called *monotonicity*—of the arm joint space and configuration space (C-space or $\mathcal{C}_f$). The essence of the monotonicity property is that for any point on the surface of a C-space obstacle, there exists at least one direction in C-space that corresponds to one of the joint axes, such that no other points in space along this direction can be reached by the arm. The monotonicity property allows the arm to infer some global information about obstacles based on local sensory data. It thus becomes an important component in sensor-based motion planning algorithms. We concluded that motion planning for a three-dimensional *XXP* arm can be done on a 2D compact surface, $\mathcal{B}_f$, which presents a deformation retract of the free configuration space $\mathcal{C}_f$.

We have further shown that any convergent 2D motion planning algorithm for moving a point on a compact surface (torus, in particular) can be "lifted" into 3D for motion planning for three-joint *XXP* robot arms. The strategy is based on the monotonicity properties of C-space.

Given the arm's start and target points $j_s, j_t \in \mathcal{C}_f$ and the notions "above" and "below" as defined in this section, the general motion planning strategy for an *XXP* arm can be summarized as consisting of these three steps:

1. Move from j_s to j_s', where $j_s' \in \mathcal{B}_f$ is directly above or below j_s;
2. find a path between j_s' and j_t' within $\mathcal{B}_f$, where $j_t' \in \mathcal{B}_f$ is directly above or below j_t; and
3. move from j_t' to j_t.

Because of the monotonicity property, motion in Steps 1 and 3 can be achieved via straight line segments. In reality, Step 2 does not have to be limited to the plane: It can be "lifted" into 3D by modifying the 2D algorithm respectively, thus resulting in local optimization and shorter paths. With the presented theory, and with various specific algorithms presented in this and previous chapters, one should have no difficulty constructing one's own sensor-based motion planning algorithms for specific *XXP* arm manipulators.

6.4   OTHER *XXX* ARMS

One question about motion planning for 3D arm manipulators that still remains unanswered in this chapter is, How can one carry out sensor-based motion planning for *XXR* arm manipulators—that is, arms whose third joint is of revolute type? At this time, no algorithms with a solid theoretical foundation and with guaranteed convergence can be offered for this group. This exciting area of research, of much theoretical as well as practical importance, still awaits for its courageous explorers.

In engineering terms, one kinematic linkage from the *XXR* group, namely *RRR*, is of much importance among industrial robot manipulators. On a better side, the *RRR* linkage is only one out of the five 3D linkages shown in Figure 6.1, Section 6.1, which together comprise the overwhelming majority of robot arm manipulators that one finds in practice. Still, *RRR* is a popular arm, and knowing how to do sensor-based motion planning for it would be of much interest. Judging by our analysis of the *RR* arm in Chapter 5, it is also likely the most difficult arm for sensor-based motion planning.

Conceptually, the difficulty with the *RRR* arm, and to some extent with other *XXR* arms, is of the same kind that we discussed in the Section 6.1, when describing a fly moving around an object in three-dimensional space. The fly has an infinite number of ways to go around the object. Theoretically, it may need to try all those ways in order to guarantee getting "on the other side" of the object.

We have shown in this chapter that, thanks to special properties of monotonicity of the corresponding configuration space, no infinite motion will ever be necessary for any *XXP* arm manipulator in order to guarantee convergence. No matter how complex are the 3D objects in the arm's workspace, the motion planning algorithm guarantees a finite (and usually quick) path solution. No such properties have been found so far for the *XXR* arm manipulators. On the other hand, similar to *XXP* arms considered in this chapter, motion planning for *XXR* arms seems to be reducible to motion along curves that are similar to curves we have used for *XXP* algorithms (such as an intersection curve between an obstacle and an M-plane or V-plane, etc.). Even in the worst case, this would require a search of a relatively small graph.

Provided that the arm has the right whole-body sensing, in practice one can handle an *XXR* arm by using the motion planning schemes developed in this chapter for *XXP* arms, perhaps with some heuristic modifications. Some such attempts, including physical experiments with industrial *RRR* arm manipulators, are in described Chapter 8 (see also Ref. 115).

CHAPTER 7

Human Performance in Motion Planning

> I ... do not direct myself so badly. If it looks ugly on the right, I take the left ...
> Have I left something unseen behind me? I go back; it is still on my road. I trace
> no fixed line, either straight or crooked.
>
> — *Michel de Montaigne (1533–1592), The Essays*

7.1 INTRODUCTION

It is time to admit that we will not be able to completely fulfill the promise contained in this book's subtitle—explain *how* humans plan their motion. This would be good to do—such knowledge would help us in many areas—but we are not in a position to do so. Today we know precious little about how human motion decision-making works, certainly not on the level of algorithmic detail comparable to what we know about robot motion planning. To be sure, in the literature on psychophysical and cognitive science analysis of human motor skills one will find speculations about the nature of human motion planning strategies. One can even come up with experimental tests designed to elucidate such strategies. The fact is, however, that the sum of this knowledge tells us only what those human strategies *might be*, not what they are.

Whatever those unknown strategies that humans use to move around, we can, however, study those strategies' *performance*. By using special tests, adhering to carefully calibrated test protocols designed to elucidate the right questions, and by carrying out those tests on statistically significant groups of human subjects, we can resolve *how good* we humans are at planning our motion. Furthermore, we can (and will) subject robot sensor-based motion planning algorithms to the same tests—making sure we keep the same test conditions—and make far-reaching conclusions that can be used in the design of complex systems involving human operators.

Clearly, the process of testing human subjects has to be very different from the process of designing and testing robot algorithms that we undertook in prior

Sensing, Intelligence, Motion, by Vladimir J. Lumelsky
Copyright © 2006 John Wiley & Sons, Inc.

chapters. This dictates a dramatic change in language and methodology. So far, as we dealt with algorithms, concepts have been specific and well-defined, statements have been proven, and algorithms were designed based on robust analysis. We had definitions, lemmas, theorems, and formal algorithms. We talked about algorithm convergence and about numerical bounds on the algorithm performance.

All such concepts become elusive when one turns to studying human motion planning. This is not a fault of ours but the essence of the topic. One way to compensate for the fuzziness is the *black box* approach, which is often used in physics, cybernetics, and artificial intelligence: The observer administers to the object of study—here a human subject—a test with a well-controlled input, observes the results at the output, and attempts to uncover the law (or the algorithm) that transfers one into the other.

With an object as complex as a human being, it would not be realistic to expect from this approach a precise description of motion planning strategies that humans use. What we expect instead from such experiments is a measure of human *performance*, of human skills in motion planning. By using techniques common in cognitive sciences and psychology, we should be able to arrive at crisp comparisons and solid conclusions. Why do we want to do this? What are the expected scientific and practical uses of this study?

One use is in the design of teleoperated systems—that is, systems with remotely controlled moving machinery and with a human operator being a part of the control and decision-making loop. In this interesting domain the issues of human and robot performance intersect. More often than not, such systems are very complex, very expensive, and very important. Typical examples include control of the arm manipulator at the Space Shuttle, control of arms at the International Space Station, and robot systems used for repair and maintenance in nuclear reactors.

The common view on the subject is that in order to efficiently integrate the human operator into the teleoperated system's decision-making and control, the following two components are needed: (1) a data gathering and preprocessing system that provides the operator with qualitatively and quantitatively adequate input information; this can be done using fixed or moving TV cameras and monitors looking at the scene from different directions, and possibly other sensors; and (2) a high-quality master–slave system that allows the operator to easily enter control commands and to efficiently translate them into the slave manipulator (which is the actual robot) motion.

Consequently, designers of teleoperation systems concentrate on issues immediately related to these two components (see, e.g., Refs. 116–119). The implicit assumption in such focus on technology is that one component that can be fully trusted is the human operator: As long as the right hardware is there, the operator is believed to deliver the expected results. It is only when one closely observes the operation of some such highly sophisticated and accurate systems that one perceives their low overall efficiency and the awkwardness of interactions between the operator and the system. One is left with the feeling that while the two components above are necessary, they are far from being sufficient.

Even in simple teleoperation tasks that would be trivial for a human child, like building a tower out of a few toy blocks or executing a collision-free motion between a few obstacles, the result is far from perfect: Unwanted collisions do occur, and the robot arm's motion is far from confident. The operator will likely move the arm maddeningly slowly and tentatively, stopping often to assess the situation. One becomes convinced that these difficulties are not merely a result of a (potentially improvable) inferior mechanical structure or control system, but are instead related to cognitive difficulties on the part of the operator. This is an exciting topic for a cognitive scientist, with important practical consequences.

To summarize, here are a few reasons for attempting a comparison between human and robot performance in motion planning:

- *Algorithm Quality.* Sensor-based motion planning algorithms developed in the preceding chapters leave a question unanswered: How good are they? If they produced optimal solutions, they would be easy to praise. But in a situation with limited input information the solutions are usually far from optimal, and assessing them is difficult. One way to assess those solutions is in comparison with human performance. After all, humans are used to solving motion planning problems under uncertainty and therefore must be a good benchmark.

- *Improving Algorithms.* If robot performance turns out to be inferior to human performance, this fact will provide a good incentive to try to understand which additional algorithmic resources could be brought to bear to improve robot motion planning strategies.

- *Synergistic Teleoperation Systems.* If, on the other hand, human performance can be inferior to robot performance—which we will observe to be so in some motion planning tasks—this will present a serious challenge for designers of practical teleoperation systems. It would then make sense to shift to robots some motion planning tasks that have been hitherto the sole responsibility of humans. We will observe that humans have difficulty guiding arm manipulators in a crowded space, resulting in mistakes or, more often, in a drastic reduction of the robot's speed to accommodate human "thinking." Complementing human intelligence with appropriate robot intelligence may become a way of dramatically improving the performance of teleoperated systems.

- *Cognitive Science.* Human performance in motion planning is of much interest to cognitive scientists who study human motor skills and the interface between human sensory apparatus and motion. The performance comparison with robot algorithms in tasks that require motion planning might shed light on the nature of human cognitive processes related to motion in space.

To be meaningful, a comparison between human and robot performance must take place under exactly the same conditions. This is very important: It makes no sense, for example, to compare the performance of a human who moves around blindfolded with the performance of a robot that has a full use of its vision sensor. Other conditions may be more subtle: For instance, how do we make sure that in the same scene the robot and the human have access to exactly the same information? While one can never be absolutely sure that the conditions under which human and robot performance are compared are indeed equal, every effort has been made to ascertain this in our study.

To formulate the right questions, we will start in the next section with observations from a few experiments, and then move in the following sections to a consistent study with more representative tests and statistics. Most of those limited experiments have been done by the author in the late 1980s while at Yale University.[1]

The surprising, sometimes seemingly bizarre results from these experiments helped prompt discussion and sharpen our questions, but also indicated a need for a more consistent study. The larger, better designed, and much more consistent studies discussed in Sections 7.4 and 7.5 were undertaken in the mid-1990s at the University of Wisconsin—Madison, within a joint project between two groups: on the robotics side, by the author and graduate student Fei Liu, and on the cognitive science side, by Dr. Sheena Rogers and graduate student Jeffrey Watson, both from the University of Wisconsin Psychology Department's Center for Human Performance in Complex Systems.

7.2 PRELIMINARY OBSERVATIONS

We will start with a task that is relatively intuitive for a human—walking in a labyrinth (a maze)—and will then proceed to the less intuitive task of moving a simple planar two-link arm manipulator, of the kind that we considered in Section 5.2 (see Figures 5.2 and 5.15). It is important to realize that in some formal sense, both tasks are of the same difficulty: Moving in a maze amounts to controlling a combination of two variables, x and y (horizontal and vertical displacement), and moving a two-link arm also requires control of two variables, representing angular displacement (call these angles θ_1 and θ_2).

7.2.1 Moving in a Maze

Many of us have tried to walk in a labyrinth (a maze). Some medieval monasteries and churches have had labyrinths on the premises, or even indoors, to entertain its visitors. Today labyrinths appear in public and amusement parks. The labyrinth corridors are often made of live bushes cut neatly to make straight-line and curved walls. The wall may be low, to allow one to see the surrounding walls; in a more challenging labyrinth the walls are tall, so that at any given moment

---

[1]Much of the software for this first stage and many tests were produced by my graduate students, especially by Timothy Skewis. The human subjects used were whoever passed through the Yale Robotics Laboratory—graduate students, secretaries, unsuspecting scientists coming to Yale for a seminar, and even faculty's children.

the walker can only see the surrounding walls. The visitor may start inside or outside of the labyrinth and attempt to reach the center, or locate a "treasure" inside the labyrinth, or find an exit from it.

Moving with Complete Information. If one has a bird's-eye view of the whole labyrinth, this makes the task much easier. To study human performance in motion planning consistently, we start with this simpler task. Consider the bird's-eye view of a labyrinth shown in Figure 7.1. Imagine you are handed this picture and asked to produce in it a collision-free path from the position S (Start) to the target point T. One way to accomplish this test is with the help of computer. You sit in front of the computer screen, which shows the labyrinth, Figure 7.1. Starting at point S, you move the cursor on the screen using the computer mouse, trying to get to T while not banging into the labyrinth walls. At all times you see the labyrinth, points S and T, and your own position in the labyrinth as shown by the cursor. For future analysis, your whole path is stored in the computer's memory.

If you are a typical labyrinth explorer, you will likely study the labyrinth for 10–15 seconds, and think of a more or less complete path even before you start walking. Then you quickly execute the path on the screen. Your path will

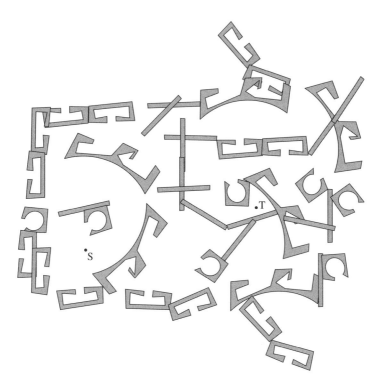

Figure 7.1   A two-dimensional labyrinth. The goal is to proceed from point S to point T.

likely be something akin to the three examples shown in Figure 7.2. All examples demonstrate exemplary performance; in fact, these paths are close to the optimal—that is, the shortest—path from S to T.[2] In the terminology used in this text, what you have done here and what the paths in Figure 7.2 demonstrate is *motion planning with complete information.*

If one tried to program a computer to do the same job, one would first preprocess the labyrinth to describe it formally—perhaps segment the labyrinth walls into small pieces, approximate those pieces by straight lines and polynomials for a more efficient description, and so on, and eventually feed this description into a special database. Then this information could be processed, for example, with one or another motion planning algorithm that deals with complete information about the task.

The level of detail and the respectable amount of information that this database would encompass suggests that this method differs significantly from the one you just used. It is safe to propose that you have paid no attention to small details when attempting your solution, and did not try to take into account the exact shapes and dimensions of every nook and cranny. More likely you concentrated on some general properties of the labyrinth, such as openings in walls and where those openings led and whether an opening led to a dead end. In other words, you limited your attention to the wall connectivity and ignored exact geometry, thus dramatically simplifying the problem. Someone observing you—call this person the tester—would likely conclude that you possess some powerful motion planning algorithm that you applied quickly and with no hesitation. Since it is very likely that you never had a crash course on labyrinth traversal, the source and nature of your powerful algorithm would present an interesting puzzle for the tester.

Today we have no motion planning algorithms that, given complete information about the scene, will know from the start which information can be safely ignored and that will solve the task with the effectiveness you have demonstrated a minute ago. The existing planning algorithms with complete information will grind through the whole database and come up with the solution, which is likely to be almost identical to the one you have produced using much less information about the scene. The common dogma that humans are smarter than computers is self-evident in this example.

Moving with Incomplete Information. What about a more realistic labyrinth walk, where at any given moment the walker can see only the surrounding walls? To test this case, let us use the same labyrinth that we used above (Figure 7.1), except that we modify the user interface to reflect the new situation. As before, you are sitting in front of the computer screen. You see on it only points S and T and your own position in the labyrinth (the cursor). The whole labyrinth is there, but it is invisible. As before, you start at S, moving the cursor with the

---

[2]Of course, in a more complex labyrinth a quick look may not be sufficient to see the solution; then one's performance may deteriorate. For the point that we are to make in this section, this fact is not essential.

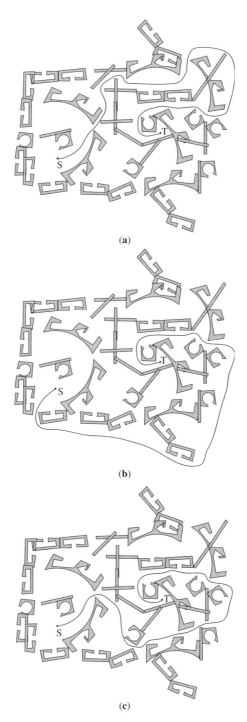

Figure 7.2 Three paths produced by human subjects in the labyrinth of Figure 7.1, if given the complete information, the bird's-eye view of the scene.

computer mouse. Every time the cursor approaches a labyrinth wall within some small distance—that is your "radius of vision"—the part of the wall within this radius becomes visible, and so you can decide where to turn to continue the motion. Once you step back from the wall, that piece of the wall disappears from the screen.

Your performance in this new setting will of course deteriorate compared to the case with complete information above. You will likely wander around, hitting dead ends and passing some segments of the path more than once. Because you cannot now see the whole labyrinth, there will be no hope of producing a near-optimal solution; you will struggle just to get somehow to point T. This is demonstrated in two examples of tests with human subjects shown in Figure 7.3. Among the many such samples with human subjects that were obtained in the course of this study (see the following sections), these two are closest to the best and worst performance, respectively. Most subjects fell somewhere in between.

While this performance is far from what we saw in the test with complete information, it is nothing to be ashamed of—the test is far from trivial. Those who had a chance to participate in youth wilderness training know how hard one has to work to find a specific spot in the forest, with or without a map. And many of us know the frustration of looking for a specific room in a large unfamiliar building, in spite of its well-structured design.

Human Versus Computer Performance in a Labyrinth. How about comparing the human performance we just observed with the performance of a decent motion planning algorithm? The computer clearly wins. For example, the Bug2 algorithm developed in Section 3.3.2, operating under the same conditions as for the human subjects, in the version with incomplete information produces elegant solutions shown in Figure 7.4: In case (a) the "robot" uses tactile information, and in case (b) it uses vision, with a limited radius of vision r_v, as shown.

Notice the remarkable performance of the algorithm in Figure 7.4b: The path produced by algorithm Bug2, using very limited input information—in fact, a fraction of complete information—almost matches the nearly optimal solution in Figure 7.2a that was obtained with complete information.

We can only speculate about the nature of the inferior performance of humans in motion planning with incomplete information. The examples above suggest that humans tend to be inconsistent (one might say, lacking discipline): Some new idea catches the eye of the subject, and he or she proceeds to try it, without thinking much about what this change will mean for the overall outcome.

The good news is that it is quite easy to teach human subjects how to use a good algorithm, and hence acquire consistency and discipline. With a little practice with the Bug2 algorithm, for example, the subjects started producing paths very similar to those shown in Figure 7.4.

This last point—that humans can easily master motion planning algorithms for moving in a labyrinth—is particularly important. As we will see in the next section, the situation changes dramatically when human subjects attempt motion planning for arm manipulators. We will want to return to this comparison when

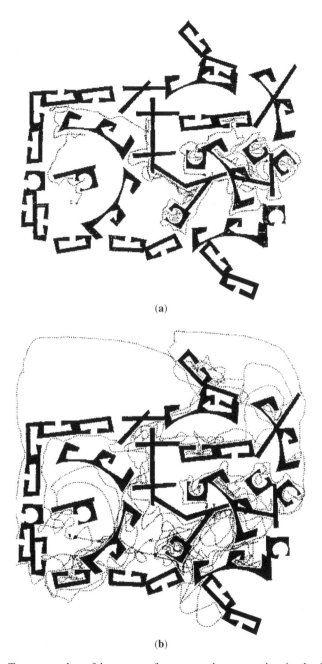

(a)

(b)

Figure 7.3   Two examples of human performance when operating in the labyrinth of Figure 7.1 with incomplete information about the scene. Sample **(a)** is closer to the best performance, while sample **(b)** is closer to the worst performance observed in this study.

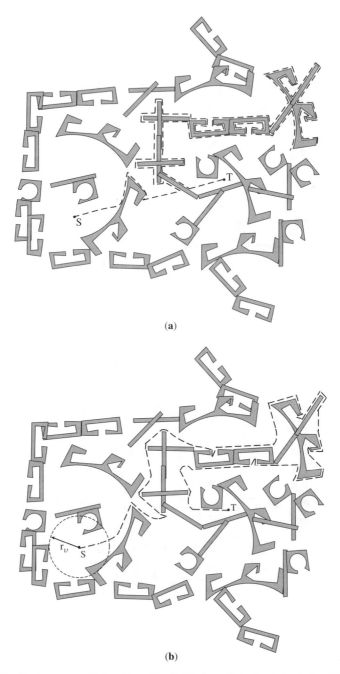

(a)

(b)

Figure 7.4 Performance of algorithm Bug2 (Chapter 3) in the labyrinth of Figure 7.1. (a) With tactile sensing and (b) with vision that is limited to radius r_v.

discussing the corresponding tests, so let us repeat the conclusion from the above discussion:

When operating in a labyrinth, humans have no difficulty learning and using motion planning algorithms with incomplete information.

7.2.2 Moving an Arm Manipulator

Operating with Complete Information. We are now approaching the main point of this discussion. There was nothing surprising about the human performance in a labyrinth; by and large, the examples of maze exploration above agree with our intuition. We expected that humans would be good at moving in a labyrinth when seeing all of it (moving with complete information), not so good when moving in a labyrinth "in the dark" (moving with incomplete information), and quite good at mastering a motion planning algorithm, and this is what happened. We can use these examples as a kind of a benchmark for assessing human performance in motion planning.

We now turn to testing human performance in moving a simple two-link revolute–revolute arm, shown in Figure 7.5. As before, the subject is sitting in

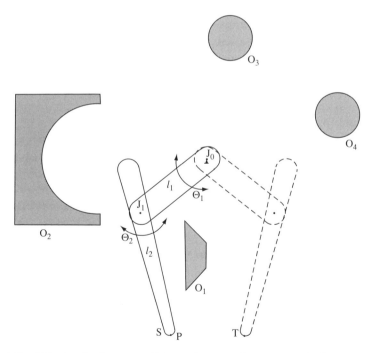

Figure 7.5   This simple planar two-link revolute–revolute arm manipulator was used to test human performance in motion planning for a kinematic structure: l_1 and l_2 are two links; J_0 and J_1 are two revolute joints; θ_1 and θ_2 are joint angles; S and T are start and target positions in the test; P is the arm endpoint in its current position; O_1, O_2, O_3, and O_4 are obstacles.

front of the computer screen, and controls the arm motion using the computer mouse. The first link, l_1, of the arm rotates about its joint J_0 located at the fixed base of the arm. The joint of the second link, J_1, is attached to the first link, and the link rotates about point J_1, which moves together with link l_1. Overall, the arm looks like a human arm, except that the second link, l_2, has a piece that extends outside the "elbow" J_1. (This kinematics is quite common in industrial and other manipulators.) And, of course, the arm moves only in the plane of the screen.

How does one control the arm motion in this setup? By positioning the cursor on link l_1 and holding down the mouse button, the subject will make the link rotate about joint J_0 and follow the cursor. At this time link l_2 will be "frozen" relative to link l_1 and hence move with it. Similarly, positioning the cursor on link l_2 and holding down the mouse button will make the second link rotate about joint J_1, with link l_1 being "frozen" (and hence not moving at all). Each such motion causes the appropriate link endpoint to rotate on a circular arc.

Or—this is another way to control the arm motion—one can position the cursor at the endpoint P of link l_2 and drag it to whatever position in the arm workspace one desires, instantaneously or in a smooth motion. The arm endpoint will follow the cursor motion, with both links moving accordingly. During this motion the corresponding positions of both links are computed automatically in real time, using the inverse kinematics equations. (Subjects are not told about this mechanism, they just see that the arm moves as they expect.) This second option allows one to control both links motion simultaneously. It is as if someone moves your hand on the table—your arm will follow the motion.

We will assume that, unlike in the human arm, there are no limits to the motion of each joint in Figure 7.5. That is, each link can in principle rotate clockwise or counterclockwise indefinitely. Of course, after every 2π each link returns to its initial position, so one may or may not want to use this capability. [Looking ahead, sometimes this property comes in handy. When struggling with moving around an obstacle, a subject may produce more than one rotation of a link. Whether or not the same motion could be done without the more-than-2π link rotation, not having to deal with a constraint on joint angle limits makes the test psychologically easier for the subject.]

The difficulty of the test is, of course, that the arm workspace contains obstacles. When attempting to move the arm to a specified target position, the subjects will need to maneuver around those obstacles. In Figure 7.5 there are four obstacles. One can safely guess, for example, that obstacle O_1 may interfere with the motion of link l_1 and that the other three obstacles may interfere with the motion of link l_2.

Similar to the test with a labyrinth, in the arm manipulator test with complete information the subject is given the equivalent of the bird's-eye view: One has a complete view of the arm and the obstacles, as shown in Figure 7.5. Imagine you are that subject. You are asked to move the arm, collision-free, from its starting position S to the target position T. The arm may touch an obstacle, but the system will not let you move the arm "through" an obstacle. Take your time—time is not a consideration in this test.

Three examples of performance by human subjects in controlled experiments are shown in Figure 7.6.[3] Shown are the arm's starting and target positions S and T, along with the trajectory (dotted line) of the arm endpoint on its way from S to T. The examples represent what one might call an "average" performance by human subjects.[4]

The reader will likely be surprised by these samples. Why is human performance so unimpressive? After all, the subjects had complete information about the scene, and the problem was formally of the same (rather low) complexity as in the labyrinth test. The difference between the two sets of tests is indeed dramatic: Under similar conditions the human subjects produced almost optimal paths in the labyrinth (Figure 7.2) but produced rather mediocre results in the test with the arm (Figure 7.6).

Why, in spite of seeing the whole scene with the arm and obstacles (Figure 7.5), the subjects exhibited such low skills and such little understanding of the task. Is there perhaps something wrong with the test protocol, or with control means of the human interface—or is it indeed real human skills that are represented here? Would the subjects improve with practice? Given enough time, would they perhaps be able to work out a consistent strategy? Can they learn an existing algorithm if offered this opportunity? Finally, subjects themselves might comment that whereas the arm's work space seemed relatively uncluttered with obstacles, in the test they had a sense that the space was very crowded and "left no room for maneuvering."

The situation becomes clearer in the arm's configuration space (C-space, Figure 7.7). As explained in Section 5.2.1, the C-space of this revolute–revolute arm is a common torus (see Figure 5.5). Figure 7.7 is obtained by flattening the torus by cutting it at point T along the axes θ_1 and θ_2. This produces four points T in the resulting square, all identified, and two pairs of identified C-space boundaries, each pair corresponding to the opposite sides of the C-space square. For reference, four "shortest" paths (M-lines) between points S and T are shown (they also appear in Figure 5.5; see the discussion on this in Section 5.2.1). The dark areas in Figure 7.7 are C-space obstacles that correspond to the four obstacles in Figure 7.5.

Note that the C-space is quite crowded, much more than one would think when looking at Figure 7.5. By mentally following in Figure 7.7 obstacle outlines across the C-space square boundaries, one will note that all four workspace obstacles actually form a single obstacle in C-space. This simply means that when touching one obstacle in work space, the arm may also touch some other

---

[3]The experimental setup used in Figure 7.6c slightly differs from the other two; this played no visible role in the test outcomes.

[4]The term "average" here has no formal meaning: It signifies only that some subjects did better and some did worse. A more formal analysis of human performance in this task will be given in the next section. A few subjects did not finish the test and gave up, citing tiredness or hopelessness ("There is no solution here", "You cannot move from S to T here"...).

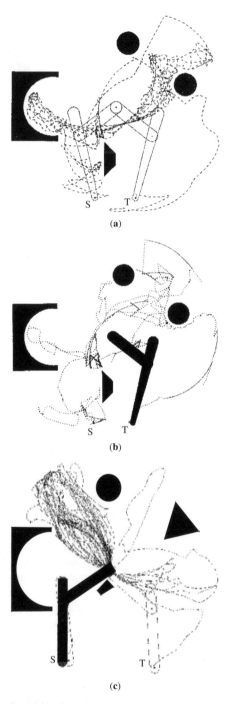

Figure 7.6 Paths produced by three human subjects with the arm shown in Figure 7.5, given complete information about the scene.

Figure 7.7 C-space of the arm and obstacles shown in Figure 7.5.

obstacle, and this is true sequentially, for pairs (O_1, O_2), (O_2, O_3), (O_3, O_4), (O_4, O_1). No wonder the subjects found the task difficult. In real-world tasks, such interaction happens all the time; and the difficulties only increase with more complex multilink arms and in three-dimensional space.

Operating the Arm with Incomplete Information. Similar to the test with incomplete information in the labyrinth, here a subject would at all times see points S and T, along with the arm in its current positions. Obstacles would be hidden. Thus the subject moves the arm "in the dark": When during its motion the arm comes in contact with an obstacle—or, in the second version of the test, some parts of the obstacle come within a given "radius of vision" r_v from some arm's points—those obstacle parts become temporarily visible. Once the contact is lost—or, in the second version, once the arm-to-obstacle distance increases beyond r_v—the obstacle is again invisible.

The puzzling observation in such tests is that, unlike in the tests with the labyrinth, the subjects' performance in moving the arm "in the dark" is on average indistinguishable from the test with complete information. In fact, some subjects performed better when operating with complete information, while others performed better when operating "in the dark." One subject did quite well "in the dark," then was not even able to finish the task when operating with a completely visible scene, and refused to accept that in both cases he had dealt with the same scene: "This one [with complete information] is much harder; I think it has no solution." It seems that extra information doesn't help. What's going on?

Human Versus Computer Performance with the Arm. As we did above with the labyrinth, we can attempt a comparison between the human and computer performance when moving the arm manipulator, under the same conditions. Since in previous examples human performance was similar in tests with complete and incomplete information, it is not important which to consider: For example, the performance shown in Figure 7.6 is representative enough for our informal comparison. On the algorithm side, however, the input information factor makes a tremendous difference—as it should. The comparison becomes interesting when the computer algorithm operates with incomplete ("sensing") information.

Shown in Figure 7.8 is the path generated in the same work space of Figure 7.5 by the motion planning algorithm developed in Section 5.2.2. The algorithm operates under the model with incomplete information. To repeat, its sole input information comes from the arm sensing; known at all times are only the arm

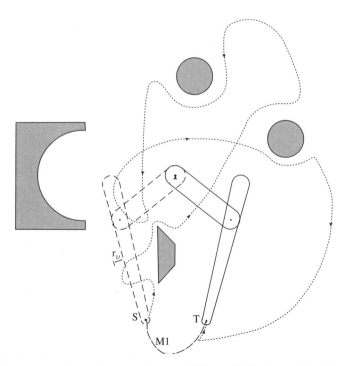

Figure 7.8 Path produced in the work space of Figure 7.5 by the motion planning algorithm from Section 5.2.2; $M1$ is the shortest (in C-space) path that would be produced if there were no obstacles in the workspace.

positions S and T and its current position. The arm's sensing is assumed to allow the arm to sense surrounding objects at every point of its body, within some modest distance r_v from that point. In Figure 7.8, radius r_v is equal to about half of the link l_1 thickness; such sensing is readily achievable today in practice (see Chapter 8).

Similar to Figure 7.6, the resulting path in Figure 7.8 (dotted line) is the path traversed by the arm endpoint when moving from position S to position T. Recall that the algorithm takes as its base path (called M-line) one of the four possible "shortest" straight lines in the arm's C-space (see lines M_1, M_2, M_3, M_4 in Figure 5.5); distances and path lengths are measured in C-space in radians. In the example in Figure 7.8, the shortest of these four is chosen (it is shown as line $M1$, a dashed line). In other words, if no obstacles were present, under the algorithm the arm endpoint would have moved along the curve $M1$; given the obstacles, it went along the dotted line path.

The elegant algorithm-generated path in Figure 7.8 is not only shorter than those generated by human subjects (Figure 7.6). Notice the dramatic difference between the corresponding (human versus computer) arm test and the labyrinth test. While a path produced in the labyrinth by the computer algorithm (Figure 7.4) presents no conceptual difficulty for an average human subject, they find the path in Figure 7.8 incomprehensible. What is the logic behind those sweeping curves? Is this a good way to move the arm from S to T? The best way? Consequently, while human subjects can easily master the algorithm in the labyrinth case, they find it hard—in fact, seemingly impossible—to make use of the algorithm for the arm manipulator.

7.2.3 Conclusions and Plan for Experiment Design

We will now summarize the observations made in the previous section, and will pose a few questions that will help us design a more comprehensive study of human cognitive skills in space reasoning and motion planning:

1. The labyrinth test is a good easy-case benchmark for testing one's general space reasoning abilities, and it should be included in the battery of tests. There are a few reasons for this: (a) If a person finds it difficult to move in the labyrinth—which happens rarely—he or she will be unlikely to handle the arm manipulator test. (b) The labyrinth test prepares a subject for the test with an easier task, making the switch to the arm test more gradual. (c) A subject's successful operation in the labyrinth test suggests that whatever difficulty the subject may have with the arm test, it likely relates to the subject's cognitive difficulties rather than to the test design or test protocol.

2. When moving the arm, subjects exhibit different tastes for control means: Some subjects, for example, prefer to change both joint angles simultaneously, "pulling" the arm endpoint in the direction they desire, whereas other subjects prefer to move one joint at the time, thus producing circular

arcs in the path; see Figure 7.6. Because neither technique seems inherently better or easier than the other, for subjects' convenience both types of control should be available to them during the test.

3. Since working with a bird's-eye view (complete information) as opposed to "in the dark" (incomplete information) makes a difference—clearly so in the labyrinth test and seemingly less so in the arm manipulator test—this dichotomy should be consistently checked out in the comprehensive study.

4. In the arm manipulator test it has been observed that the direction of arm motion may have a consistent effect on the subjects' performance. Obviously, in the labyrinth test this effect appears only when operating with incomplete information ("moving in the dark"). This effect is, however, quite pronounced in either test with the arm manipulator, with complete or with incomplete information. Namely, in the setting of Figure 7.5, the generated path and the time to finish were noted to be consistently longer when moving from position T to S than when moving from S to T. This suggests that it is worthwhile to include the direction of motion as a factor in the overall test battery. (And, the test protocol should be set up so that the order of subtests has no effect on the test results.) One possible reason for this peculiar phenomenon is a psychological effect of one's paying more attention to route alternatives that are closer to the direction of the intended route than to those in other directions. Consider the simple "labyrinth" shown in Figure 7.9: The task is to reach one point (S or T) from the other while moving "in the dark." When walking from S to T, most subjects will be less inclined to explore the dead-end corridor A because it leads in a direction almost opposite to the direction toward T, and they will on average produce shorter paths. On the other hand, when walking from T to S, more subjects will perceive corridor A as a promising direction and will, on average, produce longer paths. Such considerations are harder to pinpoint for the arm test, but they do seem to play a role.[5]

5. The less-than-ideal performance of the subjects in the arm manipulator tests makes one wonder if something else is at work here. Can it be that the human–computer interface offered to the subjects is somewhat "unnatural" for them—and this fact, rather than their cognitive abilities, is to blame for their poor performance? Some subjects did indeed blame the computer interface for their poor performance.[6] Some subjects believed that their performance would improve dramatically if they had a chance to operate a physical arm rather than a virtual arm on the computer screen ("if I had a real thing to grab and move in physical space, I would do much better...."). This is a serious argument; it suggests that adding a physical test to the overall test battery might provide interesting results.

---

[5]Of course, no such effect can be expected for the computer algorithm.

[6]An "unscientific" observation made here was that older subjects, such as visiting professors who graciously agreed to participate in the experiment, were more critical of the human–computer interface than younger subjects. The latter were more willing than the former to accept the test results as measuring their real spatial reasoning abilities.

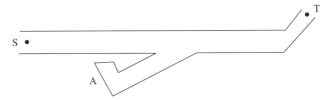

Figure 7.9 In this simplistic maze, the subjects seem less inclined to explore the dead-end corridor A when walking from S to T than when walking from T to S.

6. Based on standard practice for cognitive tests, along with some subjects' comments, it is worthwhile to explore human motion planning skills along some demographic lines. For example,

   - Performance as a function of gender (consider the proverbial proficiency of men in handling maps).
   - Performance as a function of age: For example, are children better than adults in spatial reasoning tasks (as they seem to be in some computer games or with the Rubik's Cube)?
   - Performance as a function of educational level and professional orientation: For example, wouldn't we expect students majoring in mechanical engineering to do better in our tests than students majoring in comparative literature?

7. Finally, there is an important question of training and practice. We all know that with proper training, people achieve miracles in motion planning; just think of an acrobat on a high trapeze. In the examples above, subjects were given a chance to get used to the task before a formal test was carried out, but no attempt was made to consistently study the effect of practice on human performance. The effect of training is especially serious in the case of arm operation, in view of the growing area of teleoperation tasks (consider the arm operator on the Space Shuttle, or a partially disabled person commanding an arm manipulator to take food from the refrigerator). This suggests that the training factor must be a part of the larger study.

This list covers a good number of issues and consequently calls for a rather ambitious study. In the specific study described below, not all questions on the list have been addressed thoroughly enough, due to the difficulty of arranging a statistically representative group of subjects. Some questions were addressed only cursorily. For example, attempts to enlist in the experiment a local kindergarten or a primary school had a limited success, and so was an attempt to round up enough subjects over the age of 60.

The very limited number of tests carried out for these insufficiently studied issues provide these observations: (a) Children do not seem to do better than adults in our tests. (b) Subjects aged 60 and over seem to have significantly more difficulty carrying out the tests: in the arm test, in particular, they would give

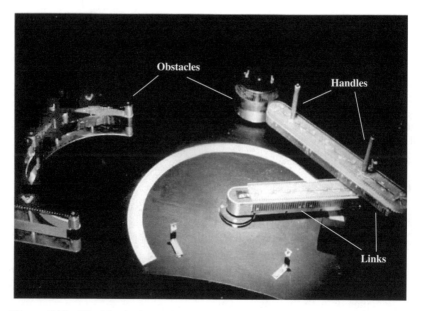

Figure 7.10 The physical two-link arm used in the tests of human performance.

up more often than younger subjects before reaching the solution. (c) The level of one's education and professional orientation seems to play an insignificant role: Secretaries do as well or as poorly as mechanical engineering PhDs or professional pilots, who pride themselves in their spatial reasoning.

The Physical Arm Test Setup. This experimental system has been set up in a special booth, with about 5 ft by 5 ft floor area, enough to accommodate a table with the two-link arm and obstacles, and a standing subject. The inside of the booth is painted black, to help with the "move in the dark" test. For a valid comparison of subjects' performance with the virtual environment test, the physical arm and obstacles (Figure 7.10) are proportionally similar to those in Figure 7.5. (Only two obstacles can be clearly seen in Figure 7.10; obstacle O_1 of Figure 7.5 was replaced for technical reasons by two stops; see Figure 7.10.)

For the subjects' convenience the arm is positioned on a slightly slanted table. Each arm link is about 2 ft long. A subject moves one or both arm links using the handles shown. During the test the arm positions are sampled by potentiometers mounted on the joint axes, and they are documented in the host computer for further analysis, together with the corresponding timing information.[7]

Special features have been added for testing the scene visibility factor. Opening the booth doors and turning on its light produces the visible scene; closing the

---

[7]The physical arm and the booth system, including hardware, electronics, and related software, have been designed by Branimir Stankovic and Steve Seaney at the University of Wisconsin Robotics Laboratory [120].

door shut and turning off the light makes it an invisible environment. For the latter test, the side surfaces of the arm links and of the obstacles are equipped with densely spaced contacts and LED elements located along perimeters of both links (Figure 7.10). There are 117 such LEDs on the inner link (link 1) and 173 LEDs on the outer link (link 2). When a link touches an obstacle, one or more LEDs light up, informing the subject of a collision and giving its exact location. Visually, the effect is similar to how a contact is shown in the virtual arm test.[8]

7.3 EXPERIMENT DESIGN

7.3.1 The Setup

Two batteries of tests, called *Experiment One* and *Experiment Two*, have been carried out to address the issues listed in the previous section. Experiment One addresses the effect of three *factors* on human performance: *interface factor*, which focuses on the effect of a virtual versus physical interface; *visibility factor*, which relates to the subject's seeing the whole scene versus the subject's "moving in the dark"; and *direction factor*, which deals with the effect of the direction of motion in the same scene. Each factor is therefore a dichotomy with two *levels*. We are especially interested in the effects of interface and visibility, since these affect most directly one's performance in motion planning tasks. The direction of motion is a secondary factor, added to help clarify the effect of the other two factors.

Experiment Two is devoted specifically to the effect of training on one's performance. The effect is studied in the context of the factors described above. One additional factor here, serving an auxiliary role, is the *object-to-move* factor, which distinguishes between moving a point robot in a labyrinth versus moving a two-link arm manipulator among obstacles. The arm test is the primary focus of this study; the labyrinth test is used only as a benchmark, to introduce the human subjects to the tests' objectives.

The complete list of factors, each with two levels (settings), is therefore as follows[9]:

A. Object-to-move factor, with two levels:
1. Moving a point robot in a labyrinth, as in Figure 7.1.
2. Moving a two-link revolute-revolute arm manipulator in a planar work-space with obstacles, as in Figure 7.5.

B. Interface factor, with two levels:
1. In this test, called the *virtual test*, the subject operates on the computer screen, moving the arm links with the computer mouse; all necessary help

---

[8]In addition to this arm, a wooden mockup of the arm, of the same dimensions as the test arm, was built and installed outside the booth, to help subjects practice their motor skills in the task.

[9]More details on the experiment design and test conditions can be found in Ref. 121.

is done by the underlying software. Both the labyrinth test (Figure 7.1) and the arm test (Figure 7.5) are done in this version.

2. In this test, called the *physical test*, the subject works in the booth, moving the physical arm (Figure 7.10). Only the arm tests, and no labyrinth tests, were done in this version.

C. Visibility factor, with two levels:

1. Visible environment: The object (one of those in factor A) and its environment are fully visible.

2. Invisible environment: Obstacles cannot be seen by the subject, except when the robot (in case of the point robot) or a part of its body (in case of the arm) is close enough to an obstacle, in which case a small part of the obstacle near the contact point becomes visible for the duration of contact. The arm is visible at all times.

D. Direction factor (for the arm manipulator test only), with two levels:

1. "Left-to-right" motion (denoted below *LtoR*), as in Figure 7.8.

2. "Right-to-left" motion (denoted *RtoL*); in Figure 7.8 this would correspond to moving the arm from position T to position S.

E. Training factor. The goal here is to study the effect of prior learning and practice on human performance. This factor is studied in combination with all prior factors and has two levels:

1. Subjects' performance with no prior training. Here the subjects are only explained the rules and controls, and are given the opportunity to try and get comfortable with the setup, before the actual test starts.

2. Subjects' performance is measured after a substantial prior training and practice.

Therefore, the focus of Experiment One is on factors B, C, and D, and the focus of Experiment Two is on factor E (with the tests based on the same factors B, C, and D). Because each factor is a dichotomy with two levels, all possible combinations of levels for factors B, C, and D produce eight tasks that each subject can be subjected to:

Task 1: Virtual, visible, left-to-right
Task 2: Virtual, visible, right-to-left
Task 3: Virtual, invisible, left-to-right
Task 4: Virtual, invisible, right-to-left
Task 5: Physical, visible, left-to-right
Task 6: Physical, visible, right-to-left
Task 7: Physical, invisible, left-to-right
Task 8: Physical, invisible, right-to-left

In addition to these tasks, a smaller study was carried out to measure the effect of three auxiliary variables:

(a) *gender*, with values "males" and "females",

(b) *specialization*, with values "engineering," "natural sciences," and "social sciences", and

(c) *age*, with values "15–24," "25–34", and "above 34".

A total of 48 subjects have been tested in Experiment One, with the following distribution between genders, specializations, and age groups:

Gender: 23 males and 25 females

Specialization: engineering, 16; natural sciences, 20; social sciences, 12

Age groups: age 15–24, 32 subjects; age 25–34, 12 subjects; age above 34, 4 subjects

From the standpoint of statistical tests, subjects have been selected randomly, and so test observations can be considered mutually independent.

One problem that must be addressed in the test protocol is avoiding the learning effect: We need to prevent the subjects from using in one test the knowledge that they acquired in a prior test. Then one's performance is independent of the order of tasks execution. This is important when randomly varying the order of tasks between subjects, a standard technique in cognitive skills tests. This has been achieved by making each subject go through 4 out of 8 tasks. For example, a subject who went through the virtual–visible–LtoR task would not be subjected to the virtual–visible–RtoL task. With this constraint, half of the subject pool did tasks 1, 4, 5, 8, while the other half did tasks 2, 3, 6, and 7 from the list above.

As a result, each of the eight tasks should have 24 related observation sets, each set including the path length, the completion time, and the actual data on the generated path for one subject. (In reality, two observation sets for the physical test, one for the visible–RtoL and the other for the invisible-LtoR combination, were documented incorrectly and were subsequently discarded, leaving 23 observations for Task 6 and Task 7 each.)

Another way to group the observed data is by the three factors (interface, visibility, direction), each with two levels: (virtual, physical), (visible, invisible), and (LtoR, RtoL). This grouping produces six data sets and is useful for studying separate effects on human performance—for example, the effect on one's performance of moving a visible arm or moving left to right (see the next section). With two lost observations mentioned above, the sizes of the six data sets are as follows:

Set 1: Interface data for "virtual"—96 observations

Set 2: Interface data for "physical"—94 observations

Set 3: Visibility data for "visible"—95 observations,

Set 4: Visibility data for "invisible"—95 observations,

Set 5: Direction data for "left to right"—95 observations,

Set 6: Direction data for "right to left"—95 observations.

A total of 12 subjects have been tested in Experiment Two.

Performance Criteria. Two criteria have been used to measure subjects' performance in the tasks:

1. The length of generated path (called Path)
2. The task completion time (called Time)

In the labyrinth tests the path length is the actual length of the path a subject generates in the labyrinth. In the arm manipulator test the path length is measured as the sum of two modulo link rotation angles in radians. *Time*, in seconds, is the time it takes a subject to complete the task.

Statistical Considerations. In statistical terms, the length of path and the completion time are *dependent variables*, and the test conditions, as represented by factors and levels, are *independent variables*. If, for example, we want to compare the effect of a visible scene versus invisible scene on the length of paths produced by the subjects, then visibility is an independent variable (with two values, visible and invisible), and the length of path is a dependent variable.

As one would expect, the dependent variables *Path* and *Time* are highly correlated: In Experiment One the correlation coefficient between the two is $r(Path, Time) = 0.74$.

A *multivariate observation* for a particular subject is the set of scores of this subject in a given task; it is thus a vector. For example, for Subject 1 the dependent variable vector (*Path, Time*) in Task 1 (virtual-visible-LtoR) happened to be (59; 175).

The concept of statistical *significance* (see e.g., [127]) is a quantitative index of reliability of a given result or statement, usually in terms of a variable in question. Specifically, the significance p-level represents the probability of an error involved in accepting an observed result (or statement) as valid, or as representative of the population. In practice, results corresponding to the significance level $p \leq 0.05$ are usually considered *significant*.

Put differently, p-level indicates the probability of error when rejecting some related *null hypothesis*. A null hypothesis, denoted as H_0, relates to making a statement about the observation data—for example, when deciding whether two sets of data came from the same population of data. If a statistical test suggests that the null hypothesis should be rejected, say with significance level $p \leq 0.01$, we can conclude that the two samples differ significantly, or that the variable of interest has a significant effect on the sample data. If the test results suggest accepting the null hypothesis, we conclude that the two samples do not differ significantly, and hence the variable of interest has no effect on the sample data.

7.3.2 Test Protocol

The salient characteristics of the test protocol can be summarized as follows (more details on the experiment design and test conditions can be found in Ref. 121):

- The primary focus in the study is on tests with the arm manipulator (see Figure 7.5). The labyrinth test is used only as a benchmark, for introducing the subjects to the tests and the study's underlying ideas.

- The bulk of the subject pool for this study was paid undergraduate students and also some graduate students. (There was no statistical difference in performance between the two groups.)

- In the first session, about one hour long, a subject would be taught how to carry out a test and would be given a pilot test to ascertain that he/she can be submitted to the test; the latter would be different from the pilot test.

- The maximum time a subject was given to finish the test with the arm manipulator was 15 minutes. Much less was allowed for the labyrinth test: As a rule, 1–2 minutes were enough. These limits were chosen as a rough estimate of the time the subjects would need to complete a test without feeling time pressure. Most subjects finished their tasks well within the time allocation; those few who didn't were not likely to finish even with significantly more time.

- Measures were taken to eliminate the effect of a subject's memory recall, or information passing, from one task to another. In the invisible version of the arm test, rotating the whole scene on the computer screen for a consequent session would practically eliminate the effect of memorization from prior sessions. This also helps from the test protocol standpoint: Using the same scene in subsequent tests allows for an "apples-with-apples" assessment of subjects' performance.

7.4 RESULTS — EXPERIMENT ONE

The basic (descriptive) statistics for motion planning tests carried out in Experiment One are given in Table 7.1. Statistics are given separately for each dependent variable, the length of path (*Path*) and the time to completion in seconds (*Time*), and within each dependent variable for each of the eight tasks listed in Section 7.3.1. Each line in Table 7.1 refers to a given task and includes the number of tested subjects ("Valid N" statistics) as well as the mean, minimum, maximum, and standard deviation of the correspondent variable.

A quick glance at the table provides a few observations that we will address in more detail later. One surprise is that the subjects' performance with the right-to-left direction of motion was significantly worse than their performance with the left-to-right direction of motion: Depending on the task, the mean length of path for the right-to-left direction is about two to five times longer than that for the left-to-right direction.[10]

---

[10]This alone would make a smart robot conclude that we humans are terribly inconsistent: What prevents one, the robot would think, from going from point B to point A along the same path one takes when going from A to B!

Another surprise is that the statistics undermines the predominant belief among subjects and among robotics and cognitive science experts that humans should be doing significantly better when moving a physical as opposed to a virtual arm. Isn't the physical arm quite similar to our own arm, which we use so efficiently? To be sure, the subjects did better with the physical arm—but only a little better, not by as much as one would expect, and only for the (easier) left-to-right direction of motion. Once the task became a bit harder, the difference disappeared: When moving the physical arm in the right-to-left direction, more often than not the subjects' performance was significantly worse than when moving the virtual arm in the left-to-right direction, and more or less comparable to moving the virtual arm in the same right-to-left direction (see Table 7.1).

In other words, letting a subject move the physical arm does not guarantee more confidence than when moving a virtual arm: Some other factors seem to play a more decisive role in the subjects' performance. In an attempt to extract the (possibly hidden) effects of our experimental factors on one's performance, two types of analysis have been undertaken for the Experiment One data:

- The first one, *Principal Components Analysis (PCA)*, has been carried out as a preliminary study, to understand the general nature of obtained observation data and to see if such factors as subjects' gender, specialization, and

TABLE 7.1. Descriptive Statistics for the Data in Experiment One

Variable/Task	Valid N	Mean	Minimum	Maximum	Std. Dev.
			Descriptive Statistics		
Length of path					
Virtual–visible–LtoR	24	58.77	18.26	147.23	31.74
Virtual–visible–RtoL	24	176.92	29.54	391.41	91.28
Virtual–invisible–LtoR	24	85.82	21.88	340.15	71.65
Virtual–invisible–RtoL	24	156.08	17.59	392.41	96.89
Physical–visible–LtoR	23	27.70	13.92	51.69	11.64
Physical–visible–RtoL	24	142.97	15.78	396.45	109.38
Physical–invisible–LtoR	24	60.57	15.17	306.13	75.78
Physical–invisible–RtoL	23	160.19	14.26	501.59	145.10
Time to completion:					
Virtual–visible–LtoR	24	265.54	82	595	163.19
Virtual–visible–RtoL	24	692.79	186	912	252.00
Virtual–invisible–LtoR	24	376.02	72	920	282.82
Virtual–invisible–RtoL	24	675.75	66	941	329.91
Physical–visible–LtoR	24	46.21	14	102	26.49
Physical–visible–RtoL	24	218.50	15	902	228.44
Physical–invisible–LtoR	24	122.88	19	612	155.37
Physical–invisible–RtoL	24	299.88	22	900	244.63

age group have a noticeable effect on the subjects' performance in motion planning tasks.

- The second, more pointed analysis addresses separate effects of individual factors on subjects' performance—the effect of interface (virtual versus physical), scene visibility, and the direction of motion. This study makes use of tools of *nonparametric analysis* and *univariate analysis of variance.* Only brief summaries of the techniques used are presented below. For more details on the techniques the reader should refer to the sources cited in the text below; for details related to this specific study see Ref. 121.

7.4.1 Principal Components Analysis

We attempt to answer the following questions:

1. To what extent are the factors used—interface (virtual or physical), scene visibility, and direction of motion—indicative of human performance?
2. Can these factors be replaced by some "hidden" factors that describe the same data in a clearer and more compact way?
3. Which factor or which part of the factor's variance is most indicative of one's performance in a motion planning task?
4. Do the patterns of subjects' performance differ as a function of their gender, college specialization, and age group?
5. Can we predict one's performance in one task based on their performance in another task?

Principal Components Analysis (PCA) addresses these questions based on analysis of the covariance matrix of the original set of independent variables [122, 123]. In our case this would be the covariance matrix of a set of two-level tasks. The analysis seeks to identify "hidden" factors—called the *principal components*—which turn out to be eigenvectors of the sample covariance matrix. The matrix's eigenvalues represent variation of the principal components; the sum of eigenvalues is the total variance of the original sample data and is equal to the sum of variances of the original variables.

With the principal components (eigenvectors) conveniently ordered from the largest to the smallest, the first component accounts for most of the total variance in the sample data, the second component accounts for the next biggest part of the total variance, and so on. The first component can thus be called the "most important hidden factor," and so on. This ordering sometimes allows the researcher to (a) drop the last few components if they account for too small a part of the total variance and (b) claim that the data can be adequately described via a smaller set of variables. Often attempts are made to interpret the "hidden factors" in physical terms, arguing that if the hidden factors could be measured directly they would allow a significantly better description of the phenomenon under discussion.

Let $\mathbf{X}$ be the column matrix of the original sample data; its element x_{ij} is the value of the jth sample variable (the jth column vector of $\mathbf{X}$) for the ith observation row vector (the ith subject, ith row of $\mathbf{X}$). Denote the covariance matrix computed from the sample matrix $\mathbf{X}$ by $\mathbf{R}$. Let matrix $\mathbf{A}$ contain as its column vectors the eigenvectors of matrix $\mathbf{R}$: The ith column of $\mathbf{A}$ is the ith principal component of $\mathbf{R}$. Both $\mathbf{R}$ and $\mathbf{A}$ are square matrices of the same rank (normally equal to the number of sample variables). Matrix $\mathbf{A}$ can then be seen as a transformation matrix that relates the original data to the principal components: Each original data point (described by a row vector of $\mathbf{X}$) can now be described in terms of new coordinates (principal components), as a row vector z; hence

$$\mathbf{Z} = \mathbf{X} * \mathbf{A} \tag{7.1}$$

Let matrix $\mathbf{\Lambda}$ be a diagonal matrix of the same rank, with eigenvalues of $\mathbf{R}$ in its diagonal positions, ordered from largest to smallest.

The matrix of principal component *loadings*, denoted by $\mathbf{L}$, is

$$\mathbf{L} = \mathbf{A} * \mathbf{\Lambda}^{1/2} \tag{7.2}$$

An element l_{ij} of the square matrix $\mathbf{L}$ is the correlation coefficient between the ith variable and the jth principal component. It informs us about the "importance," or "contribution," of the ith variable to the jth principal component. Geometrically, the loading is the projection of ith variable onto the jth component.

If only independent variables represented by the interface and visibility factors are considered in our arm test, this will include four out of the eight tasks listed in Section 7.3.1:

virtual–visible
virtual–invisible
physical–visible
physical–invisible

The loading matrix of the corresponding four principal components in the arm test is shown in Table 7.2. As seen in the table, the first PC (principal component) accounts for 36.3% of the total variance (top four numbers in column 1) and is tied primarily to the virtual–visible and physical–invisible tasks (0.679 and 0.695 loads, accordingly). The second PC accounts for the next 25.6% of the total variance and is tied primarily to the other two tasks, virtual–invisible and physical–visible (loads 0.666 and 0.739), and so on. Since the contribution of successive PCs into the total variance falls off rather smoothly, with 85.9% of the variance being accounted for by the first three PCs, we conclude (somewhat vaguely) that to a large extent the four tasks measure something different, each one bringing new information about the subjects' performance, and hence cannot be replaced by a smaller number of "hidden factors."

TABLE 7.2. Loadings for the Principal Components in the Arm Manipulator Test

Principal component number	1	2	3	4
Task				
Virtual–visible	0.679	0.082	−0.238	−0.690
Virtual–invisible	0.164	0.666	0.728	−0.011
Physical–visible	−0.170	0.739	−0.637	0.139
Physical–invisible	0.695	−0.056	−0.095	0.710
Eigenvalues	1.453	1.023	0.958	0.566
Cumulative percent of total variation	36.3%	61.9%	85.9%	100%

Using Eq. (7.1), the scores on all PCs can now be calculated for all subjects and plotted accordingly. The scores have been obtained and plotted in this study in various forms—for example, in three-dimensional space of the first three PCs and in two-dimensional plots for different pairs of PCs (e.g., a plot in plane PC1 versus PC2, etc.). By labeling the subjects (which become points in such plots) with additional information categories, such as their specialization majors, gender, and age, score plots regarding those categories have been also obtained.

These plots (see Ref. [121]) happen to provide no interesting conclusions about the importance of principal components or of their correlations with the subjects' specialization, gender, or age. Namely, we conclude that contrary to the common wisdom, engineering and computer science students, whose specialities can be expected to give them an edge in handling spatial reasoning tasks, have done no better than students with majors in the arts and social sciences. Also, men did no better than women.

This does not give us a right, however, to make sweeping conclusions of one sort or another. The Principal Component Analysis (PCA) is designed to study the input variables as a pack, and in particular to uncover the biggest sources of variation between independent variables of the original test data. Our "variables" in this study are, however, tasks, not individual variables. Each task is a combination of variables: For example, Task 1—that is, virtual–visible–LtoR—is a combination of three variables: interface, visibility, and direction of motion. Within the PCA framework it is hard to associate the test results with individual variables.

We may do better if we switch to other statistical techniques, those that lend themselves to studying specific effects in sample distributions. They can also yield conclusions about the effect of individual factors on dependent variables. For example, statistical tests may be a better tool for determining to what extent the visibility factor affects a specific side of human performance, say the length of generated paths in motion planning tasks. We will consider such techniques next.

7.4.2 Nonparametric Statistics

Brief Review. Both parametric and nonparametric statistical techniques require that observations are drawn from the sampled population randomly and independently. Besides, parametric techniques rely on an assumption that the underlying sample data are distributed according to the normal distribution. Nonparametric techniques do not impose this constraint. Because the distribution of sample data in our experiments looks far from normal, nonparametric methods appear to be a more appropriate tool.

The Mann–Whitney U-test [124] is one of the more powerful nonparametric tests. It works by comparing two subgroups of sample data. For a given variable under study, the test assesses the hypothesis that two independently drawn sets of data come from two populations that differ in some respect—that is, differ not only with respect to their means but also with respect to the general shape of the distribution. Here the *null hypothesis*, H_0, is that both samples come from the same population. If the test suggests that the hypothesis should be rejected, say with the significance level $p \le 0.01$, we will conclude that the samples differ significantly, and hence the variable of interest has a significant effect on the sample data. If the test results suggest accepting the null hypothesis, we will conclude that the two sample sets do not differ significantly, and hence the variable of interest has no effect on the sample data.

Order statistics is an ordering of the set X_i into a set $X_{(i)}$ such that

$$X_{(1)} \le \cdots \le X_{(m)}$$

Rank, referred to as R_i^*, is the new indexing of the set $X_{(i)}$, such that $X_i = X_{(R_i^*)}$.

Let $X_1, \ldots, X_m$ and $Y_1, \ldots, Y_n$ be independent random samples from continuous distributions with distribution functions $F(x)$ and $G(x) = F(x - \Delta)$, respectively, where Δ is an unknown shift parameter. The hypotheses of interest are:

- H_0: $\Delta = 0$.
- H_1: $\Delta \le 0$.

Let Q_i, $i = 1, \ldots, m$, and R_j, $j = 1, \ldots, n$, be the ranks of X_i and Y_j, respectively, among the $N = (m + n)$ combined X and Y observations. That is, R_j is the rank of Y_j among the m Xs and n Ys, combined and treated as a single set of observations. Similarly for Q_i. This implies that the rank vector $R^* = (Q_1, \ldots, Q_m, R_1, \ldots, R_n)$ is simply a permutation of sequence $(1, \ldots, N)$; although random, it hence must satisfy the constraint:

$$\sum_{i=1}^{m} Q_i + \sum_{j=1}^{n} R_j = \sum_{i=1}^{N} i = \frac{N(N+1)}{2} \qquad (7.3)$$

To test the null hypothesis H_0 against the alternative hypothesis H_1, we use the rank sum statistics by Wilcoxon and by Mann and Whitney [125]. The test statistics by Wilcoxon is

$$W = \sum_{i=1}^{n} R_i \tag{7.4}$$

That is, W is the sum of ranks for the sample observations Y when ranked among all $(m + n)$ observations. The test statistics by Mann and Whitney is

$$U = \sum_{i=1}^{m} \sum_{j=1}^{n} \Psi(Y_j - X_i) \tag{7.5}$$

where $\Psi(t) = 1$ for $t > 0$, otherwise $\Psi(t) = 0$ for $t \le 0$. It represents the total number of times a Y observation is larger than an X observation. W and U are linearly related,

$$W = U + \frac{n(n + 1)}{2}$$

Therefore, the discrete distribution of W or U under the null hypothesis H_0 is something that we might know or can tabulate from permutations of the sequence $(1, \ldots, N)$. The test will be of the form

reject H_0: $\Delta = 0$ in favor of H_1: $\Delta \le 0$ if and only if $W \ge w(\alpha, m, n)$

where $w(\alpha, m, n)$ is some accepted critical value that is dependent on a desired significance level α and sample sizes m and n. In other words, the Mann–Whitney U-test is based on rank sums rather than sample means.

Implementation. As mentioned in Section 7.3.1, Experiment One includes a total of six group data sets, with 94 to 96 sample size each, related to three independent variables: direction of motion, visibility, and interface. Each variable has two levels. The data satisfy the statistical requirement that the observations appear from their populations randomly and independently. The objectives of the Mann–Whitney U-test here are as follows:

- Compare the left-to-right group data with the right-to-left group data, thereby testing the effect of the direction variable.
- Compare the visible group data with the invisible group data, testing the effect of the visibility variable.
- Compare the virtual group data with the physical group data, testing the effect of the interface variable.

The null hypothesis H_0 here is that each of the two group data were drawn from the same population distribution, for each variable test, respectively. The alternative hypothesis H_1 for the corresponding test is that the two group data were drawn from different population distributions.

If results of the Mann–Whitney U-test show that a significant difference exists between the two group data—which means that a certain independent variable has a significant effect—we will break the group data associated with that independent variable into subgroups to find possible simpler effects. In this case, interaction effects might be found.

Results

1. The results of testing the effect of direction of motion, with data groups, RtoL and LtoR, are shown in Table 7.3. "Valid N" is the valid number of observations. Given the significance level $p < 0.01$, we reject the null hypothesis (which is that the two group samples come from the same population). This means there is a statistically significant difference between the "right to left" data set and the "left to right" data set. We therefore conclude that the direction-of-motion variable has a statistically significant effect on the length of paths generated by subjects. This is surprising, and we had already a hint of this surprise from Table 7.1.

2. The results of testing the effect of visibility factor, with the visible and invisible group data sets, are shown in Table 7.4; here, Vis stands for "visible" and Invis stands for "invisible." Given the significance level $p > 0.01$, we accept the null hypothesis (which says that the two group samples came from the same population). We therefore conclude that the visibility factor has no statistically significant effect on the length of paths generated by the subjects.

This is a serious surprise: The statistical test says that observation data from the subjects' performance in motion planning tasks contradicts the common belief that seeing the scene in which one operates should help one perform in it significantly better than if one "moves in the dark." While the described cognitive tests leave no doubt about this result, its deeper understanding will require more testing with a wider range of tasks. Indeed, we know from the tests—and it

TABLE 7.3. Results of Mann–Whitney Test on the Direction-of-Motion Factor

Mann–Whitney Test	Variable: Direction. Group 1: RtoL; Group 2: LtoR					
	Rank Sum				Valid N	
Variable	RtoL	LtoR	U	p-Level	RtoL	LtoR
Path length	6250.000	11895.00	1690.000	0.000000	95	95

TABLE 7.4. Results of Mann–Whitney Test on the Visibility Factor

Mann–Whitney test	Variable: Visibility. Group 1: vis; Group 2: invis					
	Rank Sum				Valid N	
Variable	Vis	Invis	U	p-Level	Vis	Invis
Path length	8881.000	9264.000	4321.000	0.613376	95	95

agrees with our intuition—that an opposite is true in the point-in-the-labyrinth test: One performs significantly better with a bird's-eye view of the labyrinth than when seeing at each moment only a small part of the labyrinth.

Apparently, something changes dramatically when one switches from the labyrinth test to moving a kinematic structure, the arm test. With the arm, multiple points of the arm body are subject to collision, the contacts may happen simultaneously, and the relation between some such points keeps changing as the arm links move relative to each other. It is likely (and some of our tests confirm this) that in simpler tasks where the arm cannot touch more than one obstacle at the time the visibility factor will play a role similar to the labyrinth test. It is clear, however, that if our more general result holds after a sufficient training of subjects (see Experiment Two below), we cannot rely on operator's skills in more complex tasks of robot arm teleoperation. Providing the robot with more intelligence—perhaps of the kind developed in Chapters 5 and 6—will be necessary to successfully handle teleoperation tasks.

3. The results of testing the effect of interface on the simulation group data and the booth group data are shown in Table 7.5; here, Virt stands for "virtual" and Phys stands for "physical." Given the significance level $p < 0.01$, we reject the null hypothesis (which says the two group data sets came from the same population). We therefore conclude that there is a statistically significant difference between the virtual tests (tests where subjects move the arm on the computer screen) and "physical" tests (tests where subjects move the physical arm). In other words, the interface factor has a statistically significant effect on the length of paths produced by the subjects. Furthermore, this effect is present whether or not the task is implemented in a visible or invisible environment, and whether or not the direction of motion is left-to-right or right-to-left.

While the Mann–Whitney statistical test isolates the single factor we are interested in, the interface factor, its results do not reconcile easily with the observations summarized in Table 7.1. Namely, Table 7.1 shows that while in the easier (left-to-right) task the subjects performed better with the physical arm than with the virtual arm, this difference practically disappeared in the harder (right-to-left) task. This calls for more refined statistical tests, with two separate direction-of-motion data sets. These are summarized next.

4. Here the Mann–Whitney test measures the effect of the interface factor using only the left-to-right (LtoR) data sets. The results are shown in Table 7.6. Given the significance level $p < 0.01$, we reject the null hypothesis and conclude that in the left-to-right task there is a statistically significant difference between

TABLE 7.5. Results of Mann–Whitney Test on the Interface Factor

Mann–Whitney Test	Variable: Interface. Group 1: Virt; Group 2: Phys					
	Rank Sum				Valid *N*	
Variable	Virt	Phys	*U*	*p*-Level	Virt	Phys
Path length	10427.00	7718.000	3253.000	0.000895	96	94

TABLE 7.6. Results of Mann–Whitney Test on the Interface Factor for LtoR Task

Mann–Whitney Test	Variable: Interface (LtoR only). Group 1: Virt; Group 2: Phys					
	Rank Sum				Valid N	
Variable	Virt	Phys	U	p-Level	Virt	Phys
Path length	2943.000	1617.000	489.0000	0.000002	48	47

TABLE 7.7. Results of Mann–Whitney Test on the Interface Factor for the RtoL Task

Mann–Whitney Test	Variable: Interface (RtoL). Group 1: Virt; Group 2: Phys					
	Rank Sum				Valid N	
Variable	Virt	Phys	U	p-Level	virt	phys
Path length	2505.000	2055.000	927.0000	0.134619	48	47

the virtual group data and the physical group data. This result agrees with the results above obtained for the combined LtoR and RtoL data.

5. Here the Mann–Whitney U-test measures the effect of the interface factor using only the right-to-left (RtoL) task data sets. The results are shown in Table 7.7. Given the significance level $p > 0.01$, we accept the null hypothesis and thus conclude that in this more difficult motion planning task there is no statistically significant difference between the subjects' performance when moving the virtual arm and when moving the physical arm.

The last three test results (3, 4, and 5) imply a complex relationship between the subjects' performance and the type of interface used in the test. This points to a possibility of an interaction effect between the interface factor and the direction-of-motion factor. To clarify this issue, we turn in the next section to analysis of variance of sample data.

7.4.3 Univariate Analysis of Variance

Assumptions. The purpose of analysis of variance (ANOVA), which is also the name of the technique that serves this purpose, is to probe the data for significant differences between the means of sets of data, with the number of sets being at least two. The technique performs a statistical test of comparing variances (hence the name). This objective is very similar to the objective of nonparametric analysis above, except that ANOVA can sometimes be more sensitive. In addition, besides testing individual effects of independent variables, ANOVA can also test for interaction effects between variables.

To apply the analysis of variance, we need some assumptions about the data. As before, we assume that the experimental scores have been sampled randomly and independently, from a normally distributed population with a group mean and an overall constant variance. Since the assumption may be too restrictive for real data, we need to address the effect of this assumption being invalid. The statistical F test [126] used in the next section is known to be fairly robust to deviations from normality.[11]

One-Way Analysis of Variance.   The simplest data structure, called a *one-way layout*, has one or more observations at every level of a single factor. We call each level of a one-way layout a *group* or *cell*. For example, the path length scores obtained by subjecting 96 randomly selected subjects to motion planning tasks in a visible and invisible environments form a one-way layout. The single factor is visibility, and the two groups are *visible* and *invisible*. Let us denote these groups A_1 and A_2, respectively.

The one-way analysis of variance in this example will attempt to answer the following question: Do data from the visible and invisible groups really score differently on the path length, and is the difference due to the random selection of subjects? The corresponding null and alternative hypotheses relate to unknown population averages for the groups, μ_i:

- H_0: $\mu_i = \mu$ for all A_i, $i = 1, 2$
- H_1: $\mu_1 \neq \mu_2$

Suppose X_{ij} represents the jth sampled score from group A_i, $i = 1, 2 \ldots, m$; $j = 1, 2, \ldots, n_i$. Here m is the number of groups, and n_i is the number of observations in each group; N is the total number of observations, $N = m * n_i$. Then the mean of group A_i is

$$\overline{X_{i\cdot}} = \frac{1}{n_i} \sum_j X_{ij} \tag{7.6}$$

and the grand mean for all groups is

$$\overline{X_{\cdot\cdot}} = \frac{1}{N} \sum_i \sum_j X_{ij} \tag{7.7}$$

For this one-way layout there are three estimates of variance of population:

1. *Mean Square Within*, MS_w, and *Sum of Squares Within*, SS_w, with MS_w being the average of estimates of variances within individual groups,

$$MS_w = \frac{1}{m} \sum_i \frac{1}{n-1} \sum_j (X_{ij} - \overline{X_{i\cdot}})^2$$

$$SS_w = \sum_i \sum_j (X_{ij} - \overline{X_{i\cdot}})^2 \tag{7.8}$$

---

[11] Also, according to the central limit theorem, if the sample size is fairly large, deviations from normality do not matter much [126].

2. *Mean Square Between, MS_b, and Sum of Squares Between, SS_b, with MS_b being the estimated variance of means among groups,*

$$MS_b = \frac{n}{m-1} \sum_i (\overline{X}_{i.} - \overline{X}_{..})^2$$

$$SS_b = n \sum_i (\overline{X}_{i.} - \overline{X}_{..})^2$$

(7.9)

3. *Mean Square Total, MS_t, and Sum of Squares Total, SS_t, with MS_t being the estimated variance of the total mean (ignoring group membership),*

$$MS_t = \frac{1}{N-1} \sum_i \sum_j (X_{ij} - \overline{X}_{..})^2$$

$$SS_t = SS_w + SS_b$$

(7.10)

The Mean Square Within, SS_w, is usually called *error variance*. The term implies that one cannot readily account for this value in one's data. The Mean Square Between, SS_b, usually called *effect variance*, is due to the difference in means between the groups. Consider the ratio

$$F = \frac{MS_b}{MS_w} \sim F_{(m-1, N-m)}$$

(7.11)

Under our assumption, this ratio has an F distribution with $(m-1)$ degrees of freedom in the numerator and $(N-m)$ degrees of freedom in the denominator. Note that MS_b is a valid variance estimate only if the null hypothesis is true. Therefore, the ratio is distributed as F only if the null hypothesis is true. This suggests that we can test the null hypothesis by comparing the obtained ratio with that expected from the F distribution. Under the null hypothesis, variance estimated based on the within-group variability should be the same as the variance due to between-group variability. We can compare these two variance estimates using the F test, which checks if the ratio of two variance estimates is significantly greater than 1.

To summarize the basic idea of ANOVA, its purpose is to test differences in the group for statistical significance. This is accomplished by a data variance analysis, namely by partitioning the total variance into a component that is due to the true random error (i.e., SS_w) and components that are due to differences between the means (i.e., SS_b). These latter components of variance are then tested for statistical significance. If the differences are significant, we reject the null hypothesis (which expects no difference between the means) and accept the alternative hypothesis (that the means in the population differ).

7.4.4 Two-Way Analysis of Variance

Main Effects. One-way analysis of variance handles group data for a single variable. For example, ANOVA can address the effect of visibility by testing differences between the visible group and the invisible group. Nonparametric statistics (Section 7.4.2) can do this as well. Sometimes more than one independent variable (factor) has to be taken into account.

For example, in the Experiment One data, human performance may be determined by the visibility factor and also by the interface factor. One important reason for using the ANOVA method rather than the multiple two-group nonparametric U-test is the efficiency of the former: With fewer observations we can gain more information [126].

Suppose we want to analyze the data in Table 7.8. The two rows in the table correspond to the two levels of factor A, namely, A_1, and A_2; the two columns correspond to the two levels of factor B, namely, B_1 and B_2. The levels of factor A can be, for example, the visible group and the invisible group, and the levels of factor B can be the virtual group and the physical group. The cell AB_{ij} in the table relates to the level A_i of factor A and the level B_j of factor B, $i, j = 1, 2$. In general the number of levels of A does not have to be equal to that of B. For simplicity, assume that the number of observations at every level/factor is the same, n.

Here are some notations that we will need:

I —number of levels of factor A;

J —number of levels of factor B;

n —number of observations in each cell AB_{ij};

N —total number of observations in the entire experiment; hence $N = n * I * J$;

μ —unknown population means, as follows:

$\mu_i = \frac{1}{J} \sum_j \mu_{ij}$ —the mean for level i, summed over subscript j,

$\mu_j = \frac{1}{I} \sum_i \mu_{ij}$ —the mean for level j, summed over subscript i,

$\mu = \frac{1}{I} \sum_i \mu_{i.} = \frac{1}{J} \sum_j \mu_j$ —overall mean of all μ_{ij},

$\overline{X}_{ij} = \frac{1}{n} \sum_k X_{ijk}$ —average within a cell over its subjects' scores,

TABLE 7.8. An Example of a Two-Way Data Layout

	B_1	B_2
A_1	AB_{11}	AB_{12}
A_2	AB_{21}	AB_{22}

$\overline{X}_i = \frac{1}{J}\sum_j \overline{X}_{ij.}$—average for the row i of A over related subjects' scores,

$\overline{X}_j = \frac{1}{I}\sum_i \overline{X}_{ij.}$—average for the column j of B over related subjects' scores,

$\overline{X} = \frac{1}{J}\sum_i \overline{X}_{i..} = \frac{1}{I}\sum_j \overline{X}_{.j.}$—overall mean of all the scores.

With this notation, if we only test for the *main effect* of factor B (similarly for the *main effect* of factor A), the null and alternative hypotheses can be written as

- $H_0(B)$: $\mu_j = \mu$ for all j
- $H_1(B)$: $\mu_j \neq \mu$ for at least one j

The *Main Sum Between*, MS_b^B, for J cells of factor B, is the average of estimated variance of estimated column means (this ignoring factor A). That is,

$$MS_b^B = \frac{nI}{J-1}\sum_j (\overline{X}_j - \overline{X})^2 \tag{7.12}$$

The *Main Sum Within* is the same error variance MS_w considered above; it is equal to the average of (separately estimated) variances within the individual cells. Under the null hypothesis,

$$\frac{MS_b^B}{MS_w} \sim F_{(J-1, N-IJ)} \tag{7.13}$$

That is, the ratio of two main averages has an F distribution with $(J-1)$ degrees of freedom in the numerator and $(N-IJ)$ degrees of freedom in the denominator, which is the total number of observations N minus the total number of cells, IJ.

Interaction Between Factors. Unfortunately, the main effects are not sufficient to answer questions such as, "Does the effect of factor A remain the same at different levels of factor B?" For example, in some observations of Experiment One the main effect of the visibility factor is that it significantly affects the subjects' path length: The invisible task results in longer path lengths compared to the visible task. However, we notice that the subjects' scores on the visibility factor are also affected by the interface factor: Namely, in the physical test (in the booth) the visibility factor has no significant effect on the path length, whereas in the virtual test the visibility factor has a significant effect on the path length. This suggests that the tests on main effects may be missing such interaction effects. The latter can be tested by the following formulas of interaction (for details refer to Refs. 126 and 127):

$$SS^{AB} = n\sum_i\sum_j (\overline{X}_{ij.} - \overline{X}_{i..} - \overline{X}_{.j.} + \overline{X}_{...})^2$$

$$MS^{AB} = \frac{SS^{AB}}{(I-1)(J-1)} \tag{7.14}$$

$$F = \frac{MS^{AB}}{MS_w} \sim F_{[(I-1)(J-1),N-IJ]} \tag{7.15}$$

where MS^{AB} represents the *Interaction Mean Square* between factors A and B. The ratio has an F distribution with $(I - 1)(J - 1)$ degrees of freedom in the numerator and $(N - IJ)$ degrees of freedom in the denominator.

In the case considered now, the results of F test may show that the effect of visibility factor on the path length depends on the type of interface utilized in a given task. In other words, there is an interaction between the visibility factor and the interface factor. One way to express interactions is by saying that one effect *is modified (qualified) by* another effect.

When the data indicate an interaction between factors, the notion of a main effect has no meaning. In such cases, tests of simple effects can be more useful than tests of main effects. Simple effect tests are done via one-way analysis of variance across levels of one factor, performed separately at each level of the other factor. For example, even if we suspect an interaction between the visibility factor and the interface factor, we might undertake simple effect tests for the visibility factor separately at the virtual and physical level, respectively, and see what kind of conclusions can be made based on the results.

7.4.5 Implementation: Two-Way Analysis for Path Length

We are now ready to perform the analysis of variance on the Experiment One data. From other tests above, we already know that the direction factor has a significant effect on the path length. We know, further, that the left-to-right task is significantly easier for the subjects (it results in shorter paths) than the right-to-left task. We now want to analyze the combined effect of visibility and interface factors on the subjects' performance. Even though the underlying data are not known to obey the normal distribution, we justify using the ANOVA by the F test being known to be robust.

The data set has been first separated into the LtoR and RtoL data sets. The ANOVA variables are:

- Dependent variable: Path length.
- Independent variables:
  1. Visibility factor, with two levels: visible and invisible.
  2. Interface factor, with two levels: virtual and physical.

In the tables of results that appear here, the following terms are used:

df effect—degrees of freedom for a given effect, including main and interaction effects.

MS effect—*Mean Square* for an effect, including main and interaction effects.

df error—degrees of freedom for the error variance, or *Mean Square Within*.

MS error—*Mean Square* for the error variance, or *Mean Square Within*.

Rows with the effect names "1" and "2" correspond to main effects.

Rows with more than one digit in the name, such as "12" or "123," relate to the corresponding interaction effects.

Results. For the left-to-right task, the summary of ANOVA results appears in Table 7.9. The *p*-levels for the visibility and interface factors are about 0.01, and the *p*-level for the interface is much greater than 0.01. This means the main effects of both factors are slightly significant, and there is no interaction. We therefore conclude that for both visible and invisible environments, the path length is affected only slightly by the interface factor. This reconciles with our knowing that for the physical task the path length is slightly shorter than for the virtual task. And, for both physical and virtual tasks the path length is only slightly affected by the visibility factor. Again, this reconciles with our knowing that in the visible environment the path length is slightly shorter than in the invisible environment.

The summary of ANOVA results for the right-to-left task appears in Table 7.10. Here the *p*-levels for the visibility factor, the interface factor, and the interaction are all greater than 0.01. Therefore, the main effects make no significant difference for the dependent variable, and there is no interaction. The conclusion is that

TABLE 7.9. ANOVA Results for Path Length: Interface and Visibility Factors; LtoR Task

ANOVA	Effects Studied: 1—interface, 2—visibility					
Effect	*df* Effect	*MS* Effect	*df* Error	*MS* Error	*F*-Value	*p*-Level
1	1	18828.49	91	3036.650	6.200416	0.014587
2	1	21314.74	91	3036.650	7.019164	0.009508
12	1	201.26	91	3036.650	0.066277	0.797418

TABLE 7.10. ANOVA Results for Path Length: Interface and Visibility Factors; RtoL Task

ANOVA	Effects Studied: 1—interface, 2—visibility					
Effect	*df* Effect	*MS* Effect	*df* Error	*MS* Error	*F*-Value	*p*-Level
1	1	5283.137	91	12592.88	0.419534	0.518801
2	1	77.399	91	12592.88	0.006146	0.937684
12	1	8598.556	91	12592.88	0.682811	0.410782

for either the visible or invisible environments, the path length for the physical task is not significantly different from the path length in the virtual task. Also, in either of physical or virtual tasks, the path length in the visible environment does not significantly differ from the path length in the invisible environment.

7.4.6   Implementation: Two-Way Analysis for Completion Time

In the previous section we have analyzed the effects of test factors on the length of paths generated by the human subjects in Experiment One. We will now analyze how these same factors affect another performance indicator, the task completion time.

Each completion time score is random and independent (for the 48 subjects tested here); this meets the "sampling assumption" of nonparametric statistics and analysis of variance. Even though a closer look at the completion time data shows that they do not obey a normal distribution (as the ANOVA assumption requires), we still use ANOVA, counting on the F test known to be robust.

To analyze the effect of all factors on the completion time data, a three-way analysis of variance has been done. The ANOVA variables are as follows:

- Dependent variable: Completion time.
- Independent variables:
  1. Direction factor, with two levels: LtoR and RtoL.
  2. Visibility factor, with two levels: visible and invisible.
  3. Interface factor, with two levels: virtual (simulation) and physical (booth).

Second, since we are more interested in the visibility factor and interface factor, and since the performance in LtoR task significantly differs from that in RtoL task, a two-way ANOVA was implemented. The ANOVA variables are:

- Dependent variable: Completion time.
- Independent variables:
  1. Visibility factor, with two levels: visible and invisible.
  2. Interface factor, with two levels: virtual and physical.

Results. The summary of ANOVA results of analysis of variance for all three factors used in Experiment One appears in Table 7.11. The p-levels for the interface factor, the direction factor, and the interaction between them are less than 0.01. This means that these two main effects likely significantly affect the dependent variable (completion time), and there is interaction between them. The p-levels for the remaining main effect, visibility, and for interactions with this factor are greater than 0.01. This means that there is no significant difference for these effects and interactions. However, given that an interaction has been detected, we should not be forming any conclusions from the results in Table 7.11 until we separate the factor levels.

TABLE 7.11. ANOVA Results for Completion Time: Direction, Interface, and Visibility Factors

ANOVA	Effects Studied: 1—interface, 2—visibility, 3—direction					
Effect	df Effect	MS Effect	df Error	MS Error	F-Value	p-Level
1	1	5248143.0	184	51978.68	100.9672	0.000000
2	1	189719.0	184	51978.68	3.6499	0.057626
3	1	3475104.0	184	51978.68	66.8563	0.000000
12	1	12523.0	184	51978.68	0.2409	0.624127
13	1	427953.0	184	51978.68	8.2332	0.005494
23	1	45246.0	184	51978.68	0.8705	0.352049
123	1	52450.0	184	51978.68	1.0091	0.316444

TABLE 7.12. ANOVA Results for Completion Time: Interface and Visibility Factors, LtoR Task

ANOVA	Effects Studied: 1—interface, 2—visibility					
Effect	df Effect	MS Effect	df Error	MS Error	F-Value	p-Level
1	1	1339396.0	92	32864.84	40.75467	0.000000
2	1	210132.0	92	32864.84	6.39382	0.013155
12	1	6858.0	92	32864.84	0.20867	0.648886

Hence the data for the left-to-right task was analyzed, which is one of two levels of the direction factor. The summary of related ANOVA results appears in Table 7.12. The p-level for the interface factor is smaller than 0.01, the p-level for the visibility factor is about 0.01, and the p-level for the interface factor is greater than 0.01. Therefore, the main effect of interface is statistically significant, the main effect of visibility is slightly significant, and there is no interaction between them. This reconciles with our knowledge that for both visible or invisible tasks, the completion time for the physical task is significantly shorter than for the virtual task. Similarly, for both physical or virtual tasks the completion time is slightly shorter in the visible environment than in the invisible environment.

The summary of ANOVA results for the right-to-left task appears in Table 7.13. The p-level for the interface factor is smaller than 0.01, the p-levels for the visibility factor and interaction are greater than 0.01. This means the main effect of the interface factor is statistically significant, and there is no interaction. This

TABLE 7.13. ANOVA Results for Completion Time: Interface and Visibility Factors, RtoL Task

ANOVA	Effects Studied: 1 — interface, 2 — visibility					
Effect	df Effect	MS Effect	df Error	MS Error	F-Value	p-Level
1	1	4336700.0	92	71092.52	61.00080	0.000000
2	1	24833.0	92	71092.52	0.34930	0.555959
12	1	58115.0	92	71092.52	0.81746	0.368286

reconciles with test observations: For both visible or invisible tasks the completion time in the physical task was significantly shorter than in the virtual task. Similarly, for both physical and virtual tasks the completion time in the visible environment shows no significant difference from that in the invisible environment.

7.5 RESULTS — EXPERIMENT TWO

Recall that Experiment Two was designed to analyze the effect of subjects' training and the related effect of the visibility factor on human performance. A total of 12 subjects appeared in this study. In the first group, which included six subjects, on day 1 each subject was subjected to six different training tasks, plus one test task at the end, all in the visible environment. About one week later, on day 2, the same subjects performed the same six training tasks, plus the same test task, this time in the invisible environment. In the second group, the remaining six subjects did the same tasks in the opposite order—that is, tests in the invisible environment on day 1 and tests in the visible environment on day 2. The specific task was right-to-left movement of the arm, the same as in Experiment One (recall that this is a more difficult task compared to the left-to-right task).

We therefore have a training factor *Day*, with two levels, day 1 and day 2. Subjects were expected to learn the motion planning skill through a repeated exercise.

Similar to Experiment One, human performance was measured by the path length and completion time for each of the tasks *Path* and *Time*. Path length is the measure of motion generated by the arm manipulator during the task. Completion time is the time it takes the subject to complete the task. Both measure the subjects' proficiency in carrying out motion planning. We suppose that both the path length and the completion time may be affected by such factors as training and visibility of the scene, and we would like to quantify those effects.

In statistical terms, the training and visibility factors are independent variables, whereas the path length and completion time are dependent variables. The objective of data analysis is to test whether the training and/or visibility factor improves the overall human performance in motion planning. If in terms of both dependent variables the improvement in subjects' performance turns out to be significant, follow-up tests on the separate effects on human performance should be conducted, to explain which specific aspects of human performance are responsible for such effects. Multivariate analysis of variance (MANOVA) is a good technique for data analysis of overall effects [128].

Multivariate analysis of variance is conceptually a straightforward extension of the univariate ANOVA technique described above. Their major distinction is that if in ANOVA one evaluates mean differences on a single dependent variable, in MANOVA one evaluates mean vector differences simultaneously on two or more dependent variables. In addition, the MANOVA design accounts for the fact that dependent variables may be correlated. For instance, two dependent variables in Experiment Two, the path length and completion time, are indeed relatively highly correlated, with the correlation coefficient 0.79. In this case, MANOVA should provide a distinct advantage over separate ANOVAs. In fact, performing separate ANOVA tests carries an implicit assumption that either the dependent variables are uncorrelated or such correlations are of no importance.

7.5.1 The Technique

Assumptions. The first and partly second of the three following assumptions are required by MANOVA (and are the same for the statistical tests considered above):

1. Observation scores are randomly sampled from the population of interest. Observations are statistically independent of one another.
2. Dependent variables have a multivariate normal distribution within each group of interest. This means that (a) each dependent variable is distributed normally, (b) any linear combination of the dependent variables are distributed normally as well; (c) all subsets of the variables have a multivariate normal distribution. In practice, it is unlikely that this and the next assumption are met precisely. Fortunately, similar to ANOVA, MANOVA is relatively robust to violations of these assumptions. In practice, MANOVA tends to perform well regardless of whether or not the data violate these assumptions.
3. Homogeneity of covariance matrices. That is, all groups of data are assumed to have a common within-group population covariance matrix. This can be likened to the assumption in ANOVA of homogeneity of variance for each dependent variable, or the assumption that correlation between any two dependent variables must be the same in all groups. If the number of subjects is approximately the same in the experimental groups, a violation of the assumption of covariance matrix homogeneity leads to a slight reduction in statistical power [128–130].

Multivariate Null Hypothesis. Hypotheses in MANOVA are very similar to those in univariate ANOVA, except that vectors of means are considered instead of single values (scalars) of means. For a simple example, imagine we carry out a one-way MANOVA for a visible task and invisible task groups. We would like to know if the scores of path length *and* completion time came from the same population that includes visible and invisible task data. That is, we want to compare the population mean vector for the dependent variables for one group with the population mean vector for the dependent variables for another group.

Suppose μ_{ij} represents the mean of the dependent variable i for group j, $i = 1, 2$, $j = 1, 2$. The mean vector for group j can be written as

$$\vec{\mu}_j = \begin{bmatrix} \mu_{1j} \\ \mu_{2j} \end{bmatrix}$$

Then the multivariate null hypothesis H_0 can be written as an equality of vectors:

$$H_0: \vec{\mu}_1 = \vec{\mu}_2 = \vec{\mu}$$

The alternative hypothesis H_1 in this case says that for at least one variable there is at least one group with a population mean different from that in the other group(s):

$$H_1: \vec{\mu}_1 \neq \vec{\mu}_2$$

Calculating MANOVA Test Statistics. Derivation of the MANOVA test statistics is similar to that in ANOVA but involves relatively cumbersome matrix operations and equations. Hence we will limit the discussion to a conceptual level (see Ref. 130 for more detail).

Recall that the ANOVA attempts to test if the amount of variance explained by the independent variable (namely, SS_b, see Section 7.4.3) exceeds significantly the variance that has not been explained (namely, SS_w). The variance here is a function of the sum of squares of deviations from the mean for an entire group (the latter being called the *sum of squares, SS*). The ANOVA's F statistics is a ratio of the mean square between, MS_b, to the mean square within, MS_w.

Instead of scalars of dependent variables, MANOVA employs a vector of dependent variables. A single sum of squares is replaced with a complete (*total*) matrix of sums of squares and cross-products, SP_t. Along its diagonal the matrix has the sums of squares that represent variances for all dependent variables, and in its off-diagonal elements it has cross-products that represent covariances of variables. Just as a univariate ANOVA, MANOVA divides matrix SP_t into the within-group matrix, SP_w, and the between-group matrix, SP_b. From algebra, the matrix determinant expresses the amount of generalized variance, or the total variability that is present in the underlying data and is expressed through the dependent variables. One can hence compare the generalized variance of one matrix with another.

Wilks' lambda test is perhaps the most widely used statistical test of multivariate mean differences [130]. It derives from the following idea. Since matrix SP_b

represents the amount of explained variance and covariance, and matrix SP_w represents the remaining variance and covariance, in the case of a significant effect one would expect matrix SP_b to have a larger generalized variance compared to matrix SP_w. Wilks' lambda index, Λ, is defined as a ratio of determinants of the two matrices:

$$\Lambda = \frac{|SP_w|}{|SP_t|} = \frac{|SP_w|}{|SP_w + SP_b|} \qquad (7.16)$$

where SP_t, SP_w, and SP_b are the total, within-group, and between-group SP matrices, respectively.

We associate the value of Λ with the effect's significance. The value can also be interpreted as the proportion of unexplained variance. The main effects and interaction effects in multiple-way MANOVA are conceptually the same as those in ANOVA. While computations are more complex in MANOVA, their underlying logic is the same as in ANOVA.

If an overall significant multivariate effect is found, the next natural step is to submit the data to further testing, to see whether all dependent variables or some specific dependent variables are affected by the independent variables. Performing multiple univariate ANOVAs for each of the dependent variables is a common method for interpreting the respective effects. One attempts to identify specific dependent variables that contributed to the overall significant effect.

Repeated Measures MANOVA. In our statistical tests so far, all independent variables involved in ANOVA and MANOVA were also *between-subjects variables* (or factors); we were interested in differences between means or mean vectors of several distinct groups of subjects. The observed scores were independent of each other at different levels of the between-subjects variables.

However, in Experiment Two we also want to study the difference in responses of the *same subjects* before and after treatment; in our case, treatment is training. This variable is called *repeated measures*, and its analysis is called *repeated measures MANOVA*. In a repeated measures design the several response variables are results of the same test carried out by the same subjects, applied a number of times or under more than one experimental condition. For example, in Experiment Two each subject was assessed as to their path length and completion time on day 1 and again on day 2. The variable "day" is a repeated measures variable, as well as a within-subjects variable.

In other words, a between-subjects variable is a grouping variable—similar to the visibility or interface in our study—whereas a within-subjects variable refers to the measurements for every level of the within-subjects variable. For example, a within-subjects variable may be "time," or "day," or "training factor." A study can involve both within- and between-subjects independent variables. Our Experiment Two analysis constitutes a 2 (days) by 2 (visibility levels) repeated measures MANOVA, or repeated measures ANOVA. The first independent variable, day, is a within-subjects (repeated measures) variable, and the last independent variable, visibility, is a between-subjects variable.

Repeated measures MANOVA is an extension of the standard MANOVA. The underlying principles of both are almost the same. In the standard MANOVA, vectors of *means* are compared across the levels of independent variables. In the repeated measures MANOVA, vectors of *mean differences* are compared across the levels of independent variables.

Mean differences are the differences in values of dependent measures between levels of the within-subjects variable. These can be seen as new independent variables. If, for example, the dependent variables were measured for each subject at four different time moments, say at times T1 through T4, these original four variables would be transformed to three alternative derived difference variables, denoted (T1–T2), (T2–T3), and (T3–T4). These three new variables directly address the questions of interest. The repeated measures MANOVA, therefore, compares the vectors of means across the new transformed variables, not the original scores.

When conducting a repeated measures MANOVA, a *sphericity assumption* must be met. It requires that the covariance matrix for the transformed variables be a diagonal matrix. That is, the values (variances) along the diagonal of the transformed covariance matrix should be equal, and all the off-diagonal elements (correlation coefficients) should be zeros. The purpose of the sphericity assumption is to ensure the homogeneity of covariance matrices for the new transformed variables [131, 132].

7.5.2 Implementation Scheme

Experiment One. Recall that in Experiment One the observation scores in each task were measured on two dependent variables, path length and completion time. Subjects have been randomly selected, and sets of scores were mutually independent. Further, the two dependent variables were correlated, with the correlation coefficient 0.74. We take this correlation into account when performing the significance test, since the overall set of dependent variables may contain more information than each of the individual variables. This suggests that the Experiment One data can be a candidate for a multivariate analysis of variance, MANOVA.

Since, as discussed in the previous section, the effect of direction factor in Experiment One is statistically significant, we separately perform two sets of MANOVAs—one for the left-to-right task and the other for the right-to-left task. When performing MANOVA for the left-to-right task, the data set forms a two-way array, 2 (visibility) × 2 (interface). For the right-to-left task, the data set also forms a two-way array, 2 (visibility) × 2 (interface). The results of analysis should answer questions such as: (1) does human performance improve in the visible environment compared to the invisible environment? (2) Does human performance improve in a test with the physical arm manipulator as compared to the virtual arm manipulator? (3) Does the effect of the visibility factor work across the levels of the interface factor?

TABLE 7.14. Descriptive Statistics for the Data in Experiment Two

Variable/Task	Descriptive Statistics				
	Valid N	Mean	Minimum	Maximum	Std. Dev.
day1-path	12	96.68	24.39	232.55	66.59
day1-time	12	432.67	65.00	900.00	333.66
day2-path	12	129.04	15.13	393.90	107.99
day2-time	12	432.42	36.00	900.00	365.89
vis-path	12	88.83	15.13	181.92	62.20
vis-time	12	360.25	36.00	900.00	620.83
invis-path	12	136.89	27.04	393.90	107.42
invis-time	12	504.83	90.00	900.00	361.76

Experiment Two. Table 7.14 lists basic descriptive statistics for the Experiment Two data: the number N of valid observations in each group; and means, minimums, maximums, and standard deviations in each group. In the table, "vis" means visible, "invis" means invisible, "path" means path length, and "time" means task completion time.

Similar to Experiment One, the Experiment Two data for each task were recorded for two dependent variables, path length and completion time. Subjects were randomly selected, and the sets of scores were independent of each other. The correlation coefficient of the two dependent variables is 0.79. This correlation suggests that each dependent variable contains some new information as well as some information overlapping with the other dependent variables. Accounting for this correlation allows us to test the significance of dependent variables in human performance. Since the data in Experiment Two correspond to the same subjects on day 1 and day 2, the day factor is a repeated measures variable with two levels, day 1 and day 2. These data call for a repeated measures MANOVA.

The Experiment Two data form a two-way array, 2 (day) × 2 (visibility). If any main effects or interaction effects are identified, multiple univariate ANOVA would be performed, in order to observe the effects on each dependent variable. In our data analysis we are interested in these questions: (1) Is there an improvement in human performance across day 1 and day 2? (2) Is there a statistically significant difference in human performance in the visible as opposed to invisible environment? and (3) Does the effect of one independent variable change over the levels of another independent variable?

Combined Experiment One and Two. There is another data set that we can use to test the effects of training and visibility. The first half of the data (12 subjects) in this new combined data set was extracted from the Experiment One. Six of these were randomly picked among the virt–vis–RtoL data, and another six were randomly picked among the virt–invis–RtoL data. The second half (12 subjects) of the data in the new combined set are the day 2 data (12 subjects) from Experiment Two.

The purpose of MANOVA or ANOVA analysis on this combined data set is to test whether human performance would show improvement from day 0, when the subjects executed tasks without any training (in Experiment One), to day 2, when subjects had a benefit of several training trials (in Experiment Two). The effect of visibility would also be tested here. Note, however, that the day factor in this analysis is not a repeated measure variable any longer but instead a between-subjects variable. This is because there are no pairs of data for day 0 and day 2 coming from the same subjects (which would be required by the definition of repeated measure variable). The data for day 0 are independent with respect to the data for day 2. Therefore, the new combined data form a two-way array, 2 (day) × 2 (visibility).

7.5.3   Results and Interpretation

1. The MANOVA scheme was applied to the left-to-right data in Experiment One. The variables involved are as follows:

- Dependent variables:
  1. Path length.
  2. Completion time.
- Independent variables:
  1. Interface, with 2 levels: virtual and physical.
  2. Visibility, with 2 levels: visible and invisible.

The results are shown in Table 7.15. Here df is the degrees of freedom (see Section 7.4.5). Note that the p-level for interaction between the two independent variables is significantly high. We thus conclude that there is no interaction effect. This means that the effect of one independent variable does not change across the levels of the other independent variable. The p-level for the interface factor is almost zero. We therefore reject the null hypothesis of the MANOVA, and we conclude that the interface factor has a statistically significant effect on the

TABLE 7.15. Results of MANOVA for LtoR Task, Experiment One

MANOVA	Effects Studied: 1—interface, 2—visibility			
Effect	Wilks' Lambda	df 1	df 2	p-Level
1	0.615711	2	90	0.000000
2	0.924517	2	90	0.029253
12	0.986327	2	90	0.538205

overall human performance. The p-level for the visibility factor shows that the overall human performance is only slightly improved in the visible environment compared to the invisible environment.

Given a significant effect indicated by the MANOVA results, multiple univariate ANOVAs have followed.

2. The MANOVA was applied to the right-to-left data in Experiment One. The variables are as follows:

- Dependent variables:
  1. Path length.
  2. Completion time.
- Independent variables:
  1. Interface, with 2 levels: virtual and physical.
  2. Visibility, with 2 levels: visible and invisible.

The results are shown in Table 7.16. Note that the p-level for the interaction effect between the two independent variables is large enough; we can conclude that there is no interaction effect. This means that the effect of one independent variable is not influenced by the other independent variable. The p-level for the interface factor is almost zero, hence we reject the null hypothesis of MANOVA. That is, the interface factor has a statistically significant effect on the overall subjects' performance. The p-level for the visibility factor is large; we thus conclude that the overall subjects' performance was affected by the visibility factor.

Since a significant effect was demonstrated by this MANOVA, multiple univariate ANOVAs have followed.

3. MANOVA was applied to the Experiment Two data. The variables involved are as follows:

- Dependent variables:
  1. Path length.
  2. Completion time.

TABLE 7.16. Results of MANOVA for RtoL Task, Experiment One

MANOVA	Effects Studied: 1—interface, 2—visibility			
Effect	Wilks' Lambda	df 1	df 2	p-Level
1	0.406424	2	90	0.000000
2	0.988945	2	90	0.606390
12	0.990999	2	90	0.665716

TABLE 7.17. Results of MANOVA, Experiment Two

MANOVA	Effects Studied: 1—visibility, 2—day			
Effect	Wilks' Lambda	df 1	df 2	p-Level
1	0.839068	2	9	0.454033
2	0.479769	2	9	0.036698
12	0.631314	2	9	0.126213

- Independent variables:
  1. Visibility, with 2 levels: visible and invisible.
  2. Day (repeated measures), with 2 levels: day 1 and day 2.

The results are shown in Table 7.17. The p-level for the interaction effect is bigger than the significance level (0.05), pointing to no significant interaction effect. This means the result for one independent variable is not modified by the other independent variable. The p-level for the visibility factor is also larger than the significance level, so the null hypothesis should be accepted. In other words, the overall human performance is not affected by the visibility factor. The p-level for the day factor suggests a slight effect of the day (training) factor on human performance. This reconciles with our knowing that the overall subjects' performance was better on day 2 compared to day 1.

In order to see which indicator of human performance might be affected by the day (training) factor, multiple univariate ANOVAs have been performed.

4. ANOVA was applied to the path-length-dependent variable in Experiment Two. The variables involved are as follows:

- Dependent variables:
  1. Path length.
- Independent variables:
  1. Visibility, with two levels: visible and invisible.
  2. Day (repeated measures), with two levels: day 1 and day 2.

The results are shown in Table 7.18. The p-levels for the main effects and interaction effect are larger than the significance level 0.05. Each null hypothesis for the main effects and interaction effect should hence be accepted. We conclude that there are no significant effects of the visibility factor and day (training) factor, and that these results do not change across the levels of the independent variables. In other words, surprisingly, the visibility factor has no significant effect on the subjects' path length, and the day (training) effect has no significant effect on the path length as well.

TABLE 7.18. Results of ANOVA for Path Length, Experiment Two

ANOVA	Effects studied: 1—visibility, 2—day					
Effect	df Effect	MS Effect	df Error	MS Error	F-Value	p-Level
1	1	13860.02	10	7756.985	1.786780	0.210934
2	1	6281.40	10	7292.907	0.861303	0.375232
12	1	12706.58	10	7292.907	1.742321	0.216262

TABLE 7.19. Results of ANOVA on Completion Time, Experiment Two

ANOVA	Effects studied: 1—visibility, 2—day					
Effect	df Effect	MS Effect	df Error	MS Error	F-Value	p-Level
1	1	125426.0	10	197945.9	0.633638	0.444509
2	1	0.4	10	442773.8	0.000008	0.997735
12	1	149626.0	10	44273.8	3.379561	0.095856

5. An ANOVA was applied to the independent variable of completion time in Experiment Two. The variables involved are as follows:

- Dependent variables:
  1. Completion time.
- Independent variables:
  1. Visibility, with 2 levels: visible and invisible.
  2. Day (repeated measures), with 2 levels: day 1 and day 2.

The results are shown in Table 7.19. The p-levels for the main effects and interaction effect are all larger than the accepted threshold significance level 0.05. Hence all null hypotheses for the main effects and interaction effect are accepted. We thus conclude that there are no significant effects for the visibility factor and day (training) factor, and this does not change across all levels of the independent variables. In other words, surprisingly, neither the visibility factor nor the day (training) factor has a significant effect on the completion time.

6. The MANOVA was applied to the combined data set. The variables involved are as follows:

- Dependent variables:
  1. Path length.
  2. Completion time.

TABLE 7.20. Results of MANOVA in the Combined Data Set

MANOVA	Effects Studied: 1—visibility, 2—day			
Effect	Wilks' Lambda	df 1	df 2	p-level
1	0.961175	2	19	0.686476
2	0.877370	2	19	0.288559
12	0.812598	2	19	0.139257

- Independent variables:
  1. Visibility, with 2 levels: visible and invisible.
  2. Day (repeated measures), with 2 levels: day 0 and day 2.

The results are shown in Table 7.20. Note that the p-level for the interaction effect is larger than the significance level. We therefore conclude that there is no interaction effect in these data, and hence any main effect is not modified across the levels of another main effect. The p-levels for the day factor and visibility factor are large enough, so the null hypotheses for these main effects are accepted. This means that both the day (training) factor and the visibility factor do not improve the overall human performance; that is, subjects' path lengths and completion times are not improved (decreased) by providing them with a visible environment or with the opportunity to train and practice.

7.6 DISCUSSION

The study described in this chapter focuses on experimental testing of human performance in tasks that deal with motion planning and require spatial reasoning. The experimental stage was followed with a thorough statistical analysis of the obtained test data. As said in the introduction to this chapter, the motivation for the study was two-prong. First, we wanted to use these data to compare human performance with the performance of robot sensor-based motion planning algorithms described elsewhere in this book. Second, we wanted to foresee the human performance in robot teleoperation systems, with an eye on techniques to compensate for operator deficiencies via a synergistic human–robot operation. To recap our prior discussion, assessing robot motion planning algorithms raises these questions:

- The question of quality of sensor-based robot motion planning algorithms can in principle be addressed in a number of ways. As those options are listed below, we will note that the last option—a comparison with human performance—stand out as more attractive:
  — One can compare actual generated paths with optimal paths. This comparison would make, however, little sense simply because producing

an optimal path requires complete information about the environment whereas our algorithms have only limited sensing information.

— One can assess algorithms' performance theoretically. We have done this in Chapter 3 for the case of a point robot moving in the plane. The upper bounds on the robot performance obtained there give a good idea about the worst-case performance of the algorithms, but they do not answer the direct practical "How good is it in 'normal life'?" question.

— One can attempt a comparison between paths produced by different algorithms in the same task. While some such comparisons have been done in literature, they are of limited value simply because different algorithms tend to behave differently in different tasks: An algorithm that wins in one task can easily lose in another task. That is why the task of choosing between algorithms is a hard one. And, importantly, such a comparison is not feasible for arm manipulators because today there is no competing options to the sensor-based algorithms developed in Chapters 5 and 6.

— One can compare the algorithms' performance with human performance. While we still don't know what algorithms people use in such tasks, from the practical standpoint this would be a satisfying comparison. After all, we humans do solve motion planning problems with uncertainty. We do it all the time, and so using human performance as a benchmark would be an "apples-to-apples" comparison. We tend to associate motion planning tasks with "thinking" and intelligence: If our robots perform well in such tasks, we not only can be proud of the robots' performance but can also use this fact in technical systems.

- If human performance in motion planning tasks turns out to be less than ideal—and the results described in this chapter demonstrate that this is so for tasks with arm manipulators—this conclusion should pose a serious challenge to designers of practical human-guided teleoperation systems. If robot skills in motion planning are better than human skills, and if that is still so after a substantial training by humans, this becomes a good argument for a new design approach in teleoperation systems. Namely, we should attempt to switch to human–robot synergy teams, where human intelligence is complemented with appropriate robot intelligence.

In our tests the performance of human subjects was measured in terms of two dependent variables:

- *Path length*— the length of paths a subject generates in a given task.
- *Task completion time*— the time a subject takes to complete a given task.

The experimental data appear in groups, each related to one independent variable (factor). Overall, four factors have been studied:

- *Task Interface*: Each subject operated either a virtual arm manipulator on the computer screen or a physical arm in the test booth.

- *Visibility* of the Scene: The subject either was given a bird's-eye view of the scene or was forced to "move in the dark."
- *Direction* of the arm motion in the scene (this factor had played an auxiliary role in the study): When moving the arm manipulator, the subject had to move it either from "start" to "target" position or in the opposite direction.
- *Training factor* represented by specific days on which the test was taken: Namely, on day 1 the subjects had moved the arm after only a small perfunctory practice; on day 2 they had a benefit of significant prior training.

The summary of results of statistical processing of testing data appears in Table 7.21 (here, "no" means "no effect"; the corresponding details appear in Sections 7.4.6 and 7.5.3). The table shows effects of each factor on the overall human performance and on each component of human performance, path length and completion time.

TABLE 7.21. Summary of Results in Experiments One and Two

Data Covered	Statistics Test Used	Factors Involved	Effects Found		
Experiment One	MANOVA		Effect for LtoR	Effect for RtoL	Effect for Both
		Interface	Significant	Significant	Significant
		Visibility	Slight	No	No
	ANOVA, left-to-right task		Effect on Path	Effect on Time	
		Interface	Slight	Significant	
		Visibility	Slight	Slight	
	ANOVA, right-to-left task		Effect on Path	Effect on Time	
		Interface	No	Significant	
		Visibility	No	No	
Experiment Two and combined data	MANOVA		Effect in Exp. 2	Effect in Combined Data	
		Training	Slight	No	
		Visibility	No	No	
Experiment Two	ANOVA		Effect on Path	Effect on Time	
		Training	No	No	
		Visibility	No	No	

The motion direction factor is not included in the table since it was added to the study only as a secondary factor, to shed light on the primary factors' effects.

In brief, statistical analysis of experimental data from tests with arm manipulator motion planning indicates the following:

- The interface factor has no significant effect on the length of generated paths, but has a significant effect on the task completion time.
- The visibility factor has no significant effect on human performance—neither on the path length nor on the completion time.
- Similarly, the training factor has no significant effect on the human performance.
- The motion direction factor has a statistically significant effect on human performance.

Overall, these conclusions look rather surprising. Let us discuss these findings and their implications in more detail.

Effects of the Interface Factor. The two components of this factor are the virtual (simulated) interface and the physical (arm in the booth) interface. Simple considerations and expert opinions suggest that this factor should be of much importance. After all, we humans are used to moving physical objects. Manipulating a physical object—here the arm—adds significant haptic, visual, and even auditory information about the task. Plus, the physical arm looks much like a human arm and hence adds to one's confidence. On the other hand, moving an abstract object on the screen seems far less natural. A good many observers and participants in this study had predicted that the subjects would do much better when moving the physical arm than when moving the virtual arm on the screen.

Indeed, statistical analysis of test data agrees to some extent with this intuition: For the task completion time it does show an improvement in subjects' performance. On the average, subjects moved the physical arm in a more continuous fashion, whereas in the simulation they often paused after small motions, spending extra time on figuring out what to do next.

However, an interesting result here was that the improvement was very small in the path length, and that even this small effect was erased by the motion direction factor: In Table 7.21, column "Effect on Path," observe those "slight" and "no" (effect) in the ANOVA left-to-right and in ANOVA right-to-left, respectively. The fact that no significant effect of the interface factor on the path length was found in the more difficult right-to-left task is surprising. It suggests that the importance to a human operator of the type of interface fades as the spatial tasks become harder. To put it bluntly, in nontrivial teleoperation motion planning tasks the operators will likely need help, such as from the robot intelligence; mere improvements in the control means will not go far enough.

The difference in the factor effects on the two dependent variables—path length and completion time—is not hard to explain. The length of a path generated by a human subject is, in general, independent of how quickly or slowly one moves the arm or how continuous its motion is. If the subject stops to think how to proceed, this in itself will not increase the path length, but it will increase the time to task completion. The harder the task, the more thinking the operator needs, and the more time he or she takes to think. The relation of task hardness to the length of path is hence more subtle than its relation to the time to completion. This may be a useful consideration for balancing advantages and disadvantages of virtual versus physical control means in real-world teleoperation systems.

Effects of the Visibility Factor. This factor refers to the obstacles in the environment being visible or invisible to the subjects during the test. The test subjects themselves, researchers, and practicing operators usually think that seeing the robot surroundings would significantly improve their performance.

Interestingly, our study suggests that while this common sense judgment applies to very easy tasks, it does not apply to relatively complex tasks. For the easier task in this study (which is moving the arm left-to-right), there is only a slight difference in the resulting path length and completion time. That is, seeing the environment helped a little in path length and in completion time. On the other hand, for the more difficult task (moving the arm right-to-left,) there was almost no difference in the path length and completion time. This looks puzzling, but becomes less so if one considers that many studies have demonstrated that humans are, in general, not very good in spatial reasoning based on visual data. This fact questions the large resources that are often allocated in telerobotics to help the operator see the scene. It also implies that the operator performance is affected less by the visibility factor than by the human spatial reasoning abilities.

Effects of the Training Factor. This factor has two components that refer to the day of the task execution: day 1, before training, and day 2, after training. When comparing human performance on those two days, with the other conditions fixed, any statistically significant difference should be attributed to the effect of training. Namely, a significant difference would support a common wisdom hypothesis that one's performance should improve significantly after learning from repeated exercise.

This study shows that in arm manipulator motion planning tasks, training has no significant effect on human performance, neither in terms of path length nor in the task completion time. In our tasks the subjects were unable to seriously improve their motion planning skills via training. This is no doubt very surprising. One would expect the opposite conclusion: We all know examples of tasks involving motion where, given enough training, humans become extremely adept; an acrobat on the trapeze is but one example.

There is a big difference, however: The acrobat does a once-and-for-all learned motion, whereas our tasks require constant spatial reasoning. Our test protocols do not allow a subject to simply memorize a task. We want our subjects to learn how to do a class of tasks; we want them to improve their spatial reasoning skills, rather than memorize a specific motion. Examples of positive effect of training in tasks that involve spatial reasoning are harder to come up with. Note that since the tasks given to subjects in this study are quite close to teleoperation tasks, the reported results should be taken seriously by designers of teleoperated systems.

There was one exception from this pattern: The training factor had a slight positive effect on the subjects' overall performance in Experiment Two (see the box "Experiment Two and combined data," Table 7.21; the extent of improvement is only 3.67%). The meaning of this exception is not clear. Given that the effect disappears for the combined Experiments One + Two data (further in the same box in Table 7.21), the small effect of the training factor for the Experiment Two data might be an artifact due to the insufficient data or measurement errors. Or, training may indeed improve—though only a little bit—human performance in motion planning tasks such as ours.

Effects of the Motion Direction Factor.   This factor has two components, left-to-right direction and right-to-left direction of motion. These two tasks took place in the same scene and with the same two-link arm manipulator. The only difference was that in the first task one was asked to move the arm from position S (start) to position T (target) (Figure 7.5, Section 7.2.2), and in the second task one would go from T to S. In this study the motion direction factor happened to have a significant effect on the subjects' performance. In fact, the effect has been stronger than other effects observed. Hence the motion direction factor was included in the study, to help assess the effect of the task difficulty on one's performance, with or without other factors involved.

Using the same scene and the same arm in both tasks has an added advantage that the perceived difficulty of one task over the other is then known to be "in one's head only." After all, a human subject could in principle produce exactly the same path in both tasks, which is what a robot algorithm would do.[12] The unequal difficulty of the two tasks as perceived by the subjects is very interesting. It suggests that human performance is limited by human motion planning skills no less than by the task's objective complexity. As a minimum, it demonstrates a profound qualitative difference between the human and robot algorithms.

Why do human subjects perceive the above two tasks as completely different? It is as if changing the direction of motion to right-to-left produces some additional, if unclear, difficulties; perhaps it adds more possibilities for motion planning or more ways to make mistakes. In Section 7.2.3 we made an attempt to speculate about the reasons affecting human performance in these two tasks (see design comment No. 4 and Figure 7.9).

---

[12]Formally, this is not exactly so. For example, in the example in Figure 3.5, Section 3.3.2, the path shown from point S to point T is produced using the local direction "left." If the same algorithm (here Bug2) now starts from T toward S, using the same local direction, the resulting path will be different from the one shown: It will be complementary to the shown path in that it will pass around parts of obstacles that were not passed when moving from S to T. The same is true for the arm manipulator algorithms discussed in Chapters 5 and 6. The nature of this difference is, however, not the same as in human performance. By simply switching the algorithm's local direction to its opposite, we will obtain a path identical to the one shown in Figure 3.5. Whatever rules guide human motion planning strategies, they must be very different.

To summarize, in tasks that involve spatial reasoning and motion planning for a multilink kinematic structures, human performance changes surprisingly little with the change in external conditions, such as interface, environment visibility, and training factors:

- Common wisdom suggests that seeing the complete scene should significantly help the operator. This study shows that it is not so, at least for more or less complex tasks.
- Common wisdom suggests that moving a physical device in physical space, rather than moving an equivalent abstract object on the computer screen, should significantly help the operator. This study shows that it is not the case. As the only exception, a slightly shorter completion time was achieved with the physical arm manipulator compared to the simulated manipulator. In other words, the subjects made decisions faster with the physical arm, but this did not improve the paths they produced.
- Common wisdom suggests that training should help operators in improving their performance. This study shows that at least within the training protocol of this study (which is a typical protocol in cognitive science experiments), human performance changed very little with training.

The study suggests that unlike factors that common wisdom expects to be influential in human performance (visibility, interface, training), some other "strange" factors, such as a change in the required direction of motion, may have a much stronger effect. One is left with a thought that we humans have not been designed by nature to handle motion planning tasks with objects that have lengths and joints. That, for example, finding one's place in a forest or in an unknown terrain was so essential for one's survival (note a similarity with the point-robot-in-the-maze problem) that the evolution has built appropriate intelligent strategies in our genes for this task. And that the need for moving kinematic structures—objects with lengths and joints—was not nearly as essential.

Thus today not only don't we have such skills, but we find it very difficult to acquire them. This should not be much of a surprise: We know, for example, that evolution made us capable of learning languages with unbelievable ease at a young age but made this same task excruciatingly difficult at an older age. Well, one may say, at least in languages there is the right age. True, there seems to be no right age for learning how to move kinematic structures. Then, we may want to delegate this job to robots, especially if robots can do it better, as shown in prior chapters.

The results described in this chapter look too strange to leave them unchallenged. We need more controlled experiments with human subjects. We need to understand which kind of tasks are harder and which are easier for human operators. As was done in our study, this work requires a collaboration between system designers and cognitive scientists. This will be especially important if projects of critical importance are entertained. As an example, the announced (in 2004) NASA program to create technology that will allow bringing humans to the moon and later to Mars and beyond has a provision for many highly roboticized tasks, in both autonomous and teleoperation robotic settings. Elucidation of the phenomena addressed in this chapter—in particular, the humans' cognition difficulties with spatial reasoning—will likely have a serious effect on design of robotic systems.

One likely practical design strategy coming out of such studies calls for dividing responsibilities between human and robot intelligence, so that both types of intelligence will control the system simultaneously and in a synergistic manner (see, e.g., Ref. 133).

CHAPTER 8

Sensitive Skin — Designing an All-Sensitive Robot Arm Manipulator

I'm tired of all this nonsense about beauty being only skin-deep. That's deep enough.

— Jean Kerr, writer

8.1 INTRODUCTION

Similar to Chapter 7, the present chapter may strike some readers as being some-what out of place in this book. The author agrees: As a confession, the chapter was added to the book after a serious hesitation on the author's part. After all, the primary topic of the book is strategies for sensor-based robot motion planning. Indeed, Chapters 3 to 6 do concentrate on algorithms and software. So what is the logic of suddenly switching now to hardware?

We have a problem, though. On the one hand, yes, strategies and algorithms are what we set out to study in this text. But, once the readers go through those chapters, they will be right to take us to task: How can we be sure that those strategies can be implemented in real systems? After all, those chapters imply that an important prerequisite of sensor-based motion planning algorithms is a whole-body sensing ability—an ability by the robot to sense surrounding objects at every point of its body. This is a hardware component, and such hardware is hardly an off-the-shelf item. Is it even feasible?

We are therefore obliged to convince the reader that the whole-body sensing is indeed feasible. In fact, this is exactly how the research had proceeded: Once some sensor-based algorithms appeared, the work started in earnest on appropri-ate sensing hardware. It became soon clear that the appropriate sensing device should look like a sensitive skin covering the whole robot body. Research in this area started in the late 1980s and has expanded since. A number of sensi-tive skin design projects, very much tied to the issues of robot motion planning with uncertainty, were undertaken in the author's laboratories at Yale Univer-sity in the late 1980s [106, 134] and at the University of Wisconsin—Madison

Sensing, Intelligence, Motion, by Vladimir J. Lumelsky
Copyright © 2006 John Wiley & Sons, Inc.

in the 1990s [135]. These included our own work on whole-sensitive robot manipulators, as well as joint projects with industry and government laboratories (in particular, with Hitachi Corp. [136]—an attempt to develop sensitive-skin-clad robot toys—and with Sandia National Laboratory, US Department of Energy—work on large sensitive-skin based robot manipulators for cleaning chemical and nuclear waste dumps).

The first Sensitive Skin Workshop, convened in November 1999 in Washington, DC under the aegis of the National Science Foundation (NSF) and the Defense Advanced Research Projects Agency (DARPA), brought together about 60 researchers from academia, industry, and government agencies. One surprise at the Workshop was the interest to sensitive-skin-based devices in a number of large Fortune 500 companies. The Workshop created a community of like-minded people, along with an opportunity for discussions and joint work. Details about the Workshop, along with information about some ongoing work by its members, can be found in the book called *Sensitive Skin* [137].

The following years saw an increased activity in this area. Specifically, we should mention works on development of suitable materials for the skin base. While some of this work is lacking explicit references to robotics, its relevance to systems discussed in this book is obvious. In recent years the annual meetings of the Materials Research Society have featured one or more sessions devoted to "smart skins," "smart textiles," and similar topics of obvious relevance to the robot sensitive skin. Interesting works have been reported on stretchable materials (see the discussion on this in the next section) capable of holding (stretchable) wires, and on sensitive skin electronics [138, 139].

An overall diagram of the flow of information in a robot system equipped with the sensitive skin is shown in Figure 8.1; in the figure, software blocks are shown by straight rectangles, and hardware items are shown by curved rectangles. As the robot moves in the course of executing its task, it is expected to avoid obstacles on its way. It is the sensitive skin covering its body that will be informing the robot of the approaching obstacles. (Which is not to say that other sensing means, such as vision, would not be of use in this process.) To secure this information, the robot control system continuously interrogates the whole array of skin sensors.[1] The sensor sampling rate and the corresponding hardware are designed so as to fit comfortably within the update rate of the robot actuators—which for the robot system described in this chapter is 50 *position points*, or sampling cycles, per second, or 50 Hz. (This figure is quite typical for today's complex robots.)

A position point is a set of n numbers that describe increments in the robot actuators (motors, degrees of freedom) that have to be executed in order for the robot to arrive at the desired position at the next step. For our robot, $n = 6$ (which is, again, a rather typical number in today's arm manipulators). As the robot simultaneously executes those six increments, a step motion along the robot path is generated, and so on, 50 steps per second, producing a continuous motion.

---

[1]This is not unlike the human control system, which monitors the sensing information from the skin sensors.

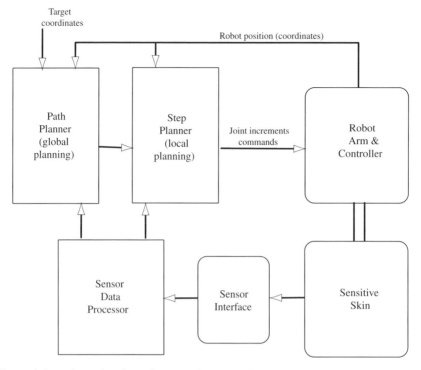

Figure 8.1   Information flow diagram of the sensitive skin-based robot control system.

In each sample cycle, information obtained from the skin sensors is passed to the Step Planner, a control unit responsible for local planning—that is, planning of individual motion steps. As common sense would dictate, only information from sensors that sense something in front of them is passed to the Step Planner. A step that will be made based on this information should (a) be such as to help the robot avoid collision with sensed obstacles, and (b) be reasonable from the standpoint of the robot's overall motion plan. The latter function is done by the Path Planner unit (see Figure 8.1). The Path Planner makes sure that each step is implemented according to the sensor-based motion planning algorithm used. (More detail on the overall scheme can be found in Ref. 115.)

As discussed in prior chapters, motion planning algorithms' requirements to the whole-body sensing include two major properties:

(a) *full coverage*, which refers to the robot's ability to detect a contact between the robot and a close-by object at every point of the robot body, and

(b) *locality identification*, an ability to infer which specific points on the robot body are involved in the contact.

Here "a close-by object" means that the distance between an object and the robot is small enough so as to require the robot to act on it in order to avoid a collision. How close is close enough depends on a number of variables, such as the robot mass, velocity, and the agility of the robot sensors. The term "contact" will thus cover a range of distances from a physical contact (a zero distance) to distant sensing, and it can therefore include tactile as well as various proximal sensors—infrared, ultrasound, capacitance, vision, and so on. Infrared sensing has been used in the system described in Sections 8.3 and 8.4.

The full coverage requirement makes one think of a sensing hardware that would be akin to a *sensitive skin*, enveloping the whole robot body the way a human skin envelops our bodies. Technically, this may be a real skin or a "virtual skin"—that is, a set of sensors that together provide information about the space around the robot's whole body.

Besides the full coverage and locality identification properties of robot sensing, the designer of a sensitive skin system has to worry about its other characteristics, such as these:

- Reliability
- Accuracy
- Resolution
- Tactile or proximal sensing?
- Ability to measure distances
- Physical principle of action: force, vision, infrared, capacitance, ultrasound, etc
- Sensors' physical shape and dimensions
- Control electronics

In the following sections we will consider details and implications of these characteristics for the robot-sensitive skin system, followed by a brief description of one such system developed and installed on a large arm manipulator, along with examples of its operation.

8.2 SALIENT CHARACTERISTICS OF A SENSITIVE SKIN

Reliability. Robot sensors must be reliable: We do not want our robot to bump into an object that its sensors "did not notice." To provide the full coverage, many sensors will likely be needed, thousands or even millions of them. (Later we will address the question, Why is having many sensors on the skin better than having fewer sensors, even if fewer sensors could do the coverage?)

Each sensor is a single device. The common wisdom says that the more devices, the bigger the chance that some of them will misbehave or die. Notice that the latter does not necessarily mean worse reliability. If sensors on the skin do their work in parallel, and if more than one sensor can functionally cover every point of the robot body (thus providing a *system redundancy*), then more misbehaving elements does not necessarily mean a less reliable system.

The system will be more reliable if it is capable of *self-diagnostics*, *self-healing*, and *graceful degradation*. Self-diagnostics is a well-developed discipline in system design. Our system would continuously poll all its sensors and inform the control unit of detected deviations from the normal. Self-healing implies an automatic repair of the failed hardware. On the level of sensors today, this feature is available in live nature but not in technology. The purpose of graceful degradation property is to avoid the need to shut off the whole system in case of losing a few elements. In our case, graceful degradation refers to the system's capacity to direct sensors in the vicinity of a broken sensor to take over its job (see, e.g., Ref. 140). Doing this will effectively increase the distance between well-functioning sensors; while slightly degraded, the system will still function. As more sensors die, the system will degrade "gracefully."

Accuracy. Sensing information has to be accurate. Sensor accuracy is defined by the difference between the sensor measurement and the actual value that the sensor attempts to measure. How accurate is accurate enough? The answer depends on many details, including the sensor types and density on the skin, and on the robot's mass, kinematics, and maximum speed. Imagine that our robot arm manipulator is equipped with short-range proximity sensors capable of detecting an object when the distance to it from the robot falls below 20 cm. The arm is a heavy body made of steel; its inertia is high. Imagine that at its maximum speed the robot control system can guarantee a successful collision-avoiding maneuver only if the distance to an object on the robot's way at the time of detection is no less than 15 cm. This means that we would not be happy with a sensor whose accuracy is ±5 cm, because with it we would run a risk of the arm colliding with surrounding objects.

Resolution. This characteristic refers to precision with which the robot can pinpoint the location of an object that appears in the vicinity of the robot body. The better the robot knows the location of an object that it tries to avoid, the better the *dexterity* of its motion and, hence, its chance to accomplish its task in a workspace filled with obstacles. Sensor resolution is tied to the sensor accuracy, but it characterizes the whole sensor system rather than a single sensor.

As an example, consider a sensitive skin that is based on proximity infrared sensing. Unlike passive sensors like a vision camera or a temperature sensor, an infrared sensor is an active sensor: It contains (a) a light-emitting diode (LED) that sends a ray of infrared light in space and (b) a detector that sits right next to the LED and detects the light's reflection from an object in front of the sensor. For better resolution we are interested in a sensor that would send light in a narrow cone. Such sensors, each with a tiny lens in front of the LED in order to produce a cone-like light ray, are common (see Figure 8.2a). They may have a limited sensitivity distance, say 20 or 30 cm. For full coverage, sensors are spaced on the skin so that their sensitivity cones overlap, forming a continuous sensitivity cushion around the robot. The fact of detecting a reflected signal thus means that the object in question is wholly or partially within the sensitivity cone and sensitivity distance of a given sensor.

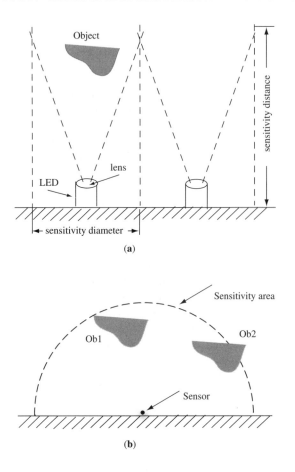

Figure 8.2   (**a**) Scheme with an infrared sensor. (**b**) Scheme with a capacitance sensor.

Assume for simplicity that the sensor sensitivity area is not a cone but a cylinder (Figure 8.2a). Let us say, this cylinder is of diameter 2 cm. Since the robot can easily know each sensor's exact location on its body, it will know the location of the obstacle relative to its body (recall the locality identification feature, Section 8.1) and will be able to use this information to maneuver around the obstacle.

To provide a full coverage of the robot body, a high enough density of sensors on the skin is needed. In our example the density of sensors is one sensor per 2 cm, which is the distance between the neighboring sensors, and also the skin resolution. (Actually, sensor placement should be here a bit more dense for this resolution, but for the sake of simplicity let us keep it at 2 cm). Providing this density will result in a large number of sensors on the skin. For a typical industrial robot manipulator (roughly of the size of a human) this density will require 1200 to 1500 infrared sensors. An example of the sensitive skin described later in

Section 8.4 has roughly this density of sensors. With this sensitive skin, a round object 10 cm in diameter will be perceived by the robot as about 13–14 cm in diameter. This also means that a robot equipped with this skin will be able to move past the obstacle at about 2 cm distance from it, but not closer—otherwise it risks a collision.

What this resolution signifies is that two objects located at a distance about 2 cm from each other may be perceived by the robot as one obstacle. Imagine that the robot arm contemplates moving between two obstacles: The diameter of the arm links is d; the distance between the obstacles is 2 cm $+ d$. Then, based on the information from the skin, the robot will not attempt to pass between the two obstacles, although it actually could. It is insufficient skin resolution that will make the robot miss the passage.

In real life the 2-cm resolution would be quite good; cases with obstacles requiring motion as tight as in the example above would be rare. The point here is that the skin resolution affects the dexterity of robot motion in the most direct way.

Note also the quadratic dependence between the skin resolution and the number of sensors on the skin: For example, decreasing in half the distance between neighboring sensors will increase fourfold the number of sensors on the skin.

Consider now another example, a skin with capacitance sensors. A capacitance sensor is a passive sensor: It works by measuring properties of the electric field that the sensor's two electrodes create. An object entering the field changes the field characteristics, and this allows the sensor to detect it. As with any electric field, the detection effect depends on the object's material and its distance to the sensor. The sensor's sensitivity area can take different shapes depending on the shape and mutual positions of the sensor's electrodes. The sensitivity area of the sensor shown in Figure 8.2b is a hemisphere that extends outwards from the robot body, with the sensor at its center.

With the sensitivity area sphere of radius, say, 10 cm (about 4 in.), a detection signal from the sensor will tell the robot that an object has entered a hemisphere of diameter 20 cm centered at the sensor. In Figure 8.2b, the robot will not know where within the sensitivity area the object is because an object in the position Ob1 or Ob2 or anywhere else within the sensitivity area may generate the same signal. To be on the safe side, the robot will have to conclude that the object occupies the whole sensitivity area. That is, an obstacle of 2 cm in diameter will become in the robot's "mind" an "obstacle" of over 20 cm in diameter. When planning its motion past the 2-cm obstacle, the robot will have to leave a large margin between itself and the obstacle, which is equivalent to suddenly increasing the dimension of every obstacle by 20 cm in every direction. That is, this capacitance-sensitive skin will effectively make the robot workspace dramatically more crowded than it actually is. Someone diving into a dirty pool without a diving mask will likely have a better vision resolution. The robot's bad maneuverability here is not the robot's fault—it is just that its sensors have unacceptably low resolution.

Compared to the 1200 to 1500 infrared sensors above, achieving a full coverage for the same large arm manipulator with our capacitance sensors will require

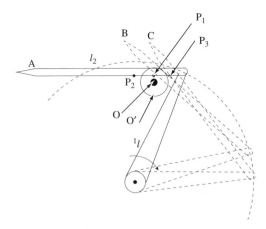

Figure 8.3   A higher density of sensors on the robot arm body translates into the better dexterity of the arm's motion.

roughly 10 sensors—perhaps a little more due to specifics of the robot shape. This is a great savings in cost and design simplicity compared to the infrared sensitive skin—but the resolution of this skin is much poorer, only 20 cm compared with the 2 cm resolution in the infrared skin.[2]

Consider the planar two-link arm manipulator shown in Figure 8.3. The body of this arm is covered with proximity sensors.[3] In its current position A (shown by solid lines), link l_2 is positioned near the obstacle O (a solid black circle). Assume that, following its motion planning algorithm, the arm intends to slide along the obstacle, with link l_1 rotating clockwise as indicated by the arrow. If an arm's sensor tells it (correctly) that the point on its body closest to obstacle O is P_1, then the next position of link l_2 would be position B, with the link l_1 position calculated accordingly (dotted lines in Figure 8.3). If, however, because of its sensors' limited resolution the arm concludes that the point of its body closest to obstacle O is somewhere at point P_3, safety considerations will make the arm move its link l_2 in the position C instead of B, which is equivalent to its reacting to a much bigger obstacle O'. Increasing the number of sensors and placing them, for example, in each of the points P_1, P_2, and P_3 would improve the skin resolution and be effectively equivalent to making the arm workspace less crowded.

Tactile or Proximal Sensing? When moving around, humans and legged animals prefer to use proximal sensing, usually vision. (More rarely we also use auditory or olfactory (smell) sensing information to guide our motion planning.)

---

[2]In theory, the resolution can be improved with extra processing of signals obtained from a few neighboring sensors. This would likely require additional assumptions about the shapes and orientations of obstacles, along with a more complex data processing scheme.

[3]In Figure 8.3 sensors are distributed along the robot's one-dimensional contour; in a real three-dimensional arm, sensors are of course distributed along the robot body's two-dimensional surfaces.

Imagine you walk into a room. It is an evening. Suddenly the lights go out. It is pitch dark. You stop for a moment, reconsider your plans, and then resume your motion—except now your movement pattern changes dramatically: You move much slower, keeping your knees and the whole body slightly bent; perhaps you outstretch your hands forward and slightly to the sides. Your motion is now guided by your tactile sensors. Your slow speed is a clear demonstration that the efficiency of our tactile sensors is not as good as of our vision.

But our tactile sensors play a much more important role in our lives than this example suggests. We use them almost continuously, sometimes in parallel with our vision and sometimes as a sole source of information. We touch objects; we keep turning in the chair depending on what the tactile sensors in the back tell us; we measure the comfort of our shoes by what our foot tactile sensors feel; we take pleasure at stroking a child's head or a fur coat. Our tactile sensing is an important component of that pleasure. We often use our tactile sensors in situations where vision would be of little help, as in the example with shoes.

In fact, while we all know that people can live without vision—we see blind people having productive lives—science tells us that humans cannot live without at least some capacity for tactile sensing. A person with no tactile sensing cannot even stand: Tactile sensing is actively used in maintaining the standing balance. Diabetes patients—who often lose partially their tactile sensing—are warned by their doctors that they should be extremely careful in their interaction with the environment.

Turning again to the moving-in-the-dark example, the reason you moved so slowly in the dark has to do with an important side effect of tactile sensing: You cannot know of an impending collision up until the moment the collision takes place. Once your hand bumps into some object on your way, you will stop, think it over, and modify the direction of your movement as you see fit.

But your body has a mass—you cannot stop instantaneously. Regardless of how slowly you are moving at the moment of collision, the "stop" will still cause a sharp deceleration of your body and forces at the point of collision. For a tiny fraction of time, your body will continue moving in the direction of your prior motion. This residual motion will be absorbed by the soft tissue of your hand, and so the collision will cause no serious harm to your body. In fact, the speed you have chosen for moving in the dark was "calculated" by an algorithm that has been refined by many bumps and pain in your childhood: Experience teaches us how slowly we should move under the guidance of tactile sensors in order to prevent a serious harm from possible collision.

Today's robot arm manipulators have no similar soft tissue to absorb forces. Their bodies are made of steel or aluminum, sometimes of hard plastic. If the robot body were to move under the guidance of tactile sensors, bumping into a suddenly discovered obstacle would spell a disaster. Once the arm collided with an obstacle, it would be too late to carry out an avoiding maneuver: Infinite accelerations would develop, and an accident would ensue. A theoretical alternative—move so slowly that those forces would be contained—is simply not realistic.

Then, why not use vision instead? The short answer is that since an arm manipulator operates in a workspace that is comparable in size to the arm itself, vision will be less effective for motion planning than proximity sensing that covers the whole arm. By and large, humans and animals use whole-body (tactile) sensing rather than vision for motion planning at small distances—to sit down comfortably in a chair, to delicately avoid an overactive next-chair neighbor on an aircraft flight, and so on. See more on this below.

This discussion suggests that except for some specific tasks that require a physical contact of the robot with other objects—such as the robot assembly, where the contact occurs in the robot wrist—tactile sensing is not a good sensing media for robot motion planning. Proximity sensing is a better sensing candidate for the robot sensitive skin.

Ability to Measure Distances. When the robot's proximal sensor detects an obstacle that has to be dealt with to avoid collision, it is useful to know not only which point(s) of the robot body is in danger, but also how far from that spot the obstacle is. In Figure 8.4, if in addition to learning from sensor P about a nearby obstacle the arm would also know the obstacle's distance from it—for example, that the obstacle is in position O and not O'—its collision-avoiding maneuver could be much more precise. Similar to a higher sensor resolution, an ability to measure distances to obstacles can improve the dexterity of robot motion. In mobile robots this property is common, with stereo vision and laser ranger sensors being popular choices. For robot arms, given the full coverage requirement, realizing this ability is much harder.

For example, at the robot-to-obstacle distances that we are interested in, 5 to 20 cm, the time-of-flight techniques used in mobile robot sensors are hardly practical for infrared sensors: The light's time of flight is too short to detect

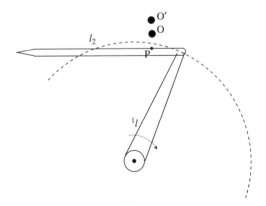

Figure 8.4 Knowing the distance between the robot and a potential obstacle translates into better dexterity of the arm's motion.

it. Ultrasound sensors can do this measurement easily, but their resolution is not good.

One possible strategy is to adhere to a binary "yes–no" measurement. In a sensor with limited sensitivity range, say 20 cm, the "yes" signal will tell the robot that at the time of detection the object was at a distance of 20 cm from the robot body. The technique can be improved by replacing a single sensor by a small cluster of sensors, with each sensor in the cluster adjusted to a different turn-on sensitivity range. The cluster will then provide a crude measurement of distance to the object.

Sensors' Physical Principle of Action. Vision sensing being as powerful as we know it, it is tempting to think of vision as the best candidate for the robot whole-body sensing. The following discussion shows that this is not so: Vision is very useful, but not universally so. Here are two practical rules of thumb:

1. When the size of the workspace in which the robot operates is significantly larger than the robot's own dimensions—as, for example, in the case of mobile robot vehicles—vision (or a laser ranger sensor) is very useful for motion planning.
2. When the size of the robot workspace is comparable to the robot dimensions—as in the case of robot arm manipulators—proximal sensing other than vision will play the primary role. Vision may be useful as well—for example, for the task execution by the arm end effector.

Let us start with mobile robot vehicles. When planning its path, a mobile robot's motion control unit will benefit from seeing relatively far in the direction of intended motion. If the robot is, say, about a meter in diameter and standing about a meter tall, with sensors on its top, seeing the scene at 10–20 meters would be both practical and useful for motion planning. Vision is perfect for that: Similar to the use of human vision, a single camera or, better, a two-camera stereo pair will provide enough information for motion planning. On the other hand, remember, the full coverage requirement prescribes an ability to foresee potential collisions at every point of the robot body, at all times. If the mobile robot moves in a scene with many small obstacles, possibly occluding each other and possibly not visible from afar, so that they can appear underneath and at the sides, even a few additional cameras would not suffice to notice those details.

The need for sensing in the vicinity of the robot becomes even stronger for arm manipulators. The reason is simple: Since the arm's base is fixed, it can reach only a limited volume defined by its own dimensions. Thinking of vision as a candidate, where would we attach vision cameras to guarantee the full coverage? Should they be attached to the robot, or put on the walls of the robot work cell, or both?

A simple drawing would show that in any of these options even a large number of cameras—which is impractical anyway—would not guarantee the full sensing coverage. Occlusion of one robot link by another link, or by cables that carry power and communication lines to the end effector, or by some other objects is hard to avoid when sensing the robot surface from a distance. Robot links often have nonconvex concavities, indentations, holes, bolts, or other pieces sticking out of them. Seeing behind every nook and cranny at all times is simply not practical. Sooner or later, some object will be hidden from all those cameras. Short of unreasonable, no number of cameras on the surrounding walls or on the robot body will do the job.

Another consideration is that for a single-step motion planning decision (and there will be 20 to 50 such steps per second) the robot needs information on all nearby obstacles simultaneously. Doing vision processing simultaneously for a significant number of cameras is too computationally expensive. We must conclude that, powerful as it is, vision is not a right solution for protecting the robot body at short distances.

Nature, of course, "noticed" this fact long ago. While supplying us with the powerful stereo vision, evolution has also supplied us with other sensors to help protect our bodies when moving in space. It gave us, in particular, the tactile sensing of our skin. The nature "concluded," in other words, that vision is not a good sensor to protect one's own body at short distances. In combination with vision and with the effect of soft tissue force absorption discussed above, tactile skin provides a rather universal protective sensor.

If so, one may ask indignantly, why hasn't the evolution been gracious enough to supply us with something better than tactile sensors—covering our skin, for example, with some proximal sensors? Then our life would be so much safer, and we would be able to move so much faster in the dark than we do now with our tactile sensing.

Unfortunately, proximal sensors that we find in nature do not fit our purpose. A bat's sonar is one example: Acting as a substitute for vision, at distances much larger than the bat's body, sonar does not protect the bat's body at very small distances. For this purpose, bats have sensitive skin. Cat's whiskers are another example: While whiskers work on a physical contact, they supply the cat with input information far enough from its body to allow for motion planning decisions typical of a proximity sensor performance. (And again, cats still need their tactile sensitive skin.) We humans have proximal sensing as well: Besides vision, we have hearing, smelling, and temperature sensing. Of these, temperature sensing is the only type of proximal sensing that appears in one's whole body and hence satisfies our requirement of full coverage. It also operates at a range of temperatures and distances: We sense a hot cup at a few centimeters' distance, and we can sense volcano lava from a distance of many meters. Unfortunately, the range of temperatures in the world around us makes temperature sensing of a limited use.

The list of sensors provided to us by technology is much bigger. Engineering progress moves in ways very different from nature. The proverbial inability of the evolution to invent a wheel does not stop there: Engineers have a whole panoply of proximity sensors that are not available in nature. Many of these—infrared, capacitance, and ultrasound sensors are but a few examples—can be used for a skin-like full coverage for robots.

Sensors Physical Shape, Dimensions, and other Physical Properties. The diameter of a link of a typical robot manipulator ranges from a few centimeters to 20–25 cm. The link diameter of the NASA Shuttle Remote Manipulator System (SRMS), likely the biggest robot arm built so far, is about 40 cm. Some arm links are short, and some are very long. Proximity sensors that we choose to cover the arm should satisfy some reasonable physical properties:

1. The sensing skin should not add significantly to the robot link diameter. What is or is not "significant"? A skin that is 1–2 mm in thickness will likely be acceptable for most arm manipulators. Many existing sensors and other necessary electronic components fall into this range. Today's surface mounting technology allows one to put those components on the skin board with only a tiny addition to the skin thickness. Future large-area electronics technology will allow printing skin sheets in a manner we produce today newspaper or wallpaper sheets.

2. If the skin base is to be a continuous medium—which is highly desirable for a high-density skin—it should be designed on a flexible carrier, so that the skin can be wrapped around robot surfaces of various shapes. To make it scalable and easier to install, the skin can be designed on separate more or less self-contained circuit board modules. Each module can include, for example, n-by-n sensors plus the related control electronics. The skin could then be extended functionally and spatially by tiling the modules to cover large surfaces.

3. Look at your own arm. When you bend it, the skin on the elbow stretches. When you stretch the arm, the elbow skin shrinks and forms wrinkles. Having the stretchability property is as important for the robot sensitive skin as it is for the human skin. In a skin built on unstretchable plastic material, every time a robot joint makes the adjoining links bend (similar to the human elbow), a gap will appear between the parts of the skin belonging to both links. The exposed part of the robot body will then lose its sensing ability and become vulnerable to the dangers of the surrounding unstructured world.

Note that having a stretchable sensing module implies stretchable wires in it, which is quite a difficult technical problem in itself. No materials fitting the needs of a stretchable sensitive skin exist today. The sensitive skin sample described later in this chapter does not have the stretchability property: Less "natural" means, such as parts of the skin that slide over each other as the robot links move, are used to compensate for the unstretchable skin material. A new and very interesting area of research in stretchable materials for sensitive skins belongs mostly to the disciplines of material science and chemical engineering (see, e.g., Ref. 137).

4. Attaching a flexible skin board to some surface may require cutting off pieces of the board. For example, if a part of the robot surface happen to be of spherical shape, a planar skin board cannot be attached to it without cutting off some portions. The board design should allow such cutting, at least to some limited degree. One problem with this is that while sensors cover the whole

board, the local control electronics occupies only a small physical area in it. We obviously don't want to cut off pieces of the board that contain control electronics. This suggests that the control electronics should be put on the board so as to simplify the cutting for typical surfaces. Another related problem is that, electrically, sensors present a load on control electronics. Cutting off some sensors changes that load, and so the control electronics should be able to handle this.

5. The arm's interaction with its environment brings additional constraints. Consider an environment where the robot arm may be hit by sharp hard objects. Without extra precautions, this environment will likely rule out an infrared-sensitive skin: Whereas these sensors have enviably high resolution and accuracy, the tiny optical lenses sitting in front of every sensor make them brittle. A better option then may be capacitance sensors: While not particularly accurate, they are quite rugged.

On the other hand, covering the infrared sensitive skin with a layer of transparent epoxy or a similar compound may still warrant its use in a harsh environment. The epoxy will pass sensors' optical beams while mechanically protecting the skin from the environment. This measure would also help in tasks where the arm is periodically covered with dirt and has to be washed, such as in cleaning chemical and nuclear dump sites. Because the content of such sites presents a danger for human workers, robots are good candidates for the cleaning job.[4]

Often the material that is to be evacuated from cleanup sites is inside large metal or concrete tanks. The robot arm has to enter the tank through a relatively small opening. Careful motion planning for the whole body of the arm is very important: A small deviation from the opening's center can spell a collision, and this may happen at various points of the robot body, depending on how deep into the tank opening the arm has to move. The operation calls for dextrous motion, which in turn requires a good resolution of the sensitive skin. Infrared sensors provide the requisite characteristics; the problem is, however, that sensors on the skin will be quickly covered with dirt. A transparent layer of protective epoxy will allow one to quickly wash off the dirt from the arm.

6. Specific applications can add their own constraints on the choice of sensitive skin components. Given their decent accuracy and physical ruggedness, arrays with tiny sonar sensors may be a good candidate for the skin. A sonar-studded sensitive skin cannot be used, however, in space applications, for the simple reason that sound does not spread outside the atmosphere.

The above need to wash off dirt from the skin is also such a constraint. Another example is applications with unusual levels of radiation. Space robots must be able to withstand space radiation. Hence only radiation-hardened components will do the job for a sensitive skin intended for space applications.

Control Electronics. Depending on the physical principle of sensors chosen for the sensitive skin, appropriate control schemes must be chosen. Ordinarily, skin

---

[4]The multi-billion cleanup Superfund project in the United States in the mid-1990s had a provision for utilizing robotics.

sensors produce analog electric signals. Before those signals are passed to the robot computer and used by motion planning algorithms, they have to be cleaned of noise, perhaps brought to some standard form, and turned into the digital form using an analog-to-digital transformation. This is done by the skin control electronics. Ideally, this could be done by an appropriate tiny control unit built into each sensor.

Today an electronic control unit will likely handle a group of sensors, say an n-by-n sensor subarray, thereby allowing an easy scaling up of the skin device. The unit also takes care of polling the whole subarray, identifying sensors that sense something in front of them, collecting information about their physical coordinates on the robot body, and passing this information to the robot "brain" for making decisions on collision-free motion. How often the polling is done depends on the robot joint motors sampling rate: 20 to 50 times per second are typical polling frequencies for large arm manipulators. Larger groups of sensors and control components are united under the control of local computer micro-processors, forming a hierarchical control system. Such architecture frees the "brain" computer for more intelligent work, and it allows scaling up the system to practically any number of sensors on the skin.

We now turn to an example of implementation of the sensitive skin concept. Space shortage will not allow us to cover all the questions that an electronics professional may have. Appropriate references will be given. The intent here is to give an idea of how the sensing skin hardware can be approached.

8.3 SKIN DESIGN

The large-area skin versions built so far are all based on optical (infrared, IR) sensors; other sensors are still waiting for their implementation in a sensitive skin. The main reason for choosing infrared sensors is the best resolution one can get with them compared to other sensors. This advantage may overweigh the drawbacks of IR sensors, such as their mechanical brittleness or their inability to measure distances at a short range. Other than this similarity, the projects carried out so far have differed in the specifications of sensors and other electronic components, in overall physical and electrical architecture of skin sections, implementation of the control scheme and robot intelligence, the mechanical installation of components on the skin (such as direct soldering or surface mounting), and so on. (For details, see references in Section 8.1 and citations therein.)

As mentioned above, an infrared sensor is an active sensing device. Each sensor presents a pair consisting of a light-emitting diode (LED) and a light detector. When initiated, the LED sends in space in front of it a beam of directed infrared light. The associated light detector detects the reflected light. If a noticeable amount of reflected light has been detected, the system assumes it was reflected from an object located in front of the sensor.[5] The LED light beam is of a conical

---

[5]In principle, a signal detected by the detector in the sensor pair X can be the light sent by an LED of some other sensor pair Y and reflected "in a wrong direction" by an object positioned in front

shape, formed by a tiny lens on the top of the LED (Figure 8.2a). The beam cones of neighboring LEDs must overlap, forming a continuous detection cushion in the space around the robot.

To increase the skin reliability, it is desirable to decrease the amount of wiring running within sensor modules, between modules, and especially between modules attached to different robot links (because these wires will have to run over robot joints). This requirement is in conflict with a desire to control every sensor independently. The latter requires parallel addressing of sensors, hence many wires, whereas a serial addressing scheme allows one to minimize the number of interconnecting wires. Another advantage of a parallel scheme is that sensing information it produces in each cycle is known to correspond to the same time moment, hence the same position of all robot links. With the serial polling scheme, the sensing information obtained from polling sensors corresponds to the robot links being in slightly different positions. The motion within one serial polling cycle is usually insignificant: The actual uncertainty depends on the serial scheme implementation and the robot speed.

A fully parallel scheme with n sensors requires roughly $\log_n$ wires. In a fully serial addressing scheme, only one wire will be sufficient to do the job. In the system described here, this conflict is resolved via a compromise parallel–serial system: The system is divided into modules that are run in parallel, whereas sensors in each module are divided into rows and columns and addressed serially.

Sensor Interface. The purpose of the sensor interface circuit (Figure 8.1) is to realize computer access to the skin sensor. The circuit's two major components are an analog-to-digital converter and a number of one-shots that control sensor addressing. In each sensor module, sensors are addressed in a serial fashion. The entire skin is reset regularly, synchronizing address counters of the sensor modules. (More information on a version of this unit appears in Ref. 134.)

Sensor Circuit Module. A sensor circuit module contains a group of sensors that, from the standpoint of control and mechanical design, are handled as a unit. A number of sensor modules makes the whole skin. The skin system described in Ref. 134 and shown in Figure 8.6 included three sensor modules, each with a different geometric shape and with an unequal number of sensors, totaling about 500 sensors. A later system described in Ref. 135 and shown in Figure 8.7 featured smaller standardized modules, each about 23 by 23 cm in size and with 8 by 8 sensors, with the whole system totaling over 1200 sensors. Each module is wrapped around and fastened to the robot arm. Neighboring modules are connected physically, using appropriate fasteners—such as Velcro fasteners—and electrically, through appropriate connectors.

Besides sensors, each module contains all necessary control electronics. The latter can be divided into two parts. The first part is a sensor addressing circuit,

---

of the pair Y. This scenario suggests an interesting hardware and processing schemes that would be checking for various combinatorial possibilities, to determine which object actually triggered the signal. No such attempts have been done so far, to my knowledge.

which decodes the order of sensor addressing. The second part is a sensor detection circuit, which amplifies and filters signals from the light detectors.

The addressing scheme is organized as follows. Each sensor module has a counter that keeps track of which sensor is being addressed currently. The counter is incremented by a clock, causing selection of a new sensor. When needed, the counter is set to zero by a long pulse from a pulse discriminator. In the earlier system, pulses longer than 10 μs are considered zero reset pulses; pulses shorter than 10 μs increment the counter. This addressing scheme allows one signal line to address a practically unlimited number of sensors.

Besides its serial nature, an obvious drawback of this scheme is that it does not allow random addressing. When picking a particular sensor, all sensors with addresses lower than this sensor will be selected. Note, however, that this is not a serious drawback, because by the nature of the skin all sensors must be addressed in turn in each cycle of sensor polling. The order in which sensors are addressed is immaterial, and so the advantages of serial addressing outweigh its disadvantages.

The sensor module circuit implemented in Ref. 134 is shown in Figure 8.5. In brief, it operates as follows. The Sensor Select signal from the Sensor Interface is first "cleaned up" by triggers IC8b and IC8c and is then passed to the "Clk" input of the 8-bit counter that keeps track of selected sensors. The function of the pulse discriminator IC6 (a dual one-shot) is to choose the time of resetting the counter.

In the pulse discriminator, when the Sensor Select line is "low," the one-shots' outputs "Q" are low, and the 8-bit counter is not reset. As a pulse arrives on the Sensor Select line at time T_a, the output "Q" of the one-shots IC6a is triggered high. If the Sensor Select line stays high longer than 10 μs, IC6a will time out, causing its output to go low at time T_b. This triggers IC6b, and its output "Q" goes high, resetting the counter. If, on the other hand, Sensor Select signal goes low before IC6a times out, no reset pulse is generated and the counter increments normally.

The infrared diode (LED) light is amplitude-modulated and then synchronously detected, to increase the system immunity to other light sources. This scheme allows operation on several "channels": For example, light transmitted by an LED on the robot link X will not be sensed by a detector on a link Y even if directly illuminated by it.

The output byte 'Out' of IC7 controls analog multiplexers that switch optical components in the sensor circuit. The least significant four bits are connected to the analog multiplexer IC2, which selects signals among the 12 preamplifiers on the skin. The analog signal is first high-pass filtered by IC1a to remove noise due to the ambient (room) light, then passed to the synchronous detector ICb, which demodulates the transmitted signal, and then low-pass filtered by a three-pole Butterworth filter composed of IC1c and IC1d. The IC1d output is then passed to one of the input channels of the Sensor Interface Board via a resistor, which provides short-circuit protection for the IC1d's output.

The setting time of the Butterworth filter is about 0.25 m, which determines the overall scheme's response time. A higher bandwidth filter would settle in less

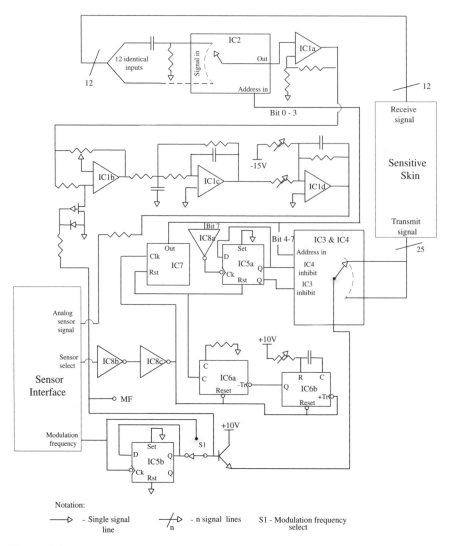

Figure 8.5   Sensor circuit module. Most notations are conventional notations found in electronics literature. Some details on components appear in this section; more detail can be found in Ref. 134.

time, but would also make the circuit more noise prone. Since the skin sensors on the arm links are polled in parallel, the limiting factor in the sensor update rate is the polling of the largest skin sections, which in this skin version happens to be on the robot arm link l_3. The resulting sensor polling rate makes the whole skin polled every sixteenth of a second. A bit faster rate—for example, equal to the arm's 25-Hz sampling rate—may be desirable, but it would require a more complex design and was deemed unessential.

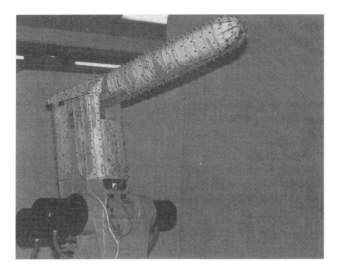

Figure 8.6 The first sensitive skin (mid-1980s). Skin sheets have been custom designed to fit this specific General Electric robot arm. The system has over 500 infrared proximity sensors.

Switch S1 selects the operating frequency of a particular sensor module. By selecting the output of IC5b, the circuit operates at 67.5 kHz, or one-half the frequency of the signal (modulation frequency) from the sensor interface circuit. The lower frequency (67.5 kHz) is used for the link l_2 skin section, and the higher frequency (135 kHz) is used on the skin of link l_3. Note that from the standpoint of motion planning these two links comprise practically the whole arm (Figure 8.6).

The Skin. The skin base is manufactured from a plastic material called Kapton, by Dupont Corporation, and is 0.26 mm (0.0085 in.) thick. (See Section 8.2 for considerations affecting the choice of base material for the skin.) Both sides of the material are copper-clad, resembling the (much thicker) inflexible material commonly used for regular printed circuits. After processing, the board provides both the necessary structural support and electrical interconnection for electronic components.

Once the skin design concept has been finalized based on preliminary experiments, skin sections have been designed using common CAD-CAM software. The actual production was done in a shop that had expertise in producing complete circuit boards on flexible materials like Kapton; a number of such shops have appeared in the United States in recent years.

8.4 EXAMPLES

The skin that covers the industrial robot arm shown in Figure 8.6 was built in 1985–1987. It is the first robot sensitive skin system ever built, and the robot system equipped with it was the first whole-body motion planning system. It is a custom-built skin, designed specifically for the robot shown. The skin consists of three large sections covering the second and third link of the robot arm—which are by far the robot's biggest links, so covering them effectively takes care of all robot parts that are subject to collision. (Motion planning still has to be done for the whole robot major linkage, in three-dimensional space.) Common components and simple soldering was used to produce the skin; its most protruding parts, LEDs, make the skin rather thick, about 6 mm.

The robot arm "under the skin" is a General Electric industrial arm manipulator. The skin resolution (distance between neighboring sensors on the skin) is about 5 cm, resulting in slightly more than 500 sensors total on the skin. The cuts in the front part of the skin were necessary to cover the robot's roundish endpoint. Note also that no hand or other end effector appears in the figure. As discussed in prior chapters, motion planning for the end effector (that is, the robot's minor linkage) is treated as a separate task, which is usually much easier due to the hand's small size.

A more advanced version of the sensitive skin system (built in 1993–1996) is shown in Figure 8.7. The skin consists of standardized sensor modules (they can be seen in the figure), each about 20 by 20 cm in size and including $8 \times 8 = 64$ sensors and necessary local control electronics. Module dimensions are chosen so that the distance between border sensors of two neighboring modules is equal to the normal between-sensor distance within a single module. This, along with

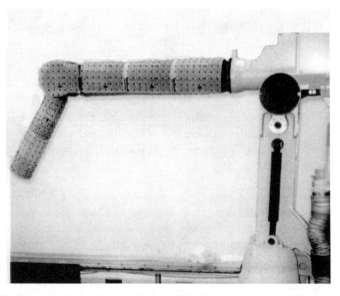

Figure 8.7 This robot is equipped with the 1993–1996 skin version. The skin is made of standardized sensor modules, each 23 by 23 cm in size and includes about 1200 infrared proximity sensors.

the fact that each module includes all the control electronics its own operation requires, makes for a universal module design. Covering the whole arm amounts to tiling it with such modules, perhaps with some modules cut to size for better coverage. The skin resolution (distance between neighboring sensors on a module) is 2.5 cm, twice as good as that of the earlier skin shown in Figure 8.6. Consequently, the skin has about 1200 sensors. Modules are electrically connected with each other by special connectors built into all four sides of each module. The skin's material base is again Kapton, but it is slightly different from the type used in Figure 8.6. The arm used in this system is a Hitachi P60 industrial arm manipulator.

One such sensor module is shown in Figure 8.8. Unlike the skin shown in Figure 8.6, surface mounting technology was used for the skin component installation. Combined with smaller components available at the time, the skin has about 2-mm maximum thickness, three times less than in the skin of Figure 8.6. Black dots in the figure are sensor detectors.

Both of the described systems went through a variety of tests using the software packages based on motion planning algorithms discussed in prior chapters. A good number of tests covered these three settings:

1. Tests with a Fully Autonomous Operation of the Robot Arm Manipulator. Typically the arm would be requested to go from some position A to a position B, while automatically avoiding collisions with previously unknown obstacles that it encounters on its way.

2. Tests with Teleoperation. In real time the operator shows the arm a rough idea of the desired trajectory, moving a master arm quickly and without any account for possible obstacles, and the main (slave) arm is expected to

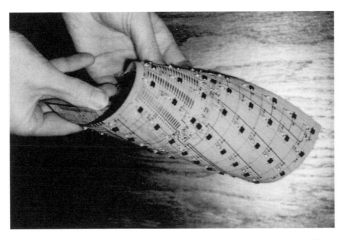

Figure 8.8 A single sensor module, one of those used in the skin on Figure 8.7. The module is 23 by 23 cm in size and includes 8 by 8 = 64 infrared sensor pairs, spaced at 2.5 cm between neighboring sensors.

move as close as possible to the shown trajectory while avoiding collisions with previously unknown obstacles.

3. Tests with a fully autonomous operation. The robot is expected to interact with the human (who shares workspace with the robot) by moving next to him or her. Whether or not the human partner is watching the robot, the robot is expected to be "out of the way" of the human. One variation of this test is when the arm "follows" the human, by keeping at some distance from him/her while avoiding collisions.

When entertaining such experiments, one is reminded of the Isaak Asimov's three laws of robotics (from his short story *I, Robot*, 1950). In tests 3 above, since the robots used were heavy industrial manipulators, Asimov's first law ("A robot may not injure a human being...") suddenly took on a literal meaning.

Manufacturer technical manuals of industrial robot arm manipulators prohibits, in no uncertain terms, humans from sharing space with such robots. Industrial robots are large and heavy and fast—they can hurt. Factory work cells with robots are often equipped with automatic means for stopping the robot if the rule is somehow neglected (e.g., this can be a laser or wire sensor guarding the robot perimeter, doors that are automatically locked when the robot is turned on, etc.).

In the described experiments, however, the whole point of the sensitive skin and collision-avoidance software is that no such protection is necessary anymore. A whole-sensitive robot is expected to never hurt a human: It should gently move aside or back or around the human. And, it should do it no matter whether the human noticed the robot. Aside from Asimov's dramatic formulations, common sense says that without such ability the robot simply cannot be used as a, say, astronaut assistant. The assistant's function may be to hand the astronaut tools and to take them back for storage, or, say, help the astronaut move and rotate bulky objects. In this team the astronaut must share space with the robot, and he or she cannot afford to always be aware of the robot presence. No less important is of course the robot's own safety; the skin should protect it from hurting itself by banging into surrounding objects. (Recall Asimov's third law: "A robot must protect its own existence...").

This author and his students, and later people who did not know much about robotics, have spent much time next to the robot, testing its "gentleness" provided by its sensitive skin and its intelligence. Nothing bad ever happened. This, of course, is not surprising: Multiple protective layers often appear in engineering systems. If worse comes to worse, the system should stop (we never came to this point in our experiments).

The pictures in Figures 8.9 to 8.12 show a few frames from videos taken in the laboratory during some of those tests.[6] The pictures in Figure 8.9 correspond to the test setting 1 above: The robot was instructed to start from some position on the left side of the scene and finish at some position on the right. Along the

---

[6]A note to the reader: videos that supplied pictures for Figures 8.9 to 8.11 can be seen in full on the web, http://aaaprod.gsfc.nasa.gov/Project/public_html-NASA/LaRue-Lumelsky.htm.

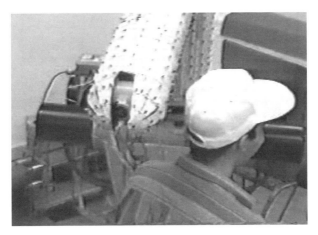

Figure 8.9   An autonomous operation of the whole-sensitive robot arm. The robot moves from the starting to the final position while maneuvering around the chess players and other obstacles as it senses them.

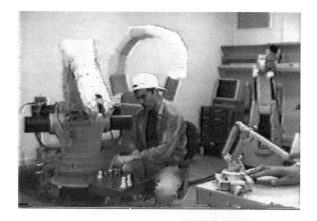

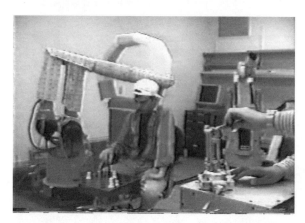

Figure 8.10 Robot teleoperation. The operator controls the main (slave) robot arm by moving a small master arm seen on the right. The robot is expected to roughly follow the path indicated in real time by the operator, while using its own sensing and intelligence to avoid a chess player in its work space. The result is a path that is as close as possible to the path shown by the operator, short of collisions.

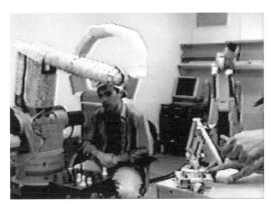

Figure 8.11 Robot teleoperation. Here the operator has attempted "to hurt" a student sitting in the robot work space. After bringing the (slave) robot right above the student's head (second picture), the operator moves the master arm sharply down, "expecting" the robot to repeat this motion. The robot starts in this direction, but then hesitates and stops. Note the difference between the robot and master arms' positions (second picture); the latter shows a clear intention to bring the robot much lower than it went. Eventually the operator moves the master arm toward us, and the robot follows (third picture). *Note*: The videos that supplied pictures for Figures 8.9 to 8.11 can be seen in full on the web, http://aaaprod.gsfc.nasa.gov/Project/public_html-NASA/LaRue-Lumelsky.htm.

Figure 8.12   A robot dancing with a ballerina (frames from a video). At all times the robot moves so as to be close to the ballerina while avoiding collisions with her. Note that at times the ballerina moves while not looking at the robot. She expects the robot to behave the same way as she would expect from a human partner—stepping in and out as necessary and reacting in a gentle "human" way. This expectation relates to the whole bodies of both partners. There is no preprogrammed motion on the robot part.

way the robot encountered obstacles—two students "oblivious" to the robot and bending slightly over the chess board; it went around and above the students' heads and continued on its trajectory.

The experiment in Figure 8.10 relates to the teleoperation setting 2 above. The robot is controlled by an operator using a small master arm that can be seen on the right, and also by its own motion planning algorithms. Unlike a common teleoperation, here we have a human–machine team, where each partner does what he/she/it is best at. The operation is thus is a real-time synergy of human and robot intelligence. The operator would move the master arm in a rather cavalier fashion, without regard for a human inside the robot workspace and without attempting a precise trajectory. The goal here was for the main (slave) robot to reach the right side of the scene and into the large loop. The operator quickly moves the master arm ahead of the robot, paying no attention to the human "obstacle" in the scene. When encountering the previously unknown obstacle—a person playing chess—the robot goes around the player's head; lowers slightly, trying to get to the operator's path; and then continues on the operator-dictated path.

The pictures in Figure 8.11 also relate to the teleoperation setting 2. Along the way the operator attempts to make the robot "hit" the human present in the robot workspace. This scene surely looks more dramatic in a videotape: The viewer sees that the operator, after first bringing the robot right above the human head, decisively pushes the master arm straight down; the viewer also sees how, after starting in this direction, the robot freezes in the air, "refusing" to hurt the human. Notice the difference between the positions of the master arm and the slave arm: The operator clearly intended the arm to go much lower than it did.[7]

The value of this experiment is more than a joke exercise. Imagine a robot assistant working next to an astronaut, controlled via teleoperation by a remote operator. If the operator is on the ground, a signal transmission delay due to large distance will introduce a bias into the operator's perception of the scene compared to the actual scene at the moment. Imagine that the operator makes a mistake, guiding the robot directly onto the astronaut. The robot behavior demonstrated in the Figure 8.11 experiment would save the situation.

Finally, the pictures shown in Figure 8.12 relate to the autonomous robot operation in the setting 3 above. A good way to demonstrate robot interaction with a nearby moving human is dance. If we can make the robot behave the way two human partners expect each other to move in a dance, we can count on the robot's adequate behavior in a human–robot crew. During the dance the partners stay close to each other. They continuously react to each other's movement. One partner does not have to look intently at the other partner at all times; he or she is confident that their partner will make sure that no collision takes place. For each of them it is not enough to know that his/her head or hand is safe: There is an expectation that one's whole body is safe from unpredictable collisions. Hence the interaction involves their whole bodies. A demonstration of this kind of interaction is a demonstration of a highly coupled robot–human team operation.

With these ideas in mind, we have carried out a special project between the University of Wisconsin (UW) Robotics Laboratory, on the one hand, and the UW Dance Department, on the other hand. Professor Tibor Zana from the UW Dance Department, who is also Artistic Director of the Wisconsin Dance Ensemble, choreographed the dance. The video frames shown in Figure 8.12 are from the resulting videos. Again, still pictures are not a good medium for showing motion: A color video looks much more interesting than these black-and-white still pictures.

The robot motion planning shown in these pictures was fully autonomous. The robot was not programmed for any specific paths. (Tests with predefined paths, which the robot would modify on the fly when reacting to the ballerina's movement, have also been tried.) The robot was only programmed to stay out of the ballerina's way and to move toward her when losing the proximity contact with her. In other words, the actual motion was in response to the ballerina movement. In a typical pair dance (e.g., waltz, tango, foxtrot, swing), one partner is the leader and the other partner is the follower. In our robot–ballerina dance the ballerina was the leader. This is admittedly not a typical dance convention today, but aren't robots the sign of the future!

The robot behavior in these experiments looks convincing and somehow "alive." We humans are not used to seeing machines behave like humans or

---

[7]For those romantically inclined after reading Isaac Asimov's robotic laws, the same would happen if the obstacle was not a human but a chair.

animals—following us and staying out of our way the way a dog or a cat do. The last thing one thinks of when first seeing these arm manipulators, large and heavy chunks of steel, would be the word "gentle." And yet this is what one thinks when watching this dance. The robot follows the ballerina, steps back when she changes her direction of motion, and hesitates when the ballerina puts her arms around the robot (as if thinking, "Where should I move now?"). The dance looks fascinating, for now at least: There are many such scenes in our future, with robots much smarter and much more sensitive and also capable of doing what we want them to do.

CHAPTER 9

Suggested Course Projects

The projects can start 3–4 weeks into the course—when the students are already familiar with the problem, notation, and terminology, and can read additional literature on their own. In this author's experience, the student would be required to write an email report once a week, outlying the work done on the project, issues to be tackled in the next period, and asking questions, if any. The work status would be also discussed in the regular meetings with the instructor. If the class is too big for that, other means of following the projects' progress are needed.

It may happen that in order to proceed with their project, some students may need the knowledge of material that would be discussed in class somewhat later. In such cases I would encourage the students to proceed with their reading of the corresponding material on their own, and would be ready to provide consultation as needed.

The topics of course projects listed here are given only as examples. Very often, students themselves try to come up with a topic that is closely related to their interests, their specialty, their thesis topic, or even another course that they are presently taking or have recently completed. With most students being already quite experienced in science and engineering, this author has chosen to make it known that such connections would be welcome; he has given students much flexibility in choosing their topics, as long as the project satisfied the expected level of effort and was closely related to the course's material. As a result, projects may vary widely, and the defense of projects at the final exam likely turns into an interesting mini-lectures series. Not rarely, course projects would have continuations in the following years (a student may continue the project for another year, or change his/her PhD topic, or prepare a conference or journal paper, etc.) Once the topic is chosen and worked on for a little while, changing the topic should be discouraged because this will not leave sufficient time for work.

One will note that the list of suggested projects below involves topics that, while related to the material taught in the course, go beyond the specific material studied. One such example is motion planning for multiple mobile robots. The idea here is that the course has prepared the student to use the material they have already learned for more advanced work. A topic like this would encourage the student to read literature and to apply creatively the ideas and techniques taught in the course to produce new knowledge. The instructor may want to recommend appropriate literature in such cases.

Suggested course topics can be roughly divided into three categories: theory and algorithms; computer simulation; experimental (hardware) work. A project may fall into more than one category; for example, one may be assigned to develop an algorithm and validate it by theoretical analysis and a computer simulation. Examples of topics in each category are given below.

Theory and Algorithms for Sensor-Based Motion Planning

1. Sensor-based motion planning for multiple mobile robots:
   - Centralized motion planning: All or most commands to all robots come from one center.
   - Decentralized motion planning: Each robot makes its motion planning decisions on its own, without coordinating it with other robots. Various options can be considered here, for example: A robot knows nothing about other robots' decisions except by watching them move; or, robots may distribute their next step decisions to other robots; or, in a given pair of robots, one is required to report its planning decisions to the other but not vice versa. A mathematically interesting problem is an interception problem: One robot tries to escape from a robot that attempts to intercept it; different assumptions about sensing means, speeds, and knowledge about the other robot abilities gives rise to different schemes.
   - Collective behavior of multiple robots. Again, different behavioristic schemes addressing motion planning can be considered here, such as making and changing formations, covering a maximum area (the Mushroom Pickers problem), etc.
   - Motion planning for tethered robots. Such problems appear in some industrial settings—for example, a Robot World system where tethered robots are floating above an assembly table (a two-dimensional setting); or underwater robot probes connected by tethers to the mother ship (a three-dimensional setting). Options include the tether being able, or not being able, to sense surrounding objects or other robots and tethers. One may want to consult related literature.
   - Computational geometry: efficient (static or dynamic) division of work space between a few robots (such as when vacuum cleaning a supermarket floor); the rendezvous problem for two robots. The performance criteria may include minimizing the total path (equivalent to minimizing energy), or minimizing total time spent on the job, or minimizing wasted walks over areas that have been cleaned already or walks to charging stations.
2. Hierarchical motion planning systems: combining prior knowledge (e.g., a map of the area) with on-line sensor data. For example, inspection of a nuclear plant after a major disaster: Some areas may be still as they were before the event, and some may be damaged by an explosion, so the robot must be able to deviate from the map using its sensing and intelligence.

3. VLSI routing as a motion planning problem (designing a printed board is a variation of this problem). Assume that after a VLSI chip has been already designed, a need appears to add another few wires. With the traditional tools, the whole design must be recalculated from scratch; this is expensive and not always possible. Address the task as a motion planning problem. The new wires cannot intersect those already on the chip, so the old wires and other electronics on the chip present obstacles to new wires. Options may include motion only in one plane or between planes, in which case a wire may jump from plane to plane only in designated points.

4. Motion planning amongst moving obstacles. The topic can be formulated for either a mobile robot or an arm manipulator. Investigate the constraints, if any, to be imposed on the velocities and directions of motion of the obstacles (relative to the robot) to make the problem tractable.

5. Mutual collision avoidance of two arm manipulators operating in shared space.

6. Effect of arm kinematics on workspace accessibility in the presence of obstacles. As we learned in Chapters 5 and 6, an interaction between the arm and an obstacle creates *shadows*, which is effectively a part of workspace that is not accessible to the arm. For the same size and shape obstacle, the amount of workspace lost due to this effect depends on the arm kinematics and relative dimensions of its links. Investigate this phenomenon.

7. Algorithm for coordinating information from different sensors—for example, in the context of sensitive skin sensors. As presented in Chapter 8, each sensor of the skin looks exactly in front of it, catching the reflection of its light from an obstacle in this direction. If, for example, the obstacle's surface is at an angle to this sensor, much of reflected light will go in a different direction, possibly to another sensor. Therefore, the fact of the object detection by both sensors may provide additional information about the object's shape, location, and distance from the robot body. Investigate the coordinated use of sensor data for such and other useful inferences about the obstacles.

8. Effect of robot dynamics on motion planning. This set of topics is good for students with a strong background in control theory. The study of effects of robot dynamics in Chapter 4 is done for a mobile robot that is either a point, or a very symmetrical body.
   - Investigate effects of dynamics for a mobile robot of a more complex shape, e.g., a boxy four-wheel vehicle.
   - Investigate effects of dynamics for an arm manipulator of a given kinematics.

9. Motion planning for highly redundant kinematic structures:
   - Snakes
   - Multi-finger wrists as multi-snake systems: power (whole-wrist) grasping, precision (two-point) grasping; pick-and-place operation, and so on

- Two-legged locomotion among obstacles; same with gravity
- Same with dynamically stable motion
- Multi-legged locomotion
- Advanced sensing (vision, range sensing) and redundancy
- Effect of joint limits in redundant systems; handling potential self-collisions of robot links

Computer Simulation, Real-Time Animation

10. Computation and visualization of configuration space of an arm manipulator.
11. Animation of motion planning algorithms for locomotion.
12. Real-time human-centered systems (blending human and machine intelligence in physical and virtual tasks):
    - Human-assisted interaction between multiple mobile robots (e.g., top view control of C-space or W-space); 2D and 3D tasks; same for a single mobile robot; for an arm manipulator
    - Same for advanced versus simple (tactile) sensing.
    - Human-assisted virtual part assembly/disassembly.
13. Taking advantage of advanced sensing for arm manipulator motion planning: For example, design an arm manipulator version of VisBug algorithms.
14. Simulation of 3D vision-based underwater exploration.
15. Motion of two arms sharing common space: each with two links; each with three links.
16. Implementation of motion planning algorithms; animation of robot motion.
17. Development of modular robot motion simulation software.

Experimental Work

Projects in this area tend to vary greatly and to be highly individualized. Examples below are shown only to give a taste of possible topics.

18. Experimental validation of, using a robot available for this work, one or more sensor-based motion planning algorithms.
19. Experimental study of a few existing sensing devices for the choice of technology that satisfies in the best way a given motion planning system. Examples: infrared, ultrasound, capacitance, electromagnetic, and so on, sensors.
20. Sensor/processing systems for detecting surrounding objects and doing motion planning.
21. Experimental analysis of human performance in spatial reasoning tasks: teleoperation skills; motion planning skills; effects of training.

■■■■ REFERENCES

1. V. Lumelsky, S. Mukhopadhyay, and K. Sun. Dynamic path planning in sensor-based terrain acquisition. *IEEE Transactions on Robotics and Automation* **6**(4): 462–472, 1990.

2. F. Preparata and M. Shamos. *Computational Geometry*, Springer-Verlag, New York, 1985.

3. J. C. Latombe. *Robot Motion Planning*, Kluwer Academic Publishers, Dordrecht, 1991.

4. Y. Lajoie et al. Gait of a deafferented subject without large myelinated sensory fibers below the neck. *Neurology* **47**(1):109–115, 1996.

5. S. Rossignol et al. Peripheral and spinal mechanisms in neural control of movement. *Progress in Brain Research* **123**:297–309, 1999.

6. M. Brady, J. Hollerbach, T. Johnson, T. Lozano-Pérez, and M. Mason, eds. *Robot Motion*, MIT Press, Cambridge, MA, 1982.

7. J. Craig. *Introduction to Robotics, Mechanics and Control*, 2nd ed., Addison-Wesley, Reading, MA, 1989.

8. R. Paul. *Robot Manipulators: Mathematics, Programming, and Control*, MIT Press, Cambridge, MA, 1981.

9. A. Bejczy. Robot arm dynamics and control. Technical memorandum 33-669. Technical report, Jet Propulsion Laboratory, Pasadena, CA, 1974.

10. R. H. Taylor. Planning and execution of straight line manipulator trajectories. *IBM Journal of Research and Development*, **23**(4):424–436, 1979.

11. J. Schwartz, J. Hopcroft, M. Sharir, eds. *Planning, Geometry, and Complexity. Robot Motion Aspects*, Ablex Publishing Corporation, Norwood, NJ, 1986.

12. D. Lipski and F. Preparata. Segments, rectangles, contours. *Journal of Algorithms* **2**:63–76, 1981.

13. T. Lozano-Pérez. Spatial planning: A configuration space approach. *IEEE Transactions on Computers* **32**(3):108–120, February 1983.

14. J. Reif. A survey of advances in the theory of computational robotics. In *Adaptive and Learning Systems. Theory and Applications*, K. S. Narendra, ed., Plenum Press, New York, 1986.

15. J. Reif. Complexity of the Mover's Problem and generalizations. In *Proceedings, 20th Symposium of the Foundations of Computer Science*, 1979.

Sensing, Intelligence, Motion, by Vladimir J. Lumelsky
Copyright © 2006 John Wiley & Sons, Inc.

16. J. Schwartz and M. Sharir. On the "Piano Mover's" problem. II. General techniques for computing topological properties of real algebraic manifolds. *Advances in Applied Mathematics* **4**:298–351, 1983.

17. V. Lumelsky and A. Stepanov. Effect of uncertainty on continuous path planning for an autonomous vehicle. In *23rd IEEE Conference on Decision and Control*, Las Vegas, 1984.

18. J. Schwartz and M. Sharir. On the "Piano Mover's" problem. I. The case of a two-dimensional rigid polygonal body moving amidst polygonal barriers. *Communications on Pure and Applied Mathematics* **34**:345–398, 1983.

19. J. Hopcroft, J. Schwartz, and M. Sharir. On the complexity of motion planning for multiple independent objects: PSPACE hardness of the 'warehouseman's problem'. *International Journal of Robotics Research* **3**(4):76–88, 1984.

20. J. Hopcroft, D. Joseph, and S. Whitesides. On the movement of robot arms in 2-dimensional bounded regions. In *Proceedings, 20th IEEE Symposium on Foundations of Computer Science*, Chicago, November 1982.

21. C. Ó'Dúnlaing, M. Sharir and C. Yap. Retraction: A new approach to motion planning. In *15th ACM Symposium on the Theory of Computing*, Boston, MA, 1983.

22. T. Lozano-Pérez. Automatic planning of manupulator transfer movements. *IEEE Transactions on Systems, Man, and Cybernetics* **SMC-11**(10):681–698, 1981.

23. J. O'Rourke. Convex hulls, Voronoi diagrams, and terrain navigation. In *Proceedings of the Pecora IX Remote Sensing Symposium*, Sioux Falls, SD, 1984.

24. D. Pieper. The kinematics of manipulators under computer control. Ph.D. thesis, Mechanical Engineering Department, Stanford University, 1972.

25. S. Udupa. Collision detection and avoidance in computer controlled manipulators. In *Proceedings of 5th Joint International Conference on Artificial Intelligence*, Cambridge, MA, 1977.

26. B. Faverjon. Obstacle avoidance using an octree in the configuration space of a manipulator. In *Proceedings of the IEEE International Conference on Robotics and Automation*, Atlanta, GA, March 1984.

27. T. Lozano-Pérez and M. Wesley. An algorithm for planning collision-free paths among polyhedral obstacles. *Communications of the ACM* **22**:560–570, 1979.

28. H. Moravec. The Stanford cart and the CMU rover. *Proceedings of the IEEE* **71**(7):872–874, July 1983.

29. R. Brooks. Solving the find-path problem by good representation of free space. *IEEE Transactions on Systems, Man, and Cybernetics*, **13**(3):190–197, 1983.

30. T. Binford. Visual perception by computer. In *Proc. IEEE International Conference Systems, Science, and Cybernetics*, Miami, FL, 1971.

31. R. Paul. *Modeling trajectory calculation and servoying of a computer controlled arm*. Ph.D. thesis, Stanford University, 1972.

32. J. Canny. A new algebraic method for robot motion planning and real geometry. In *Proceedings of the 28th IEEE Symposium on Foundations of Computer Science*, Los Angeles, CA, 1987.

33. L. Meijdam and A. de Zeeuw. On expectations, information, and dynamic game equilibria. In *Dynamic Games and Applications in Economics*, T. Basar, ed., Springer-Verlag, New York, 1986.

34. E. Moore. The firing squad synchronization problem. In *Sequential Machines: Selected Papers*, E. Moore, ed., Reading, Addison-Wesley, MA, 1964.

35. M. Blum and D. Kozen. On the power of the compass (or, why mazes are easier to search than graphs). In *Proceedings of the 19th Annual Symposium on Foundation of Computer Science (FOCS)*, Ann Arbor, MI, 1978.

36. H. Abelson and A. diSessa. *Turtle Geometry*, MIT Press, Cambridge, MA, 1981.

37. J. Traub, G. Wasilkowski, and H. Wozniakowski. *Information, Uncertainty, Complexity*, Addison-Wesley, Reading, MA, 1983.

38. L. Euler. *Commentationes Arithmeticae Collectae*, St. Petersburg Academy, St. Petersburg, 1766.

39. C. Berge. *Graphs and Hypergraphs*, North-Holland, Amsterdam, 1973.

40. Hermann Kern. *Labyrinthe—Erscheinungsformen und Deutungen—5000 Jahre Gegenwart eines Urbilds*, Prestel-Verlag, Munich, 1982.

41. Hermann Kern. *Through the Labyrinth: Designs and Meanings over 5,000 Years*, Prestel Publishing, Munich, 2000.

42. O. Ore. *Theory of Graphs*, American Mathematical Society, Providence, RI, 1962.

43. E. Lucas. *Recreations Mathematique*, A. Blanchard, Paris, 1892.

44. G. Tarry. Le problem des labyrinthes. *Nouvelles Annales de Mathematiques* **14**:187–189, 1895.

45. A. Fraenkel. Economic traversal of labyrinths. *Mathematics Magazine* **44**:12, 1970.

46. A. Fraenkel. Economic traversal of labyrinths. *Mathematics Magazine* **43**:125–130, 1971.

47. B. Bullock, D. Keirsey, J. Mitchell, T. Nussmeier, and D. Tseng. Autonomous vehicle control: An overview of the Hughes project. In *Proceedings of the IEEE Computer Society Conference "Trends and Applications: Automating Intelligent Behavior,"* Gaithesburg, MD, May 1983.

48. A. M. Thompson. The navigation system of the JPL robot. In *Proceedings of 5th Joint International Conference on Artificial Intelligence*, Cambridge, MA, 1977.

49. C. Thorpe. Path relaxation: Path planning for a mobile robot. Technical report CMU-RI-TR-84-5, Carnegie-Mellon University, 1984.

50. D. Keirsey, E. Koch, J. McKisson, A. Meystel, and J. Mitchell. Algorithm for navigation of a mobile robot. In *Proceedings of the International Conference on Robotics*, Atlanta, GA, 1984.

51. R. Chatila. Path planning and environment learning in a mobile robot system. In *Proceedings European Conference on Artificial Intelligence*, Torsey, France, 1982.

52. R. Chattergy. Some heuristics for the navigation of a robot. *International Journal of Robotics Research* **4**(1):59–66, 1985.

53. J. Crowley. Navigation for an intelligent mobile robot. *IEEE Journal of Robotics and Automation* **RA-1**(1):31–41, 1985.

54. A. Petrov and I. Sirota. Control of a robot manipulator with obstacle avoidance under little information about the environment. In *Proceedings of the VIII Congress of IFAC*, Vol. XIV, Kyoto, Japan, 1981.

55. In *Proceedings of the IEEE International Conference on Intelligent Robots and Systems, IROS '2003*, Las Vegas, 2003.

56. V. Lumelsky. A comparative study on the path length performance of maze-searching and robot motion planning algorithms. *IEEE Transactions on Robotics and Automation*, **7**(1):57–66, 1991.

57. W. S. Massey. *Algebraic Topology*, Harcourt, Brace & World, New York, 1967.

58. V. Lumelsky and A. Stepanov. Path planning strategies for a point mobile automaton moving amidst unknown obstacles of arbitrary shape. *Algorithmica* **2**:403–430, 1987.

59. V. Lumelsky. Effect of kinematics on dynamic path planning for planar robot arms moving amidst unknown obstacles. *IEEE Journal of Robotics and Automation* **RA-3**(3):207–223, 1987.

60. A. Sankaranarayanan and M. Vidyasagar. Path planning for moving a point object amidst unknown obstacles in a plane: The universal lower bound on the worst path lengths and a classification of algorithms. In *Proceedings of the IEEE International Conference on Robotics and Automation*, Sacramento, CA, 1991.

61. A. Blake and A. Yuille. *Active Vision*, MIT Press, Cambridge, MA, 1992.

62. M. Crowder. *Interactive Image Processing for Machine Vision*, Springer, Berlin, 1993.

63. K. Kutulakos, V. Lumelsky, and C. Dyer. Vision-guided exploration: A step toward general motion planning in three dimensions. In *Proceedings of the IEEE International Conference on Robotics and Automation*, Atlanta, May 1993.

64. T. P. Skewis. Incorporation of vision or range information into robot motion planning: Theoretical and implementation issues. Ph.D. thesis, Department of Electrical Engineering, Yale University, 1990.

65. A. Sankaranarayanan and M. Vidyasagar. Path planning for moving a point object amidst unknown obstacles in a plane: A new algorithm and a general theory for algorithm development. In *Proceedings of the 29th IEEE International Conference on Decision and Control*, Honolulu, HI, 1990.

66. I. Kamon, E. Rivlin, and E. Rimon. A new range-sensor based globally convergent navigation algorithm for mobile robots. In *Proceedings of the IEEE International Conference on Robotics and Automation*, Minneapolis, MN, 1996.

67. E. Rivlin and A. Rosenfeld. Navigational functionalities. *Computer Vision Graphics and Image Processing—Image Understanding* (10), 232–244, 1995.

68. I. Kamon and E. Rivlin. Sensory-based motion plannning with global proofs. *IEEE Transactions on Robotics and Automation* **27**(12):108–112, 1997.

69. S. L. Laubach and J. W. Burdick. An autonomous sensor-based path-planner for planetary microrovers. In *Proceedings of the IEEE International Conference on Robotics and Automation*, Detroit, MI, 1999.

70. S. L. Laubach, J. W. Burdick, and L. Matthies. An autonomous path planner implemented on the rocky 7 prototype microrover. In *Proceedings of the IEEE International Conference on Robotics and Automation*, Minneapolis, MN, 1998.

71. H. Noborio. Several path planning algorithms of mobile robot for an uncertain workspace and their evaluation. In *Proceedings of the IEEE International Workshop on Intelligent Motion Control*, Vol. 1, 1990.

72. N. Rao, S. Iyenger, and G. deSaussure. The visit problem: visibility graph-based solution. In *Proceedings of the IEEE International Conference on Robotics and Automation*, April 1988.

73. H. Choset, I. Konukseven, and J. Burdick. Mobile robot navigation: issues in implementating the generalized voronoi graph in the plane. In *Proceedings of the IEEE/SICE/RSJ International Conference on Multisensor Fusion and Integration for Intelligent Systems*, 1996.

74. H. Choset, I. Konukseven, and J. Burdick. Sensor based planning for a planar rod robot. In *Proceedings of the 1996 International Conference on Robotics and Automation,* Minneapolis, MN, 1996.

75. N. Rao, S. Iyenger, C. Jorgensen, and C. Weisbin. On terrain acquisition by a finite-sized mobile robot in plane. In *Proceedings of the 1987 IEEE International Conference on Robotics and Automation*, Raleigh, NC, May 1987.

76. I. Kamon, E. Rimon, and E. Rivlin. Range-sensor based navigation in three dimensions. In *Proceedings of the IEEE International Conference on Robotics and Automation*, Detroit, MI, 1999.

77. V. Lumelsky and K. R. Harinarayan. Decentralized motion planning for multiple mobile robots: The cocktail party model. *Autonomous Robots Journal* **4**(1):121–135, 1997.

78. D. T. Greenwood. *Principles of Dynamics*, Prentice-Hall, New York, 1965.

79. Z. Shiller and H. H. Lu. Computation of path constrained time optimal motions along specified paths. *ASME Journal of Dynamic Systems, Measurement and Control* **114**(3):34–40, 1992.

80. J. Bobrow. Optimal robot path planning using the minimum-time criterion. *IEEE Journal of Robotics and Automation* **4**(4):443–450, August 1988.

81. B. Donald and P. Xavier. A provably good approximation algorithm for optimal-time trajectory planning. In *Proceedings of the IEEE International Conference on Robotics and Automation*, Scottsdale, AZ, May 1989.

82. Z. Shiller and S. Dubowsky. On computing the global time optimal motions of robotic manipulators in the presence of obstacles. *IEEE Transactions on Robotics and Automation* **7**(6):785–797, 1991.

83. C. O'Dunlaing. Motion planning with inertial constraints. *Algorithmica* **2**(4): 431–475, 1987.

84. J. Canny, A. Rege, and J. Reif. An exact algorithm for kinodynamic planning in the plane. In *Proceedings of the 6th Annual Symposium on Computational Geometry*, Berkeley, CA, June 1990.

85. O. Khatib. Real-time obstacle avoidance for manipulators and mobile robots. *International Journal of Robotics Research* **5**(1):90–99, 1986.

86. R. Volpe and P. Khosla. Artificial potential with elliptical isopotential contours for obstacle avoidance. In *Proceedings of the 26th IEEE International Conference on Decision and Control*, Los Angeles, 1987.

87. D. Koditschek. Exact robot navigation by means of potential functions: Some topological considerations. In *Proceedings of the IEEE International Conference on Robotics and Automation*, Raleigh, NC, May 1987.

88. J. Barraquand, B. Langlois, and J. C. Latombe. Numerical potential fields techniques for robot path planning. *IEEE Transactions on Systems, Man, and Cybernetics* **22**(2):224–241, March 1992.

89. C. De Medio and G. Oriolo. *Robot Obstacle Avoidance Using Vortex Fields.* In *Advances in Robot Kinematics*, S. Stifter and J. Lenarcic, eds, Springer-Verlag, New York, 1991.

90. T. Wikman, M. Branicky, and W. Newman. Reflexive collision avoidance: A generalized approach. In *Proceedings of the IEEE International Conference on Robotics and Automation*, Raleigh, NC, May 1993.

91. P. Jacobs, J. P. Laumond, and M. Taix. Efficient motion planners for nonholonomic mobile robots. *IEEE International Conference on Intelligent Robots and Systems (IROS)*, Osaka, Japan, August 1991.

92. J. C. Latombe. A fast path planner for a car-like indoor mobile robot. In *Proceedings of the 9th National Conference on Artificial Intelligence*, Anaheim, CA, 1991.

93. J. Barraquand and J. C. Latombe. Nonholonomic multibody mobile robots: Controllability and motion planning in the presence of obstacles. In *Proceedings of the IEEE International Conference on Robotics and Automation*, Sacramento, CA, May 1991.

94. A. De Luca and G. Oriolo. Local incremental planning for nonholonomic mobile robots. In *Proceedings of the IEEE International Conference on Robotics and Automation*, San Diego, CA, May 1994.

95. T. Fraichard and A. Scheuer. Car-like robots and moving obstacles. In *Proceedings of the IEEE International Conference on Robotics and Automation*, San Diego, CA, May 1994.

96. A. Shkel and V. Lumelsky. The role of time constraints in the design of control for the Jogger's Problem. In *34th IEEE Conference on Decision and Control*, New Orleans, 1995.

97. G. Korn and T. Korn. *Mathematical Handbook*, McGraw-Hill, New York, 1968.

98. L. Hocking. *Optimal Control*, Clarendon Press, Oxford, 1991.

99. A. Shkel and V. Lumelsky. The Jogger's Problem: Accounting for body dynamics in real-time motion planning. *Automatica* **33**(7):1219–1233, 1997.

100. L. S. Pontryagin. *The Mathematical Theory of Optimal Processes*, Interscience Publishers, New York, 1962.

101. Unece: World robotics 2003. Technical report, United Nations Economic Commission for Europe, Geneva, 2003.

102. R. Brooks. Planning collision-free motions for pick-and-place operations. *International Journal of Robotics Research* **2**(4), 1983.

103. V. Milenkovic and B. Huang. Kinematics of major robot linkages. In *Proceedings of the 13th International Symposium on Industrial Robots*, Chicago, 1983.

104. M. Mason. Compliance and force control for computer controlled manipulators. In *Robot Motion*, K. S. Narendra, ed., MIT Press, Cambridge, MA, 1982, pp. 305–322.

105. H. Behnke et al., eds. *Fundamentals of Mathematics*, Vol. II: *Geometry*, MIT Press, Cambridge, MA, 1974, Chapter 16.

106. E. Cheung and V. Lumelsky. Proximity sensing in robot manipulator motion planning: System and implementation issues. *IEEE Journal of Robotics and Automation* **5**(6):740–751, 1989.

107. V. Lumelsky and K. Sun. A unified methodology for motion planning with uncertainty for 2d and 3d two-link robot arm manipulators. *International Journal of Robotics Research* **9**(5):89–104, 1990.

108. T. Lozano-Pérez. Spatial planning: A configuration space approach. *IEEE Transactions on Computers* **32**(3):108–120, February 1983.

109. John E. Hopcroft and Gordon Wilfong. Motion of objects in contact. *International Journal of Robotics Research* **4**(4):32–46, 1986.

110. M. H. A. Newman. *Elements of the Topology of Plane Sets of Points*, Cambridge University Press, Cambridge, 1961.

111. D. McCloy and M. Harris. *Robotics: An Introduction*, Open University Press Robotics Series, Halsted Press, New York, 1986.

112. M. P. Groover, M. Weiss, R. N. Nagel, and N. G. Odrey. *Industrial Robotics: Technology, Programming, and Applications*, CAD/CAM, Robotics, and Computer Vision, McGraw-Hill, New York, 1986.

113. A. J. Critchlow. *Introduction to Robotics*, Macmillan, New York, 1985.

114. A. Aho, J. Hopcroft, and J. Ullman. *The Design and Analysis of Computer Algorithms*, Addison-Wesley, Reading, MA, 1974.

115. E. Cheung and V. Lumelsky. Real time path planning procedure for a whole-sensitive robot arm manipulator. *Robotica* **10**:339–349, 1992.

116. N. Sliwa and R. Will. A flexible telerobotic system for space operations. In *Proceedings of the Space Telerobotics Workshop*, Pasadena, CA, 1987.

117. T. Matsui and M. Tsukamoto. An integrated robot teleoperation method using multimedia display. In *Proceedings of the 5th International Symposium of Robotics Research*, Tokyo, Japan, 1989.

118. S. Hayati, T. Lee, K. Tso, P. Backes and J. Lloyd. A testbed for a unified teleoperated-autonomous dual-arm robotic system. In *Proceedings of the IEEE International Conference on Robotics and Automation*, Cincinnati, OH, 1990.

119. Thomas B. Sheridan. *Telerobotics, Automation, and Human Supervisory Control*, MIT Press, Cambridge, MA, 1992.

120. S. Seaney and B. Stankovic. Design and construction of the human tester algorithm experiment booth. Technical report, University of Wisconsin—Madison, Robotics Laboratory, 1992. Also, Technical report RL-92004.

121. Fei Liu. Multivariate analysis of human performance in motion planning. Technical report, MS thesis, University of Wisconsin—Madison, Department of Mechanical Engineering, 1997. Also Technical report RL-97003.

122. A. Basilevsky. *Statistical Factor Analysis and Related Methods*, John Wiley & Sons, New York, 1994.

123. I. T. Jolliffe. *Principal Components Analysis*, Springer-Verlag, New York, 1986.

124. J. Hajek. *A Course in Nonparametric Statistics*, Holden-Day, San Francisco, 1969.

125. R. Bradley. *Distribution-Free Statistical Tests*, Prentice-Hall, Englewood Cliffs, NJ, 1986.

126. H. Lindman. *Analysis of Variance in Experimental Design*, Springer-Verlag, New York, 1992.

127. J. Tukey D. Hoaglin, and F. Mosteller. *Fundamentals of Exploratory Analysis of Variance*, John Wiley & Sons, New York, 1991.

128. J. Bray and S. Maxwell. *Multivariate Analysis of Variance*, Sage Publications, Thousand Oaks, CA, 1985.

129. G. Dunteman. *Introduction to Multivariate Analysis*, Sage Publications, Thousand Oaks, CA, 1984.

130. I. Bernstein. *Applied Multivariate Analysis*, Springer-Verlag, New York, 1988.

131. D. Hand and C. Taylor, *Multivariate Analysis of Variance and Repeated Measures*, Chapman and Hall, New York, 1987.

132. M. Crowder. *Analysis of Repeated Measures*, Chapman and Hall, New York, 1990.

133. V. Lumelsky and E. Cheung. Real-time collision avoidance in teleoperated whole-sensitive robot arm manipulators. *IEEE Transactions on Systems, Man, and Cybernetics*, **23**(5):194–203, 1993.

134. E. Cheung and V. Lumelsky. A sensitive skin system for motion control of robot arm manipulators. *Journal of Robotics and Autonomous Systems* **10**:9–32, 1992.

135. D. Um, B. Stankovic, K. Giles, T. Hammond and V. Lumelsky. A modularized sensitive skin for motion planning in an uncertain environment. In *Proceedings of the 1998 IEEE Conference on Robotics and Automation*, Leuven, Belgium, May 1998.

136. C. Miyazaki, A. Hirai, M. Fujie and V. Lumelsky. Development of proximity sensing system for obstacle detection. In *Conference of the Japanese Society of Instrument and Control Engineers (SICE)*, Kanazawa, Japan, 1993.

137. V. Lumelsky, M. Shur, and S. Wagner. *Sensitive Skin*. World Scientific, Singapore, 2000.

138. S. Périchon-Lacour, Z. Huang, Z. Suo, and S. Wagner. Stretchable gold conductors on elastomeric substrates. *Applied Physics Letters* **82**(15):2404–2406, 2003.

139. S. Wagner, S. Périchon-Lacour, P.-H. I. Hsu, J.C. Sturm and Z. Suo. Stretchable and deformable macroelectronics. In *61st IEEE Device Research Conference Digest*, 2003.

140. D. Um and V. Lumelsky. Fault tolerance via analytic redundancy for a modularized sensitive skin. *International Journal of Robotics and Automation*, **15**(4):99–108, 2000.